ENVIRONMENTAL STANDARDS

Scientific Foundations and Rational Procedures of Regulation with Emphasis on Radiological Risk Management

ENVIRONMENTAL STANDARDS

SCIENTIFIC FOUNDATIONS AND RATIONAL PROCEDURES OF REGULATION WITH EMPHASIS ON RADIOLOGICAL RISK MANAGEMENT

A Monograph
by the Study Group: Environmental Standards

K. Pinkau (Speaker)
O. Renn (Editor of English Edition)
K. Decker
C.F. Gethmann
H.W. Levi
J. Mittelstraß
S. Peyerimhoff
G. zu Putlitz
A. Randelzhofer
C. Streffer
F.E. Weinert

A Report of the Berlin Academy of Sciences and Technology

KLUWER ACADEMIC PUBLISHERS
BOSTON / DORDRECHT / LONDON

A C.I.P. Catalogue record for this book is available from the Library of Congress.

ISBN 0-7923-8109-2

Published by Kluwer Academic Publishers,
P.O. Box 17, 3300 AA Dordrecht, The Netherlands.

Sold and distributed in the U.S.A. and Canada
by Kluwer Academic Publishers,
101 Philip Drive, Norwell, MA 02061, U.S.A.

In all other countries, sold and distributed
by Kluwer Academic Publishers,
P.O. Box 322, 3300 AH Dordrecht, The Netherlands.

Printed on acid-free paper

Translated from German by: Gisela Jäger and Sonja Kais-Heinrich

Printed in the Netherlands

TABLE OF CONTENTS

PREFACE TO THE ENGLISH EDITION

It is not uncommon that a group of scientists from many different disciplines join a working group, discuss a topic of interest and edit a volume of articles related to this topic. They may even agree on a jointly written introduction or conclusion. The study group „Environmental Standards“ established in 1987 as an expert panel of the German Academy of Sciences and Technology in Berlin broke with that common tradition and became involved in a fascinating, but also pain-staking experiment to compose a document on setting environmental standards that has been literally written and authorized by all group members. The group consisted of eleven individuals representing the following disciplines: physics, chemistry, biology, engineering, law, psychology, and sociology/economics.

The study group had two major goals: First, to provide a document that summarizes the state of the art in each discipline with respect to the topic of environmental standards. Since it is impossible to cover all environmental hazards in one book, the panel members agreed to limit the discussion of the scientific material to one major case study: the effects of ionizing radiation. This topic was selected because the scientific data base is well developed in this area and levels of anthropogenic release can be compared with natural background levels. These two conditions are rarely met by most chemical hazards. Within the subject of ionizing radiation and its impacts, the group intended to provide a complete and accurate picture of the respective health impacts and environmental problems. This task was accomplished by asking individual members of the group and outside experts to provide basic knowledge and background material on scientific issues such as exposure data, dispersion, dose-effect-relationships, and health impacts. In addition, the normative aspects of setting limits of exposure were discussed in papers on legal handling of standards, rational decision making, risk perception, and institutional and social processes for setting and justifying environmental standards.

The second goal was to provide guidelines for a desirable procedure that would guarantee the inclusion of all scientific data and reasoning but would also assure the incorporation of social costs, cultural values and individual preferences in the decision making process. The study includes a section that describes an analytic framework for collective decision making under uncertainty and a section on institutional innovations necessary to implement a procedure for integrating scientific input and rational decision making. The procedural suggestions are focused around a new environmental institution that is empowered to initiate and promulgate new standards on the basis of a present process of linking sicentific data with social values and preferences.

Meeting these two goals turned out to be an ambitious task. Starting a constructive dialogue between natural and social scientists took some time and effort, but the main difficulty was to have all eleven members agree with each sentence of the study. The organizational principle of the group was to ask one or two members to provide a first draft, a so called „seed paper" on a topic that all members regarded as relevant for the subject of the study. After the seed paper was completed, the text was discussed and reviewed by all members of the group. In a second effort the initial author or another member of the group was asked to compose a second version, which again was scrutinized and reviewed by all members. The process was iterated until all members felt comfortable with the text.

In addition to the internal review process by the panel members, assistance to improve the study was given by many external experts. The study group conducted two initial workshops in early 1988. More than 20 contributors from both the natural and the social sciences were invited to present their expertise on topics such as „necessity and function of environmental standards", „concrete examples of standard setting in Germany", „scientific criteria for evaluating technical expertise" and „social and individual risk evaluation. Furthermore, each chapter of the final study was reviewed by several experts in the respective field. These reviews were incorporated in the final version of the study. Once completed the study was presented and discussed in a Council Meeting of the Academy of Sciences and Technology in Berlin in June 1990. A final expert meeting was held on July 4, 1990 to take into account the suggestions and comments made by the Council members. By the end of 1990, the final text was completed. The study was published in 1992 by the German publishing Company de Gruyter.

Several years after its initial publication, the study is still considered one of the seminal source books on environmental standard setting among German academics, professionals and regulators. Furthermore, the book represents a rare example of integrating the natural and social sciences and developing consistent and complete procedural guidelines for standard setting in a democratic society. This alone would be a sufficient reason to make the work of the panel available to a larger audience and publish it in English. In addition, countries such as the United States and the United Kingdom face similar problems in standard setting and might benefit from the perspective and suggestions of the German expert panel. Several organizations, such as the U.S. EPA or the Washington based NGO „Resources for the Future" have taken considerable efforts in reviewing old and designing new policies for risk evaluation. They might find the ideas expressed in this volume helpful for their own discussion.

This book is intended to reach several audiences. For the interested scientist, the book provides a concise series of scientific material that conveys the state of the art on radiation and its impacts as seen from different scientific disciplines. In particular, experts in fields such as radiation physics, health physics, toxicology, epidemiology, and related areas may find the sections on environmental impacts of radiation helpful and stimulating for their own research. The social scientist may benefit from the description and analysis of German institutions and their regulatory style. Another focus of the book is on the perception of nuclear risks by the public at large. Issues such as individual risk perception, social amplification of risk, and societal responses to collective decision making have become part of an international agenda that seems relevant in almost any country of the world. Regulators and politicians may find the normative conclusions and the suggested procedures for standard setting interesting and stimulating. These suggestions are based, of course, on the conditions and traditions of the political culture in Germany and are thus not readily transferable to other countries. Many aspects of regulation, however, are universal in character and may well apply to other countries. Finally, the book is intended to appeal to the non-professional reader who may be interested in the complete picture of standard setting and would like to explore the scientific as well as the normative side of environmental regulation.

At the end of this preface, we wish to acknowledge the valuable contributions provided by many experts who helped us to complete the study. Renate Mayntz (Cologne) gave us helpful insights to section 1.3.4 by giving a lecture and sharing her lecture notes with the group. Peter Schwabe (Cologne) wrote the „seed text" to section 3.2. The ideas of Section 4.4 were developed and confined on the basis of several publications from and discussion with the former state undersecretary of environmental affairs, Manfred Popp (Wiesbaden). Susanne Hahn (Essen) participated in writing the „seed text“ to section 5.4, and Ernst Mohr (Kiel) contributed to section 6.2.1 by a lecture and a manuscript. Special thanks are also due to Ludger Honnefelder (Bonn), Albrecht Kellerer (Munich), Lutz von Rosenstiel (Munich) and Martin Weber (Kiel) who provided written comments to the first draft of the manuscript. In addition, Renate Mayntz (Cologne), Karl Aurant (Berlin), Walter H. Goldberg (Göteborg), Dietrich Henschler (Würzburg), Wolfgang Jacobi (Munich), Michael Kloepfer (Trier), Heinrich Freiherr von Lersner (Berlin) and Ernst Mohr (Kiel) attended the final workshop on July 4, 1990 and helped us to polish the final version of the manuscript. We also like to acknowledge the many experts who contributed their expertise to the two workshops and other meetings.

We are deeply indebted to Helga Knebel (Garching) for countless hours of typing and supplementing the „opus magnum“, Susanne Hahn (Essen) for editing the text, Gisela Jaeger and Sonja Kais-Heinrich for the enormous contribution in translating more than 600 pages of sophisticated German text into „readable“ English, and finally Sabine Mücke who was the final text editor for the English edition.

The Academy of Sciences and Technology in Berlin as well as the Gottlieb Daimler and Karl Benz Foundation in Ladenburg enabled and supported in various ways the interdisciplinary cooperation documented in this study. We are also grateful to the „Kollegium“ of the Academy of Sciences and Technology for providing support and approval for funding this translation. We also appreciate the financial support by the Federal Minister for Research and Technology for the study group „Environmental Standards“[1]. We also would like to express out gratitude to the editors of the Kluwer publishing company who offered their assistance in finalizing the English edition.

Munich and Stuttgart

K. Pinkau
O. Renn

ANNOTATION

[1] Support for this study was provided by the German Federal Minister for Research and Technology under the grant PLI 1395/0. The study reflects the viewpoints of the members of the Council of the Academy of Sciences and Technology in Berlin. These views are not necessarily identical with the views of the Federal Ministry

LIST OF ABBREVIATIONS

AbfG	(Abfallbeseitigungsgesetz) German law on waste disposal
Abl.	(Ablage) file
Abs.	(Absatz) paragraph
AEC	Atomic Energy Commission
AGS	(Ausschuß für Gefahrstoffe) German committee on dangerous substances
AGU	(Arbeitsgemeinschaft Umweltschutz) a German interest group on environmental protection
AKW	(Atomkraftwerk) nuclear power station
ALARA	As low as reasonably achievable
Anm.	(Anmerkung) annotation
Art.	(Artikel) article
AtG	(Atomgesetz) German law on atomic energy
AtVfV	(Atomrechtliche Verfahrensverordnung) ordinance on procedure under the German atomic energy law
BauGB	(Baugesetzbuch) German building code
BauR	(Baurecht) German building law
BEAR	Biological Effects of Atomic Radiation
BEIR	Biological Effects of Ionizing Radiation
BGA	(Bundesgesundheitsamt) German federal agency for public health (dissolved in 1995)
BGB	(Bürgerliches Gesetzbuch) German civil code
BGBl	(Bundesgesetzblatt) German Federal Law Bulletin
BGH	(Bundesgerichtshof) German Federal Supreme Court
BGHZ	(Entscheidungen des Bundesgerichtshofes in Zivilsachen) decisions of the German Federal Supreme Court in civil cases
BImSchG	(Bundes-Immissionsschutzgesetz) German federal law on immission protection
BImSchV	(Bundes-Immissionsschutzverordnung) German federal regulations on immission protection
BMU	(Bundesministerium für Umwelt, Naturschutz und Reaktorsicherheit) German Federal Ministry for the Environment, Nature Conservation and Nuclear Safety
BNatSchG	(Bundesnaturschutzgesetz) German federal law on nature conservation
BT-Drs.	(Bundestagsdrucksache) Document of the German Parliament (Lower House)
BT-Prot.	(Bundestagsprotokolle) minutes of the German Parliament (Lower House)

BVerfGE	(Entscheidungen des Bundesverfassungsgerichts) decisions of the German Supreme Constitutional Court
BVerwG	(Entscheidungen des Bundesverwaltungsgerichts) decisions of the German Federal Administrative Court
BWaldG	(Bundeswaldgesetz) German federal law on forests
CDU	German Christian Democratic Party
CFC	chlorofluorocarbon
ChemG	(Chemikaliengesetz) German chemical control act
CO_2	carbon dioxide
DFG	(Deutsche Forschungsgemeinschaft) a German institution for research promotion, similar to the National Science Foundation in the U.S.A.
DIN	(Deutsche Industrienorm) German industrial standard
DOE	U.S.-Department of Energy
D...V	(Die öffentliche Verwaltung) „Public Administration", a German juridical journal
DVBl	(Deutsches Verwaltungsblatt) German Administration Bulletin
E	(Entscheidungssammlung) collection of decisions
EAG-V	(Vertrag zur Gründung der europäischen Atomgemeinschaft) treaty on the foundation of the European Atomic Energy Community
EPA	Environmental Protection Agency (United States of America)
ET	(Energiepolitische Tagesfragen) „Current Issues of Energy Policy", a jounal on energy policies
ed	edition
ed.	editor
eds.	editors
FDP	German Liberal Democratic Party
FernStrG	(Fernstraßengesetz) German law on federal roads
GAU	(größter anzunehmender Unfall) maximum credible accident
GefStoffV	(Gefahrstoffverordnung) German act on dangerous substances
GG	(Grundgesetz) German constitution
GSF	(Gesellschaft für Strahlen- und Umweltforschung, jetzt Forschungszentrum für Umwelt und Gesundheit) Center for Research on Radiation and the Environment, now renamed in Research Center for Environment and Health
HdUR	(Handbuch des Umweltrechts) a manual of German environmental law
IAEA	International Atomic Energy Agency
ICRP	International Commission on Radiological Protection
Jungdemokraten	youth organization of the FDP
Junge Union	youth organization of the CDU
Jungsozialisten	youth organization of the SPD

JuS	(Juristische Schulung) „Juristic Education“, a German juridical journal
JZ	(Juristenzeitung) „Journal for Lawyers“, a German juridical journal
KFA	(Kernforschungsanlage Jülich) a former nuclear research facility in Jülich, Germany; now: Research Center Jülich
LAI	(Länderausschuß für Immissionsschutz) committee of the German states on immission protection
LET	linear energy transfer
LuftVG	(Luftverkehrsgesetz) German law on air traffic
MAK	(Maximale Arbeitsplatzkonzentration) maximum admissible concentration of occupational exposure
MAUT	multi-attribute utility theory
Münch-Komm.	(Münchner Kommentar zum BGB) a comment on the German civil code
NIH	National Institute of Health (USA)
NGO	Non-Governmental Organization
NJW	(Neue Juristische Wochenschrift) „New Juristic Weekly“, a German juridical journal
NOEL	No observable effect level
NOx	nitrogen oxides
Nr.	(Nummer) number
NRC	Nuclear Regulatory Commission
NuR	(Natur und Recht) „Nature and Law“, a German juridical journal
NVwZ	(Neue Zeitschrift für Verwaltungsrecht) „New Journal on Administrative Law“, a German juridical journal
n.y.	no indication of the year
OTA	Office of Technology Assessment
OVG	(Oberverwaltungsgericht) a German higher administrative court
p.c.	post conceptionem
PuO_2	plutonium dioxide
RBE	relative biological effectiveness
Rdnr.	(Randnummer) marginal number
RERF	Radiation Effects Research Foundation
RGBl	(Reichsgesetzblatt) Pre-WorldWar II Law Bulletin
RöV	(Röntgenverordnung) German ordinance on X-rays
RVO	(Rechtsverordnung) ordinance
RWE	(Rheinisch-Westfälisches Elektrizitätswerk) a German power-supply company
SO_2	sulfur dioxide
SPD	German Social Democratic Party

SSK	(Strahlenschutzkommission) German commission on radiation protection
StGB	(Strafgesetzbuch) German criminal code
StörfallVO	(Störfallverordnung) German ordinance on emergency planning for accidents in nuclear plants
st. Rspr.	(stehende Rechtsprechung) standing jurisdiction
StrlSchV	(Strahlenschutzverordnung) German dordinance on radiation protection
StrVG	(Strahlenschutzvorsorgegesetz) German law on preventive radiation protection
TA Luft	(Technische Anleitung zur Reinhaltung der Luft) German technical ordinance for the preservation of clean air
TRGS	(Technische Regel für Gefahrstoffe) German technical ordinace on dangerous substances
TRK	(Technische Richtkonzentration) German technical reference levels of concentrations
U	uranium
UBA	(Umweltbundesamt) German Federal Environmental Protection Agency
UNSCEAR	United Nations Scientific Committee on Effects of Atomic Radiation
UPR	(Umwelt- und Planungsrecht) German environmental and planning law
VCI	(Verband der chemischen Industrie) German association of chemical industry
VDI	(Verband Deutscher Ingenieure) association of German engineers
VwGO	(Verwaltungsgerichtsordnung) rules of the German administrative court
VwVfG	(Verwaltungsverfahrensgesetz) German law on administrative procedures
WEC	World Energy Conference
WHG	(Wasserhaushaltsgesetz) German law on water supply
WHO	World Health Organization
WZB	(Wissenschaftszentrum Berlin) Scientific Center of Social Research in Berlin
ZPO	(Zivilprozeßordnung) German code of civil procedure
ZZP	(Zeitschrift für Zivilistische Praxis) „Journal on Civilian Practice", a German juridical journal

SUMMARY

The study presented here has four main sections. The first section – identical with chapter 1 – defines the problem and the terms of reference. It clarifies the notion of environment and discusses the necessity and function of environmental standards.

The second section – chapters 2 and 3 – describes the present situation regarding facts and scientific evaluation problems. Chapter 2 is concerned with the scientific and medical foundations of radiation protection, chapter 3 gives an overview of the development of environmental standards concerning radiation and other noxious agents.

The third section – identical with chapter 4 – describes and analyzes the present regulatory situation in Germany. It starts with a description of risk perception by individuals and continues with an analysis of the social, political, and institutional processes that govern or influence the regulatory decision making process. Issues such as the media_s influence on public perception and the legal procedures for promulgating standards are addressed specifically in chapter 4.

The fourth section – chapters 5 and 6 – is devoted to the development of normative advice of how to include scientific reasoning and rational procedures in the political decision making process for promulgating environmental standards. These procedures need to be designed in such a way that they are also commensurable with the social and political structures of Germany as described in chapters 3 and 4. Chapter 5 introduces the technique of decision analysis and demonstrates its application for standard setting. Chapter 6 examines institutional and organizational perspectives, which, according to the aim of this study, develops strategies of coping with complex environmental problems that require governmental interventions.

Chapter 1: *Necessity and Function of Environmental Standards*

„Environmental standards" are legal and administrative regulations or civil-law rules, which help to transform yet undetermined legal concepts of environmental protection (such as ALARA = as low as reasonably achievable) into definite norms via operationalization and standardization of measurable quantities. Reaching environmental standards is a complex process, which takes into account scientific insights of different disciplines, normative beliefs and values, and the general social context.

The starting point of the study is a discourse about the notion of „environment" in the context of the environmental debate. This debate differs from the debate in the behavioral sciences, which regard environment as everything outside of a system's boundaries (cf. v. Uexküll). Following the movement known as philosophical anthropology, humans are regarded as social beings open to the world („weltoffen"), for whom environment at the same time enables and menaces life. Human beings are in a special position insofar as they are not only subject to their environment but they face the anthropological necessity of shaping it as well. Against this background, the study opposes a „naturalistic" notion of environment and favors a „culturalistic" one: environmental standards are human-made obligations of action, which societies impose on themselves as „voluntary" norms if and insofar as they want to sustain the natural resources for future generations, build community and foster social cohesion. Naturalism, however, becomes manifest when an environmental standard is thought to be directly derivable from some physically-defined threshold value. This concept cannot be sustained for the mere fact that in many relevant environmental fields (such as radiation protection) there are no physical threshold values. But even if they existed, there are often good reasons to remain below them; although, under certain circumstances it might be expedient to allow the threshold to be exceeded.

The problem of establishing environmental standards shows that only a moderate anthropocentrism, as opposed to an exploitative human egotism or a radical ecocentrism, represents an adequate philosophical framework for viewing and handling nature. This, however, implies the abandonment of a regulatory concept that would promise a comprehensive standardization of all human interventions which affect the environment. To establish environmental standards, one has to proceed by and large from an environment as humans have known it for some generations. With reference to this environment, environmental standards operate rather conservatively: they are to protect the environment from further damages and to reverse already existing damages, which are recognized as such. Even though the image of an „undisturbed" environment is culturally constructed, the human task to shape the environment in accordance with the reference image does not leave the establishment of environmental standards to political arbitrariness. The latter is – apart from general standards of rationality – constrained by scientifically ascertained facts (as, e.g., dose response relations), the numerous social and cultural objectives of the social assessment process and the rules of rationally balancing risks and opportunities.

When definite environmental standards are established, it is assumed that the definition of limits does not only depend on the probability of an expected damage but also on the benefit which the potentially harmed person gains by taking the risk. So the general function of environmental standards is to define limits of human actions in such a manner that the potential benefits associated with different options available to the acting person or institution are balanced with the potential detriments (risks) that the decision maker or somebody else may face as a result of each decision option. Looking at standards from a decision analytical perspective, setting standards is a specific case of making rational decisions under uncertainty. The notion of risk lies at the core of the discussion about establishing and justifying environmental standards. Accordingly, environmental standards are self-imposed restrictions of actions as a means to reduce potential damages when facing uncertain outcomes.

Many philosophers deny the possibility of rational action; in fact, the principles of rational action largely differ in their structure from the principles of cognitive rationality. Nevertheless, the study group made a case that, under certain conditions, one is able to reconstruct the conditions for rational actions. Such a reconstruction leads to considerable complications when actions under uncertainty are targeted. The understanding of uncertain outcomes are linked both to the understanding of probability and to a precise definition of the notion of damage. It is not easy to define a rational rule for making choices when each outcome is characterized by a mix of potential damages and probabilities. Using expected values (multiplying the damage with its concomitant probability) is often regarded as the optimal procedure for calculating comparative risk estimates. Other rationales, such as minimizing the maximum damage, deviate from the expected value approach, but can also be justified by means of rational reasoning. In addition, decisions under uncertainty include other dimensions beyond probability and damage. Issues such as fairness in distributing risks and benefits or resilience with respect to sustaining the productivity of an environment, require an multi-dimensional approach to decision making. The study group endorsed the use of multiattribute decision analysis, a technique that allows for a balancing procedure of risks and benefits based on several target dimensions. Implementing such a procedure raises difficult methodological and substantive problems (a detailed suggestion for a solution of those problems is developed in chapter 5).

With regard to environmental standards, one can observe a considerable discrepancy between (normative) acceptability and (factual) acceptance of risks in all contemporary societies. The distinction between acceptability as a normatively justified procedure to set standards and acceptance as the factual attitude of individuals and groups towards the risk taking activity is crucial for legitimizing environmental policies. Each policy must reflect both acceptability and acceptance. Without considering acceptability, the government would have no right to legally impose risks on its citizens, which in fact is inevitable when personal and collective rationality do not coincide. Without considering acceptance, the government would not be legitimized to impose these standards with at least a minimum of public consent. Both concepts need to be reconciled.

Facing considerable gaps between public perception and scientific assessment of risk, reconciliation between acceptance and acceptability has been a major problem in modern societies. A promising attempt to resolve this conflict is to define procedures of decision making that assure the public that the best scientific estimate and the salient values of the affected populations are taken into consideration without pre-determining the outcome of the decision process. It is much easier to reach a consensus on procedure than on substance. Citizens are assumed to act in a pragmatically consistent way, i.e., they have to be prepared to have risks imposed on them which they, too, expect others (and themselves) to take. This principle of pragmatic consistency is the supreme norm, on the basis of which acceptability of risks can gain public support. Therefore, the important point is that government and society organize their decision processes in such a way that the procedures include science and social values and are transparent enough to be easily understood by the citizens concerned. The public discourse, which leads to the establishment of environmental standards, must have a high „potential of value sensitivity".

Chapter 2: *Scientific-Medical Foundations of Radiation Protection*

Scientific-medical foundations of radiation protection have been explored systematically for several decades. Today, our knowledge is more comprehensive and secure than knowledge available for most other protective schemes against environmentally noxious agents. According to the two parameters which determine risks caused by environmental noxae – exposure and effect – the necessary scientific foundations for a protection concept comprise, on the one hand, procedures of exposure determination (dosimetry) and, on the other hand, knowledge and understanding of biological effects of the respective noxae.

The first aim of research into radiation dosimetry, which has essentially been achieved by now, was the development of methods to measure radioactivity, radiation doses and its effects, as far as they are directly measurable. The next step then led from measured energy doses to calculated dose equivalents, which take into consideration biological impacts of ionizing radiation as a function of radiation quality. The third step, finally, and the one to which, above all, radiation dosimetry owes its superiority to toxicological dosimetry, is calculation of the dose distribution in the human body, both following external irradiation and internal irradiation by incorporated radionuclides. The present state of radiation dosimetry thus allows the determination of the biologically effective dose at the place of its impact, which is satisfactory for most radiation protection requirements.

Apart from that, research into radiation dosimetry has led to two further important results. Mean exposure of the population to natural radiation, and its variation, is quite well known. This holds true both for exposure to external terrestrial and cosmic radiation and for exposure to internal primordial radionuclides, especially potassium-40. The dose limit for the population at large is guided by the variation of this exposure to natural radiation, which amounts to at least 0.3 milliSievert (mSv) in Germany.

With respect to the effective dose, this part of exposure to natural radiation contributes about 30% to the mean exposure of the German population. Another approximate 30% stem from radon in the air in closed rooms, which is of natural origin as well, but the impacts of which are at least partly due to civilization. They depend on constructional features, and also on personal behavior (ventilation). Mean exposure for medical purposes amounts to about 40%. For the individual, of course, this level of exposure depends on personal health conditions. Finally, exposure to artificial radiation due to the whole range of human use of nuclear fission and radioactivity contributes to radiation exposure with a long-term mean share of about 2%, where nuclear power plants in Germany are involved with a few tenths of one per cent.

One important insight about impacts, which also holds for many chemical noxae, is the fact that there are non-stochastic and stochastic effects. Non-stochastic effects are characterized by threshold doses below which they do not occur and above which they occur with increasing severity. Such threshold doses are of the order of 0.1 Sievert and higher, thus far higher than expected exposures to environmental radioactivity. Therefore, they can be neglected when environmental standards are established. The occurrence of damage is determined by the dose level. It should be taken into account, though, that even threshold doses, like all biological phenomena, vary within a certain range.

Stochastic impacts are those which lead to genetic damages or to cancer or leukemia. Here, it is not the severity of the damage which increases with increasing dose but the probability of its occurrence. There is no dose for which this probability is 100%, but probably no dose either for which it is 0%. Stochastic damages are due to changes in the DNA of a cell. As is obviously the case with the induction of hereditary deficiencies, it is assumed that, also in the case of induction of cancer, one single damaged cell already, if though with a correspondingly low probability, may cause such damage. With non-stochastic damages, many cells have to be damaged for an effect on health to become manifest. To be on the safe side with an assessment, it is therefore assumed with respect to stochastic damages that there are no threshold doses and that the dose response relation is a linear one, even for the smallest doses. Empirical evidence either for or against this assumption is not available. It must also be taken into account that there are natural repair mechanisms for radiation damages in the DNA.

To be on the safe side, it is assumed that there is no threshold dose for the radiation induction of cancer and genetic damages. It has to be presupposed as well that no radiation dose, however small it might be, is innocuous. This is taken into account by the German law on radiation protection, ruling that any exposure to radiation has to be justified by an expected benefit and has to be minimized as far as this is compatible with the expected benefits.

Another important result of radiobiological research is the quantitative assessment of the risk for contracting radiation cancer. The risk factors indicating how much larger the risk is after an exposure to radiation, compared with the normal risk of cancer, were established on the basis of epidemiological studies. The most important collection of persons for these epidemiological studies are the survivors of the atomic explosions in Hiroshima and Nagasaki. According to these studies, which have been revised in recent years, a cancer risk may reasonably be assumed to be within a range of about 5% per Sievert for sparsely ionizing radiation, at low doses and dose rates. With the German dose limit of 0.3 mSv/a for the general population (assuming that this dose limit is indeed reached), this would mean an additional lifetime risk to fall ill with cancer of about 1/1'000. This has to be compared with a normal cancer risk of 200/1'000.

Chapter 3: *Development of Environmental Standards for Radiation and Other Noxae*

An important foundation for establishing environmental standards is scientific and medical knowledge about impacts of noxious agents on the environment with its organisms, and on human health. The development of environmental standards have benefited from the improvement of this knowledge, a better description of the dose-response relationship and, linked to it, an improved measurement of exposure. In addition, the increasing sensitivity of people in industrialized countries to potential dangers caused by noxious agents, which can harm environment and health, have advanced the quest for effective protection standards. In many cases, standards for workers' protection against ionizing radiation and other noxae were established first. The experiences gained from occupational standards were then used to formulate standards for environmental protection with respect to the general public and the principles they are based on.

In developing protection standards for ionizing radiation, the focus was originally on dose limits for occupational exposure to ionizing radiation. The growing spread of technological and medical nuclear facilities led to an establishment of radiation protection standards for the population at large. In this process, comparison with exposure to natural radiation always played a significant role. Dose limits for the population have been fixed at a level far below the threshold doses for non-stochastic effects. Therefore, only stochastic effects are taken into account when risks are considered. For these effects, too, dose limits are of an order for which risks can only be calculated by extrapolation. In Germany, dose limits for individuals of the population in the surroundings of nuclear plants, at the most unfavorable places, amount to 0.3 mSv per annum. With the help of radioecological models, exposure to both external and internal radiation has been taken into account. The dose limit lies within the mean variation range of exposure to natural radiation. The rules of minimization and optimization, moreover, require that attempts are to be made to remain below the dose limits by means of „sensible" measures. Regulators in Germany have rigorously applied this minimization principle with the result that individual exposures to radiation are far below the limit of 0.3 mSV per year in Germany.

The development of standards for workers' protection or environmental protection with regard to other noxae followed similar lines. Standards for workers' protection were developed for different chemical substances by the MAK-commission (commission on maximal admissible concentrations of toxic substances for occupational exposure). This commission, however, defines limits only for non-carcinogenic substances, for which a threshold dose exists. For carcinogenic substances, technical reference levels of concentrations (Technische-Richt-Konzentrationen, TRK) are established by the Committee on Dangerous Substances (Ausschuß für Gefahrstoffe, AGS), which are published as technical rules on dangerous substances (Technische Regeln für Gefahrstoffe, TRGS) by the German Minister for Labor and Social Affairs (Bundesminister für Arbeit und Soziales). Similar to ionizing radiation, dose-response relationships without threshold doses are assumed for carcinogenic substances. The minimization rule applies here as well.

The establishment of protection standards for asbestos, for instance, followed the same general principles. As with ionizing radiation, the principles of risk extrapolation and risk assessment were applied with regard to the impact of asbestos fibers at the workplace and in the environment. The committee of the German states on emission protection (Länderausschuß für Immissionsschutz, LAI), a body of the conference of the state ministers for the environment (Umweltministerkonferenz), suggested a value of 10^{-3} for the life-time risk caused by the total of carcinogenic air substances including asbestos. This risk would be of the same order as the radiation risk for annual exposure, provided, however – and this may well be excluded – that the persons concerned are exposed to 0.3 mSv per annum during their entire lives.

The establishment of environmental standards for substances for which environmental damage is the target criterion and not human health follows a different procedure. For example, chlorofluorocarbons damage the ozone layer and sulfur dioxide affects vegetation and forests. Knowledge about the causal mechanisms of these agents are often only very fragmentary; for an assessment of environmental damages, further research is needed. Standards for substances that can damage the environment are set in a complex decision process, in which economic, ecological, and social factors are balanced against each other.

With respect to all noxious agents, processes of distribution and transportation in the environment are of great importance. The spreading mechanisms for airborne harmful substances, for instance, are similar for many noxae, including radioactive substances. They can be described with the help of basic ecological models, while the parameters for a quantitative description of the processes may be quite specific for each substance under investigation. As to the damaging impact, the following rough classification can be made:
- changes due to physical-chemical environmental damages;
- ecological damages in plants and animals;
- damages done to human health.

Although such impacts are different from those of ionizing radiation, even for substances connected with the third category, the most important effects, namely carcinogenesis and mutagenesis, are often the same. Therefore, some analogies can be established. With regard to chemical substances that demonstrate carcinogenic activities, the trend is also towards establishing limits on the basis of the acceptable risk concept taking into account costs and benefits.

Chapter 4: *Individual, Social and Governmental Risk Assessment*

Risks are ubiquitous in human life. Individuals cognitively and affectedly perceive and assess risks so that they are able to react. These individual processes are influenced, and often even determined by the dynamics of public opinion and reference group judgments. If certain risks are not acceptable according to scientific insights or if they are not accepted by influential groups of the population, governmental action is required. This may take either the form of legal regulations or of administrative measures. Chapter 4 focuses on the social forces and institutional means that create the momentum for agencies and political bodies to engage in risk assessment and standard setting. In doing so, they are confronted with the scientific and normative principles of risk assessment and with a variety of public demands ranging from risk tolerance to risk prohibition. Chapter 4 provides an empirical basis for how these conflicts about risk management evolve over time and how they are resolved, if at all.

The first section of the chapter describes the emergence of individual judgments about and attitudes towards risks. As lay persons normally have insufficient knowledge with respect to environmental risks (e.g. the radiation risk), they have to rely for their judgments on robust cognitive rules of thumb, on general value orientations and on the opinion of relevant social reference groups. Thus, risks are often drastically overestimated or underestimated depending on intuitive heuristics and processes of social amplification or attenuation of risk.

The second section is an analysis of social processes and conflicts in risk management. Its starting point is the collective awareness of the ambiguous nature of risks. This ambiguity gave rise to a new field of social conflicts, for which no approved traditional forms of conflict resolution were available. Substantially amplified by the influence of modern mass media, this situation led to a polarization of public opinion, mobilized highly active protest groups, and contributed to a credibility crisis of many risk management institutions. A rational establishment of standards as to different risks was and is, of course, not facilitated by such conditions. From this analysis of the situation, new forms of a social handling of technological risks appear to be necessary.

Among them, legal regulations play a prominent role. This is demonstrated in the third section by using the case study of German radiation protection law. The section outlines the history of regulation with respect to radiation, it explains the basic principles of German environmental law (in particular. the principles of causality, precaution and cooperation), and points out the influence of regulations both of international and European law. The different German acts and ordinances are characterized and case studies presented in which these legal regulations have been applied. Among the most prominent regulations are the German radiation protection ordinance, the German licensing procedure for nuclear facilities, the German law on preventive radiation protection and the German ordinance on X-rays. The section concludes that the legal requirements of protection meet most of the scientific as well as rational conditions for both effective health protection and efficient balancing of risks and benefits. The greater a potential damage, the greater the demands that are to be made on a reduction of its probability. Apparently lacking is a convincing procedural and institutional arrangement for implementing these legal provisions.

The last section of the fourth chapter deals, therefore, with organizational and institutional conditions for policy formulation, promulgation and implementation. The process of lawmaking in the field of environmental regulation is described and analyzed.

The section centers on the German ministerial administration responsible for managing environmental problems. These institutions often produce and execute decisions under strong external pressure. In addition, the process of decision making is often confusing to the public, is rather restrictive on public involvement, and emphasizes legal over substantive considerations. The analysis of these problems leads to some suggestions for a more efficient environmental policy. All in all; the existing reality of individual, social and governmental risk handling is complex and multi-directional. Public perception and institutional handling of risks are not without practical effectiveness and political wisdom, and yet they show great deficiencies.

Chapter 5: *Formal Criteria and Instruments for Improving the Decision Process*

Environmental standards establish an acceptable risk level, i.e., a level that is collectively regarded as tolerable considering the respective benefit. The establishment of these standards is based on scientific knowledge about the impacts of harmful substances on human beings and on the environment, but in addition it depends on a social balancing between risk-induced costs and costs of risk avoidance. Such a balancing process cannot be reduced to a simple cost-benefit-analysis. First, many impacts of environmental pollution are not commensurable with monetary units. Second, distribution of impacts among different groups and individuals is at least as important as are aggregate effects of costs and benefits to society as a whole. Third, symbolic connotations associated with nature and technology need to be incorporated since they also matter in people's perceptions. Hence, a judgment about acceptability of risks always depends on a balance among a wide variety of assessment criteria (need for multiattribute decision analysis).

In order to reach a sensible balance in establishing standards, a three-step process is required:
- establishing objectives serving both the protection of life, health and environment and allowing for a rational allocation of social resources;
- studying the possible outcomes of implementating these objectives;
- considering social costs or damages, including opportunity costs and benefits which will arise when any of the available options are not further pursued.

Rationality in establishing environmental standards does not extend to the definition or deduction of objectives for society, but to the choice of means and strategies with the help of which these objectives can be realized. Protection objectives thus have to be introduced into the process of finding standards through existing legal regulations or political processes. Decision sciences may, however, examine these objectives for coherence and internal consistency.

Once the objectives are established, the second step is to ascertain the impacts of different environmental noxae on the operationalized criteria. This task is to be fulfilled and can be fulfilled by collecting the relevant scientific knowledge. Although opinions about possible impacts differ within the scientific community, these differences in opinions are by no means arbitrary but methodologically and rationally well-founded. An inquiry aiming at a distinction between well-founded and unfounded statements in the field of science can only be carried out by the scientific system itself and not by external groups.

On the basis of scientific assessments of the impacts of different environmental noxae and the costs of their limitation, the desired degree of risk reduction and the respective costs have to be determined and then weighed against each other. Several formal procedures are suitable for this purpose:
- *risk-risk comparisons,* which consider a risk to be sufficiently reduced when it does not significantly increase the total risk of a chosen life-form, i.e., when it remains below the thresholds of already accepted risks in everyday life;
- *the cost-effectiveness procedure,* according to which a fixed budget for risk reduction measures is distributed in such a way as to minimize the total risk;
- *the cost-benefit analysis,* in which costs and benefits of the respective risk reduction are at first transformed into one homogeneous unit (usually monetary values) and are then compared in order to determine the risk reduction with the greatest net benefit to be expected;
- *decision analytic procedures,* which systematically relate utility losses due to environmental standards to utility gains, according to the preferences of the decision maker;
- *precautionary procedures,* according to which a risk is to be reduced as far as permitted by the state of technology, or only as far as this is still reasonable (ALARA: as low as reasonably achievable).

All these procedures have their advantages and disadvantages, which are discussed in detail in chapter 5. With respect to the meta-criteria „normative validity“ and „practicability“, the decision analytic approach is recommended. Since such an approach is time-consuming, risk-risk comparisons are also regarded suitable as a screening device as well as a tool for assessing the acceptability of the dimension „risk“ within the multidimensional decision framework. The precautionary principle is applicable where noxious impacts are still too poorly known to allow for a weighing of benefits and costs.

The decision analytic approach has the advantage that it structures the process of reaching a decision according to rational criteria and creates transparency to the public at the same time. It does not provide an algorithm for automatically producing correct decisions, it only defines a logical framework for action, which systematically and sensibly links factual knowledge and value-based preferences. The individual steps of this analysis are described for decisions of individual and collective decision making.

Chapter 6: *Institutional and Organizational Perspectives*

The analyses of the preceding chapters are integrated in chapter 6 in an institutional proposal. This proposal includes four crucial elements: first, the contribution of scientific expert knowledge; second, a decision making process by which one of the two major pathways of either pursuing the precautionary principle (minimizing impacts) or the causal principle (balancing costs and benefits) are selected and implemented; third, the pursuit of rational procedures for weighing costs and benefits; and fourth, the compatibility with democratic rulemaking and due processes. Integrating these four elements requires an analytical separation between scientific reason, which is capable of creating the necessary expert knowledge, and a procedure of defining objectives and evaluating decision options, which ensures that due respect is given to public opinion according to rational rules. Moreover, due to the complexities of and time requirements for setting standards, the proposal must also address the problem of priorities in establishing new standards, in view of the multitude of potential environmental impairments. Setting priorities necessarily means setting posteriorities, too. These are often hard to convey to public opinion; yet renouncing an action deliberately and reasonably may often better contribute to an efficient environment protection than some undistinguished activism.

Another problem is the establishment of exposure standards for such cases, where knowledge about a dose-response relation is still insufficient for an effect-related assessment. In this case, precautionary procedures are warranted. Such procedures enjoy high popularity but are usually very costly. In public discussion the optimization rule of weighing costs and benefits associated with the causal principle is frequently understood as a minimization rule in line with the precautionary principle. Public outrage is often observed when economic considerations enter the deliberations about standard setting. The working group, however, recommends the causal route of optimization if impacts are well-known and the remaining uncertainty has been assessed in a reasonably accurate manner. Using the precautionary principle for all environmental agents would stalemate economic development and, in its ultimate consequence, paralyze all innovative actions. This holds true both for the establishment of standards and for their implementation. Rational procedures of weighing advantages and disadvantages are necessary to maintain the social capability of collective action.

There are two major possibilities to pursue a process of weighing costs and benefits. One can enhance the existing instruments of the market economy so that environmental damages are internalized in the price system, or one can rely on legal regulations that provide binding instructions for all citizens and institutions. Chapter 6 describes the principles of market instruments in detail and outlines their advantages and disadvantages. It turns out that these instruments promise higher cost-effectiveness in environmental policy, yet in most cases they are not effective on their own but only if accompanying regulations are present. Limitations of the market mechanism also occur when the loss of health and life is at stake. Furthermore, an optimal allocation does not automatically lead to a desirable distribution of effects. Distributional aspects of fairness and social justice need to be included through democratic processes such as political negotiations, mediation or social discourse. Market mechanisms are also not very flexible if, due to new insights about the activity of harmful substances, it becomes necessary to alter existing limits (extension or reduction of pollution rights). To sum up, market instruments turn out to be of eminent importance as a supplement to environmental policy, yet they need completion and limitation by legal regulations.

Both regulations of civil law and of public law are necessary instruments to ensure effectiveness and enforcement of environmental regulation. The analysis shows, however, that public law enables a more flexible, more rapid, and hence more effective environmental protection than civil law under the existing constitutional framework in Germany.

Environmental decision making is always confronted with cognitive uncertainty and evaluative dissension. To cope with both problems, legitimacy of decisions relies mainly on the rationality of procedure rather than on the outcome of the decision. Required is an environmental discourse that takes the „state of the art in scientific knowledge“ into consideration, reflects the plurality of preference attitudes, and observes rational procedures prescribed by the normative framework of decision analysis. Such a discourse is not going to happen by itself. The present „muddling through“ approach in German policy making does not meet the criteria of scientific accuracy, rational reasoning and plural value input[1]. The study group suggests, therefore, the establishment of an independent environmental institution. This institution should be assigned the tasks to prioritize the areas in which new standards are necessary, choose the appropriate assessment principle (causal or precautionary principle), clarify the actual and operational preconditions for a rational weighing of costs and benefits, suggest a quantification for the environmental standard and introduce the quantified environmental standard as a recommendation into the government decision process.

The study groups have given such an institution the title: „Environmental Council“. The Environmental Council should be structured in analogy to the existing German Science Council. The Science Council has two decision making bodies, called chambers. The first chamber consists predominantly of scientists, the second chamber of administrators of political institutions. Suggestions for policies are first articulated in the scientific chamber and then passed over for discussion or modification to the second chamber. After the two chambers had ample opportunity to debate over the proposals, members of both chambers convene to formulate a common proposal that will be submitted to the usual legislative or administrative process. The study group would like a similar body to initiate standard setting in Germany. This Council should be complemented by expert commissions, a part of whose members will also be members of the scientific chamber. Among the (permanent) expert commissions should be the already existing commissions in the environmental field (like the commission on radiation protection) and new commissions still to be founded (e.g., a commission on hazardous chemicals). The Environment Council is supposed to emphasize, above all, the contributions of science and rational reasoning; therefore no general obligation to reach an agreement between both chambers is demanded. The administrative chamber, consisting of representatives of the German federal and state governments, is supposed to participate in the formulation of recommendations, yet it should not be mandated that compromises between scientific and political reason must be reached.

The Environmental Council cannot and should not handle all proposals for environmental policies. It is not supposed to interfere with the normal tasks of federal and state environmental authorities. Only if environmental regulation is likely to raise considerable conflict or if precedents in environmental policies are expected, should it become active. Routine environmental regulations should be left with the respective state and federal institutions. To facilitate their job, the study group has made several procedural suggestions for establishing routine environmental standards. These procedures should be applied by the environment authorities, if necessary in accordance with the Environment Council.

Potential objections against establishing a new institution such as the Environmental Council are also discussed and refuted by the study group. The chapter closes with a discussion on the necessity of risk communication as a necessary link between the work of the Environmental Council and the public at large. Public involvement is seen as a necessary element in any regulatory decision making process as a means to establish transparency about the procedure of regulation and —even more importantly— to assure the inclusion of public values in the multiattribute balancing procedure. The Environmental Council is seen as a natural institution to foster and promote risk communication.

ANNOTATION

1 The federal environmental minsitry and the state ministries are in charge of environmental regulation in Germany. In contrast to the US-EPA, the German environmental agency in Berlin is part of the envrionmental ministry and not an independent political institution. Environmental standards are often prepared by private organizations such as the association of German engineers. These associations are mandated by law to suggest standards. The process of environmental decision making is described in more detail in Chapter 4.

INTRODUCTION

The main goal of this study is to investigate tasks and functions of environmental standards, to show their scientific foundation and to describe, from the point of view of rational reasoning, the problems of decision processes in cases of conflicting objectives. To do this, the study attempts to integrate a variety of contributions from different scientific disciplines in order to present the public with proposals for handling one particular aspect of the so-called „ecological crisis", i.e. proposals to develop, promulgate and legitimate environmental standards based on the best available knowledge and rational procedures of incorporating social values and public interests.

In this study, radiation protection is taken as an example against which theoretical considerations and practical proposals are tested. Radiation protection is a good example, since exposure to radiation is easily measured and its impact on the organism can be clearly proved. There exists both exposure to natural radiation, differing within a considerable range according to the geographical location, and exposure to artificial radiation. This might, of course, also serve as an argument against the choice of radiation protection as the selected case study, since our knowledge in the field of noxious chemical agents, for instance, is substantially more limited. The question is whether the results of the study can be transferred to other environmental hazards.

Better knowledge about exposure to and impact of ionizing radiation, however, helps to undertake an analysis focused on principles, procedures and justifications of environmental standards, and thus paradigmatically shows the direction which should be taken in an evaluation of other noxious agents as well. Environmental standards have to regulate routine operations. This requires a distinction between the routine operation and an accidental release of hazardous substances. Accident situations cannot be regulated by environmental standards; the avoidance of accidents is a matter of the licensing procedure. Therefore, this study does not deal with accidental risks related to nuclear reactors or other technologies that release radiactive particles. The arguments presented in this volume are meant to reflect a procedure which allows for a reasonable establishment of environmental standards, independently of the rate of malfunctions or accidents theoretically calculated or known from experience.

The phenomena subsumed under the notion of „ecological crisis" have brought about a far-reaching mistrust of scientific expert knowledge with regard to these problems, although scientists were the first to warn against problems of exposure to radiation, of chlorofluorocarbons and of carbon dioxide (greenhouse effect). Science has often triggered an increased environmental concern. Nevertheless, it became clear in many situations that much of the knowledge and abilities required for finding a solution to actual problems is not within the range of science, because the data base is not precise enough, or not at all conclusive, or it cannot even be collected in the first place. Moreover, science is often blamed for deficiencies beyond its sphere of responsibility, for instance in the sphere of political decisions.

And yet, in spite of incompleteness, ambiguity and preliminary status of the present knowledge, environmental problems must be assessed according to procedures based on generalizable insights, if and insofar as the means of coping with a problem demand rational decision making. Coping with overcomplexities by irrational inspirations is not equivalent, let alone superior to rational procedures because the irrational remains cognitively arbitrary and cannot, therefore, claim general validity. It can, thus, be reinforced in the long term by tutelage and manipulation only. There is no alternative to practical and rational reasoning informed by science, even if this should entail the most difficult and time-consuming way of proceeding.

How difficult and time-consuming this way may be becomes clear if one realizes that 200 years ago the Enlightenment and the efforts of putting Kant's practical philosophy into practice provided an attempt to translate a rule of reason into the decision of public affairs. At that time, this rational way of proceeding was founded on the idea that the goal had to be an optimum realization of individual human dignity within the framework of the socially available options. The basis for this was the appeal to individual responsibility towards the duty expressed in the categorical imperative. The modern state was supposed to realize this objective.

History has taken another way. It was nationalism which became the „great" problem solver, and this ideology occupied Europeans till the end of World War II. Only now, after Europeans have suffered from the consequences of this blind alley, does the old idea of the Enlightenment reappear again. This allows us to hope that it has only been temporarily buried and will penetrate public thinking against still existing irrationalisms.

One of those irrationalisms is the fact that in the environmental debate people often orient themselves by ideologies and myths rather than by practical reason informed by science. As important myths may be to provide orientation and psychological security, they cannot replace practical reason as the main source of collective decision making. Practical reason alone corresponds with a scientific perception of reality and a responsible management of limited resources and opportunities.

Therefore, the concept of rationality as a principle guideline to cope with environmental problems has played the role of a regulative idea in this study. Due to the complexity of problems and the inevitable mix of facts and values in designing environmental policies, however, the postulate of rationality cannot be restricted to findings of the natural sciences. Environmental standards reflect normative rules and regulations for society; they are, hence, cultural instruments established by society. Regulations result from an evaluation process, which includes, on one hand, the insights gained by the natural sciences and, on the other hand, the values and goals of the society for which these regulations have validity. Such evaluation processes will, in the course of this study, prove to be multidimensional assessments based on facts and rational procedures of incorporating values in the decision making process. Determining and justifying environmental standards demands a logically consistent integration of the scientifically revealed dose-effect-relationships and the social preferences within a complex multidimensional decision framework. This integration demands the cooperation of scientists, decision analysts, regulators, and representatives of the public.
Considered „rational" are those arguments for which „good reasons" can be given in procedures understandable to everybody. This principle of procedural rationality knocks the bottom out of every strategy of securing beliefs by immunization and dogmatization. At the same time, the procedural rules themselves are not arbitrary. Procedural rules are the rules of rational argumentation, in accordance with the developed standard of international scientific and ethical discourse. Although procedures of rational argumentation will, in many cases, not provide unambiguous results, there is, after all, no way but to insist on the non-ambiguity of the notion of „argumentation". Such a form of rational argumentation open and accessible to all publics meets the requirements of a public discourse in a democratic society.

1

NECESSITY AND FUNCTION OF ENVIRONMENTAL STANDARDS

Achieving environmental standards is a complex process, in which scientific insights of different disciplines, normative beliefs, and general social conditions interact with each other. Environmental standards are – like other standards – social rules operating within the framework of political orders. But specific to them is their reference to a human dealing with nature, so that, for any informed way of acting, the findings of the natural sciences provide an indispensable foundation for the establishment of environmental standards. General social conditions are, in the first place, relevant for an understanding of the „environment" of humans, which is by no means a culturally variant quantity (1.1).

In view of the numerous factors affecting the formation of environmental standards, those conditions are of crucial importance which, if met, will imply a process of establishing environmental standards that may be regarded as rational. Obviously, the way rationality is understood in this context cannot simply be determined by the scientific way of gaining knowledge, yet not without it either. A more detailed analysis shows that a determination of acceptable risk limits in view of the expected collective benefit is the core function of environmental standards. If, however, we try to describe this risk problem more precisely, we are faced with considerable difficulties, which are, above all, due to the „multiattributivity" of the decisions to be made (1.2).

Environmental standards are established in the framework and according to the rules of the respective political system. In democratic societies, this poses, in particular, the question about the relation between factual acceptance and the determination of acceptability according to rational criteria, on one hand, and according to the democratic procedures of formation of opinion and will, on the other hand. In almost all problematic fields, a situation is likely to occur where a minority not ready to take a certain risk will be imposed this very risk. To settle the question whether such risk impositions are justifiable, those rules which appeal to the pragmatic consistency of the actors are of special importance. Moreover, the particular problems of deciding under risk make certain demands on institutions and procedures of decision making (1.3).

1.1 ENVIRONMENT AND ENVIRONMENTAL STANDARDS

1.1.1 The Notion of Environment

The notions of „environment" and „environmental protection" are used very vaguely in public debate. For some people, environment is *that* nature which surrounds them, yet what is „natural" in nature remains undefined. For others, environment is everything, natural or artificial, which affects them. For a third group, each animal and each plant has an environment as well, which is worth protecting. If there is to be a discussion about „environmental standards" and the different roles of politics and science in finding and justifying them, this vagueness cannot remain undiscussed, since standards must include reliable, well-defined quantities, should they rightly claim to serve a well-defined purpose.

Unlike in common use, science mostly uses the notion of *„environment"* in a more restricted sense, which in biology and behavioral research goes back to von Uexküll.[1] But von Uexküll linked the notion to a reductionist anthropology, so that it cannot consistently be used in the context of the problem of „environmental standards". According to von Uexküll, human beings have a specific environment, like the tick or the fly, a sign-world and an action-world („Merk- und Wirkwelt"), which, in its qualities and limitations, is specifically linked to them and which they can transcend as little as other creatures can transcend their respective environments. The description of a creature fitting in its environment can be understood negatively or positively. Seen negatively, the environment represents a fundamental, biologically determined limit due to fitting. Seen positively, fitting means an optimum functional relation between creature and environment.[2]

In both respects, this concept of environment cannot be applied to humans without modification. Phenomenology[3] and Philosophical Anthropology (Scheler 1928, Plessner 1928, Gehlen 1940, Hengstenberg 1957) made it clear in their criticism of von Uexküll that human beings transcend their specific environment. They do so, on the one hand, by exploring the environments of other creatures and by being able to a certain degree to put themselves in the place of others. This ability is a crucial argument in recent environmental debate, since, e.g., animal protection is based on it inasmuch as it exceeds a mere „human-egoistic" view; a creature strictly bound to its specific environment cannot care for the welfare of other creatures. At the same time, human beings can make their own environment the subject of their cognition and their preparatory planning of actions; otherwise, for instance, a protective and caring attitude towards the environment would not be possible. But also the positive interpretation of the concept of fitting must be questioned, since this would imply that the harm humans do to their environment had to be interpreted as just another specific aspect of their optimal fitting.

Any creature, as a center of its specific environment, struggles for its identity by differentiating itself from its environment, distinguishing itself from it, coping with it. The dog, the bird, the bee do this in their respective different ways, yet they are and will remain completely subject to the superiority of nature. Human beings, however, have evolved in such a way as to make nature and its forces serve their purposes to some extent. Their nature-transcending identity is characteristic of human beings, like flying is characteristic of a bird. Humans are the emancipated beings, which have been set free from nature to some degree by their more comprehensive knowledge of nature.[4]

The „Philosophical Anthropology" contrasts Uexküll's idea of an environment of humans with the concept of *„openness to the world"* („Weltoffenheit"). „Openness to the world" implies that it is characteristic of the human relation to the environment not to consist in a life-securing and at the same time constraining „fitting", but that human beings always have to create their „fitting" in interaction with their environment. Their environment is not simply given to humans as a precondition, it is given to them as the double task of coping with it and taking care for it: the environment poses problems to humans (to the point of menacing their lives), which they have to solve; yet they can survive only if they realize that they must not wipe out their environment but have to take care of it and preserve it. Their environment is at the same time *a threat to and a precondition for* a life in human dignity. Thus, human structural openness to the world results in a relation to the environment which is not simply one of fitting but one shaped by the tension of the diametrically opposed threat and precondition. A specifically human relation to the environment is reached only if humans succeed in achieving a life-securing balance between their defense against the threat by the environment and their dependence on the environment. Therefore, humans are not „fitted" into an environment but they have to produce the adequate fitting, which can neither consist in surrendering to the environment nor in subjection of the environment.

Producing that fitting should, however, not be misunderstood as freeing oneself from one_s environment. It is specific to humans that their openness to the world and their ties to the environment are permanently *interlinked*.[5] Since humans, the „cultural beings", remain „natural beings", they cannot simply dissociate from their natural environment; since they are „natural beings" only as „cultural beings", their unbreakable natural ties become their cultural task.

Although the aspect of producing is emphasized, it should not be ignored that there is, for different sectors of human agency, a different variability in establishing standards with respect to a chosen objective. The dose-response relation in radiation protection, for instance, shows a far lower variability with respect to objectives like preserving life, prolonging life, or health, than standards of film rating show with respect to the objective of protecting the young. Basically, this variability depends on the precision of the target determination and on our knowledge about causalities. The concept of „prolonging life", for instance, can be defined more precisely than the concept of „protecting the young". In those cases where we have to refer to explanations of the social sciences for establishing cause-effect relations, we have to take a broader range of variability into account, among other things, for reasons of philosophy of science. But even if – as in the case of radiation protection – a precise causal analysis is at least basically achievable, a certain variability of determining standards remains unavoidable because of the set targets. A closer look at target concepts, even at concepts like „health", reveals that their normative content results from optimizing several set targets as functions of the existing state of affairs. Accordingly, the exposure to radiation considered reasonable in medical checkups is different from the one considered reasonable in the context of nuclear facilities; similarly, a greater radiation dose is considered acceptable for occupational exposure than for the population in general.[6]

1.1.2 Special Position of Human Beings

Humans occupy a special position among living beings, due to their specific relation to the environment. Since this special position consists exactly in their interaction with their specific natural *surroundings*, an interaction which shapes culture and is determined by it, environment is more than natural surroundings. It is therefore recommendable to abandon a primarily biologically determined concept and to characterize the notion of „environment" in the sense of a specific „human environment". Air, water and soil, the natural systems for transportion of matter, are an important part of this environment, but only a part; human environment is more than nature. Understood in a broader sense, „environment" also comprises cooperation, communication, culture, traditions and institutions.

As a result of these considerations, the frequently found *„naturalistic"* understanding of environment cannot be sustained and has to be replaced by a *„culturalistic"* understanding.[7] The difference between the two concepts is directly reflected in their different understanding of status and function of „environmental standards". While in the naturalistic view, environmental standards derive their necessity from those circumstances according to which we are constrained to act (apodictic validity); in the culturalistic view, environmental standards are obligations of action, which societies impose on themselves if and insofar as they want to live together in some specified way (hypothetical validity). It should be noted that such standards are obligatory in both cases: yet in the first case, the obligation is given by the circumstances of action, in the second case by the ends of action. While in the naturalistic view standards cannot be argued against once they are found – natural phenomena can only be recognized or ignored – in the culturalistic view, environmental standards are open to discussion, they are even in need of being discussed: they are justified by their instrumentality, i.e. in relation to a socially lived or set purpose. The obligations arising from an understanding of nature as a task must, however, not be misunderstood as something *merely* set; nature as a „task" has rather to be interpreted in the sense of a non-arbitrary yet malleable configuration of requirements (according to the mentioned linkage between openness and ties). Within this configuration of requirements, differentiations are quite appropriate; for instance specific requirements are obligatory only to the degree that their fulfillment is a precondition for satisfying other desirable requirements („basic needs").

Naturalism becomes manifest where an environmental standard is thought to be deductible directly from morphological particularities of the dose-response relationship. This idea cannot be sustained, which is obvious from the mere fact that zero-exposure would have to be taken as a threshold dose for stochastic effects, for which no threshold dose can be determined.[8] But also for non-stochastic effects, for which a threshold dose can be ascertained, this threshold dose does by no means necessarily represent or at least determine the limiting value. Since often there is no threshold dose, or a threshold dose is not automatically a limiting value, it becomes clear that the limit is primarily a social restriction or construction. One reason for problems arising in connection with environmental standards is exactly the fact that threshold doses are not automatically limits.

Of course, resort to factual information must be determinative in establishing such restrictions of action. The action directive, however, never results as a conclusion from descriptions of facts („*naturalistic fallacy*“).[9] The naturalistic fallacy logically characterizes any naturalism.

Even in the culturalistic view, however, reality is not describable as the product of human agency. Instead, action always unfolds on the basis and within the framework of an existing reality, „*nature*“, which is real without any human intervention (concursus). This implies, however, that the perception of these circumstances of action, the conflict between adaptation to and resistance against them, depends on human attention and intention, on cognitive and operative attitudes in the social context. *What is interpreted as „nature“ is itself a cultural phenomenon.*[10]

1.1.3 Human Task

The environment of human beings, be it their natural or artificial surroundings, are both the result of the environment affecting humans and of human attitude affecting the environment. In this sense, each human being is the pragmatic point of reference of both the impacts directed towards him or her and those emanating from him or her. This is not to say that each human being is an autonomous, self-determined individual in full command of his or her environment. Even those who live unconscious of the determinants surrounding them have to structurally differentiate from their environment and affect it. It follows from this structural relation that nobody can choose whether to affect the environment or not. The subject matters of the human relation to the environment are largely at the disposal of the individual: profession, leisure, clothes, art; individual differences exist even with respect to such qualities as sensitivity to noise, and destructive or creative drive.

One person may attach a higher priority to the conservation of forests than another, some may like rapid highways for a less dangerous driving, while others prefer damp biotopes to runways. What is not optional, however, is the situation that humans, by emancipating themselves from nature, have created themselves surroundings which now affect them as environment, from which they are distinct entities as much as they depend on these surroundings for living and well-being. Being different from the environment is a precondition for taking actions that shape and arrange the environment according to an individual or collective will.

The ability to choose a lifestyle within a specific environment, to select desirable elements from the surroundings, to control to some extent the environmental influence on humans and to create new environment by human interventions and their manifestations of life, makes the distinction between „actors“ and „victims“ disappear for humans as a species. All people become „actors“ affecting the environment as long as they live, and at the same time they are „victims“ of the environment, due to their own existence and the existence of their fellow-beings.

In this context, H. Markl (Markl 1984) emphasized the fact that the relation between living beings and their environment cannot be conceived as a harmonic equilibrium in analogy to thermodynamics. On the contrary, each being is doomed which does not try to hold its own, confronting others in economic competition, i.e., propagating if ever possible. „And therefore, life necessarily creates a shortage in preconditions for life, exactly through the thus necessarily created shortage in products of life. Life is a process exhausting resources, since it has been an autonomous creative process from the beginning“ (Markl 1984, 10). According to Markl, it is humans that, for the first time, have the chance to interrupt the blind process of emergence and disappearance of species, by using their superiority due, above all, to their scientific-technological civilization, to free their ends of action from the natural compulsion to compete.

The present discussion about an anthropocentric versus a cosmocentric (physiocentric, ecocentric) world view turns out to be a pseudo-discussion, in view of the human relation to the environment. Humans are subject to the tension between *threat to life* and *precondition for life*.[11] Since humans must also hold their own against their environment and emancipate themselves from it, they cannot position themselves above the environments of all living beings as seemingly neutral observers and arbiters and, in a patronizing way, grant every living being the same right to live. One needs only to mention viruses or bacteria — that alone reveals such ideas to be absurd; for, even if humans conceded equal rights to those beings and thus perished themselves, the state of competition between different participants in the process of life would be restored all the same.

Humans can define their relation to the environment only according to their *specific* vital interests. This definition cannot be arbitrary because the environment is at the same time precondition for their vital interests. Humans always talk about the environment *as humans*; this does not imply, however, that they are allowed to subject the environment at will without running the risk to harm themselves.

According to the explained linkage of openness to the world and ties to the environment, anthropocentrism is therefore correctly understood only if it is „moderated“ both by the conditions arising from humans being part of nature and by the duties arising from humans transcending their environment. So humans cannot restrict themselves to meeting their specific vital interests, but they have also to employ their ability of recognizing and preserving their surrounding reality *as such*. It is, for instance, an intrinsic aspect of this cognition and preservation that certain distinctions result from „nature itself", like, e.g., the distinction of living systems according to their degree of differentiation. Without acknowledging this distinction, which humans make all the time, it could not be understood why, for instance, animal protection treats animals with subjective feelings different from those without such feelings. Nevertheless, the anthropocentric element shows here again, because the degree of differentiation does not mark an absolute border but merely a rule for human balancing considerations with respect to different values.

The structurally inevitable anthropocentrism should therefore not be confounded with an ultimately misanthropical human egotism. Nor is an altruistic relation to nature realizable, though. The call for „peace with nature"[12] goes unheard in situations where nature is threatening us; on the contrary, nature continues to produce new beings threatening human life (e.g. HIV-viruses causing AIDS). This demonstrates drastically that the category of „peace" borrowed from the field of human coexistence can at best be applied metaphorically to the relation between humans and the environment. Nature can neither be conceived as a peace partner nor as a war enemy.

This insight has an effect on legal policy, down to concrete questions like, e.g., laying down environment protection as a national objective. The German constitution establishes a federal legislative supremacy with respect to partial aspects of environment protection (e.g., Art. 74 Nr. 11a: protection against dangers resulting from the use of nuclear energy, or otherwise from ionizing radiation, and disposal of radioactive substances; Art. 74 Nr. 24: waste disposal, air pollution control and noise abatement). These legal provisions entrust the federal government with the legislative *right* in the mentioned sectors, they do not imply a *duty*. On the other hand, the constitutional court has repeatedly deduced from the constitution (Art. 2, Abs. 2, sentence 1: „Everybody has the right to life and freedom from injury") a governmental *duty* for protection against respective impairments: against dangers to body and life resulting from peaceful use of nuclear energy (BVerfGE 49, 89; 53, 30, 57ff), against air-traffic noise (BVerfGE 56, 54, 73 ff), against chemical air pollution (BVerfGE, NJW 1983, 2931f). According to the prevalent opinion, however, the quoted sentence from the second constitutional article does not grant an original right to environment protection (BVerfGE, NJW 1983, 2931 f; BVerwGE, NJW 1978, 554 ff). For it becomes clear in the cited decisions of the constitutional court that the government has a considerable scope for meeting its protection duty, i.e., about whether and how to meet it. Utilizing this scope, it must and can, in particular, weigh this protection duty against other constitutionally legitimized objectives like, e.g., national security, the safeguarding of power supply, the safeguarding of economic development etc. For this reason, the quoted sentence from the second constitutional article does not imply a general obligation for a comprehensive environment protection. Nor does such an obligation explicitly result from any other constitutional provision.

As the public has become increasingly aware of the endangering of the environment since the early seventies, a growing number of people in politics and jurisprudence have called with growing intensity for an integration of environment protection into the constitution, either as a constitutional right or as a national objective. The environment program of the then governing coalition of social democrats (SPD) and liberals (FDP) of October 14, 1971 states that it was intended to „examine how the citizen could be granted a legitimate claim to have those who cause heavy environmental damages sued by the state" (BT-Drs. VI/2710,9). In his policy statement of January 13, 1973, the late German prime minister Brandt said: „All people have an elementary right to an environment ensuring human dignity, a right, which should be laid down in the constitution." (BT-Prot. 7, 127 D). Political parties and environmental organizations made equivalent or similar proposals. There is also jurisprudential literature calling for an integration of environment protection into the constitution, mostly in the form of defining a national objective. The vast majority of jurists were skeptical, however.[13] They warned against establishing a constitutional right to environment protection, above all for the reason that such a right would create expectations it would never be able to meet.

In 1981, the German ministries of the interior and of justice convened an expert commission on „definition of national objectives/legislation tasks". In their report, published in 1983, the majority of the members of the commission suggested the introduction of environment protection into the constitution in the form of defining a national objective. Paragraph 1 of the constitutional article 20 should read: „The Federal Republic of Germany is a democratic and social confederation. *It protects and preserves culture and the natural foundations of human life*" (the text in italics denotes the suggested amendment of Art. 20 Abs. 1 GG). A respective bill submitted to the national parliament on the part of the German social democratic party was rejected by the Christian democrats (CDU) on January 16, 1986 (BT-Prot. 10, 14254 ff). In the same year, however, the then environment minister Wallmann (CDU) favored the adoption of environment protection in the constitution as a defined national objective. After the elections for the national parliament in 1987, the governing coalition of CDU, CSU and FDP also agreed on it. Prime

Minister Kohl confirmed this in his policy statement of March 18, 1987 (Bulletin No. 27 of March 19, 1987, 205, 212).

Meanwhile, respective bills have been introduced in the Lower House of the German Federal Parliament, the Bundestag, by the Upper House, the Bundesrat (BT-Drs. 11/885) and the parliamentary group of the social democratic party (BT-Drs. 11/10). According to the Bundesrat bill, the following Art. 20a should be inserted into the constitution after Art. 20 GG:

> Art. 20a (1): The natural foundations of *human* life (in italics by the authors) are protected by the state.
> (2) Federal and state governments regulate the details by law, weighing this objective against other objects of legal protection and government tasks.

The parliamentary group of the SPD suggested the following Art. 20a:

> Art. 20a: The natural foundations of life are protected by the state.

At first sight, one gets the impression that the Bundesrat starts from an anthropocentric concept of environment, the SPD, however, from a purely ecocentric one. On October 14, 1987, an expert hearing concerning both bills took place in the Bundestag committee on legal affairs.[14] The vast majority of experts then voted in favor of an anthropocentric concept of environment (should it come to an integration of environment protection into the constitution; the respective necessity was rejected by a majority).

This has to be agreed to. Yet the argument put forth most often, namely the proposition that only an anthropocentric concept of environment would match with Art. 1 Abs. 1 GG, which acknowledges human beings and their dignity as supreme values and charges the state with their protection, turns out to be not as convincing in every respect as is often assumed. It might very well suit human dignity to acknowledge that nature is worth protecting, even beyond its utility for humans, at least to the borders where nature becomes dangerous for human health and life.

It is not because of Art. 1 Abs.1 GG that the anthropocentrism of environment protection and hence of the concept of environment is inevitable; this fact is derived from far more basic contexts, it lies in the nature of things, it is not merely relevant for the legal order, by which, however, it has to be taken into consideration above all. The idea of legal protection of nature for its own sake is ultimately based on the idea of natural objects having legal capacity. In particular, it was the American jurist Christopher Stone (Stone 1972) who triggered a vivid discussion by putting forward this thesis, which found supporters in German scientific literature, too.[15]

According to him, non-human nature has interests as well, which can be translated into subjective rights and safeguarded on trust. Analogies to guardianship for minors or the mentally ill and to legal representation of juristic persons were established. If even completely insane persons were adjudged subjective rights, the latter could not be denied to non-human life.

These analogies are not suited to support the thesis. The „interests" of represented juristic persons are given as legally defined purposes, and the legal agent of minors or the mentally ill at least acts on behalf of his or her equals. But this idea of equality cannot be sustained far beyond human nature. How, for instance, could representable interests of rivers and lakes, of mountains and woods, of air and landscapes be formulated? Nobody else but humans could formulate them, and this only according to their understanding, which changes in space and time. Thus, anthropocentrism has rightly been called inescapable, and it has been warned against „re-mythologizing nature" (Hofmann 1988, 277f). Only in an anthropocentrically oriented environmental ethic can human responsibility for nature be justified in a way which stands to reason and can thus claim universal validity. A definition of environment detached from humans is therefore useless.

It has to be stressed, however, that anthropocentrism of environment protection must not at all mean, and actually is not supposed to mean in the constitutional amendment suggested by the Bundesrat, that nature is only supposed to be protected to the extent to which it is useful and utilizable for people living now. As natural foundations of human life, the foundations of the lives of future generations have to be protected and preserved, too. Anthropocentric environment protection is also posterity protection.

Laying down a national objective of environment protection in the constitution cannot aim at establishing non-human legal entities, thus throwing back legal development by centuries; instead, it is supposed to declare human responsibility for nature irrespective of its *direct* utility for human purposes.[16]

1.1.4 Objectives of Environmental Standards

By „environmental standards" we understand legal, administrative or private regulations (like, e.g., industrial standards), by means of which vague legal concepts referring to the environment (like „harmful impact", „precaution", „due diligence", „acknowledged technical rules") are translated into definite prescriptions, mandatory regulations or permits, by operationalization and standardization of measurable quantities.[17]

If we entertain a notion of „environment" referring to human beings as point of reference, environmental standards have to be rules and regulations for humans, of importance in regulating the dynamic process of life-preserving exchange between humans and human environment. „Environmental standards" have to be rules and regulations put up as „traffic signs" in this dynamic process, in order to regulate it in a way which is held beneficial. This function of environmental standards also explains why the environmental debate cannot take place on the level of humans as a species alone. Here, also the evolutionary interpretation categories are found insufficient. For a consciously acting human individual, the survival of the species is no genuinely perceived motivation of action.[18]

In addition to the above considerations, the environmental problem has also to be seen as subject to the tension between *individual and society*. The culturalistic interpretation implies that the concept of environmental standards includes a social dimension. In social systems, with increasing population density, but beginning in the individual family already, different environmental spheres grow together: the environment of one person collides with that of another person, they mutually modify each other. This collision of environmental spheres of individuals leads to aggregations and contradictions. Aggregations with respect to common efforts for solutions to common problems, contradictions with respect to impairments of one person's environmental sphere by that of another person. It is my sewage which causes problems to others, it is my use of energy which possibly requires the construction of power plants.

At the same time, however, if there were no others, my sewage would cause them no problems, if there were no energy use on the part of others, I would not have to think about the consequences of my energy use. It is exactly the understanding of this reciprocity of relations between the environmental spheres of individuals, on which the necessity for a rational discourse about environmental standards is based.

Since procedures of managing social conflict are institutionalized by a legal system of norms, the legal structures based hereupon are the framework for any coping with conflicts. So, for instance, the solution to environmental problems in the past and the future cannot be understood apart from the legal and political context. If the establishment of standards in countries of the second and third world is regarded for a comparison, it turns out that economic difficulties in certain countries require a view of environmental problems differing from that of countries with high economic efficiency. Thus it becomes clear that environmental problems are differently assessed in countries with different national objectives. Therefore the concept of what environment is and how environental standards should be established cannot be conceived without considering the interplay of its being interpreted by human beings and its being elaborated by the society in the respective prevalent national idea. Environmental standards are a cultural phenomenon since they are achieved according to the desired interpretation as result of a multidimensional decision process. Hence their establishment cannot be expected to be valid globally and for all times.

Since the interpretation of the human relation to the environment depends on the cultural self-image, the task of shaping the environment can always merely be a partial and sectorial one. This means that it cannot be the task of environmental standards to provide a comprehensive standardization of environmental matters. In the foreseeable future, the necessary information to base such a project on and the necessary procedures to legitimize it will not be at human avail. The attempt to radically redesign the environment would therefore fail because of unbridgeable dissensions and conflicts. So we have to proceed from an environment which by and large corresponds with what humans have known for some generations and are responsible for. With respect to this environment, environmental standards operate rather conservatively: they are supposed to protect the environment from damage, but to reverse already existing damages, which are recognized as such, only to a limited extent.

So, what is behind the formulation of environmental standards is not the wish to realize a forward-oriented or backward-oriented utopia. The aims of finding and justifying environmental standards are more modest. An example: notwithstanding everybody's right to imagine a reforestation of the Lüneburg Heath (a human-made environment in the North of Germany now regarded as as an area worthy of preservation) utopian model, the task of environmental standards is to protect the heath as it is now. This landscape has been developing for some generations. It cannot be the duty of scientific expert knowledge and of legislation based thereupon to indicate a target state of environmental matters, which, after its discovery or invention, could provide the standards for human agency adequate to the environment, to everybody and over a long time. The status quo (obviously deplorable states of affairs not included) has a strong normative force in environmental policy.

This partial and conservative character is expressed above all in the fact that *environmental standards* are formulated as *prohibitions*, i.e., are supposed to mainly protect human beings and their environment against noxious influences. In contrast to this, it can at best implicitly be a task of formulating environmental standards to bring about a target state of the environment. An environmental policy aiming at a comprehensive and positive new formation of the environment (including the putative restoration of a former state) must be rejected as totalitarian in its tendency and must be criticized with regard to the underlying ideologies.

From the restricted human possibilities to realize concrete environmental utopias it follows also that the need for some regulation by environmental standards has to be proved. The aim of environmental standards is not the wish for a change as such, but the intention to avert a definite damage. In addition, with a far-reaching environmental policy and a population with a strong desire for protection, the aspect of the density of the net of regulations gains importance. As is well known, the sum of all regulations may operate counterproductively if it exceeds a certain threshold. For instance, too dense a net of traffic regulations increases insecurity. It is foreseeable that in future, the problem of the density of the net of regulations in German environmental policy will be of growing importance, and a specific optimization problem will arise.

Aggregating the individual answers to the question about the meaning of life, the solutions of conflicts in the procedures of living together in a society and a state, and the elaboration of their contents in both the idea of the state and its normative stipulations, provide the essential dynamics of the formation of human political consciousness. Therefore, the definition of what environment is and what it should look like is a result of the formation of political consciousness, too. Environmental standards are, among other things, tools in the political process, meant to help find these definitions and realize them.

The political process, the social communication about establishing environmental standards as a result of an assessment of conflicting targets, must, however, not evolve in an arbitrary way, because of its far-reaching consequences. The following conditions constrain political arbitrariness, in addition to general standards of rationality:

(a) scientifically ascertained facts, as, for instance, dose response relations in different technological procedures. It is the task of science to study by means of „cause-effect research" the impacts on nature and human beings of certain chemical agents and radiations, and moreover of techniques and social behavior, which are determined by natural laws. Apart from the causal analysis, science must specify in particular the „dose-response relationships". Research on effects of human interventions and the specification of the dose-response relationship are strictly scientific tasks. In this respect, science and politics are two separate fields. If science discovers that some effect means a particular danger – achieved by systematic proceeding or in the course of basic scientific research or through specific early-warning sensors – it has thereby settled one important task which

initiates any establishing of environmental standards: an early realization of possible later impacts. History proves that it was often science which gave an early warning, e.g. in radiation protection, with respect to the impact of chlorofluorocarbons on the stratospheric ozone layer, with regard to the greenhouse effect linked to the carbon dioxide problem, and as to impacts of genetic engineering.[19]

(b) the inclusion of every dimension of different impacts and different set targets into the assessment process. Environmental standards have effects on the environment itself, on other environmental standards, on the economy, on the job market, on climate, on well-being etc. This should be emphasized especially for the reason that one particularity of the political process is its proceeding „sectorially“ and not „holistically“. There is thus a tendency to discuss individual political „issues“ without taking their effects on other sectors into consideration. An energy system may be discussed from an environmental point of view, while questions of economic efficiency or availability are of secondary importance. Perhaps this tendency exists, among other things, for the reason that one tactical trick of political debate is to succeed by focusing the discussion on only those points and limiting it to those points for which one's own objectives entail superior consequences.

(c) the necessity to achieve a rational judgment with respect to these consequences and chance-risk relations, which, on their one respective dimension, can be ordered according to dose-response relationships. The judgment which has to be brought about in the political process has to be rational in order to be conveyable in the long term, so that it can be accepted by society as a whole.

1.2 RATIONAL ESTABLISHMENT AND EVALUATION OF ENVIRONMENTAL STANDARDS

1.2.1 Environmental Standards and Acting under Risk

If environmental standards are established, it is assumed that the determination of limits does not only depend on the probability of an expected damage but also on the benefit which the potentially harmed person gains by taking the risk. Therefore higher limits are accepted for occupational activities than for the population in general, higher limits for handling accidents than for routine releases, and, to name quite a trivial example, higher limits for the strain on health caused by medical diagnoses than for that caused by buying shoes (interdiction of X-ray machines which were common in shoe-shops in the fifties). These examples are evidence for the general argument that it is the purpose of environmental standards to indicate limits for certain cases of acting under risk, *namely in a way that the question about the limiting value always leads to the counterquestion about what risks the persons concerned are willing to trade off in exchange for some benefit.*

So the more detailed analysis shows the problem of environmental standards to be part of a more comprehensive problem, namely that of specifying *criteria of rationality for acting under uncertainty.* The notion of risk is the core concept in the discussion about environmental standards. Accordingly, environmental standards are conventional restrictions of acting under uncertainty. This means that they are mandatory regulations, interdictions or permits for actions whose consequences will, with a certain probability, exceed a certain extent of damage or will remain below it. Such restrictions of action can be reconstructed as demands to do something or refrain from doing something. Hence the question arises whether such utterances are amenable at all to an examination according to criteria of rationality, i.e., whether reflections about expediency are amenable to rational criticism in a similar way as reasons about truth.[20]

Examining the rationality of prescriptive utterances, however, follows other lines than examining descriptive utterances. The truth of an assertion is proved in another way than the expediency of a norm. It is therefore appropriate to introduce a terminological differentiation. While the claims of validity of assertions and other constative utterances are substantiated, the claims of validity of norms and other regulative statements are *justified.* Accordingly, *reasoning discourses* and *justifying discourses* are to be distinguished.[21]

So we have to ask which criteria environmental standards have to meet to qualify as rationally justified. For a more precise treatment of this question and its possible answers, we have to explain how *actions can be reconstructed* so as to become possible matters of reflections on rationality.

One possible approach characterizes actions as observances of norms.[22] An advantage of this view is the fact that norms are on their part verbal actions, which share with some other verbal actions (e.g., assertions) the property of being possible conclusions of a possibly correct deduction. In other words: norms are amenable to argumentation (which is not true for many other verbal actions like congratulating, swearing, baptizing etc.). Deductions are characterized by three elements – this is agreed upon from Aristotle to modern logic. They consist of a conclusion, a class of premises (possibly empty) and a combination of premises and conclusions according to a deduction mode the correctness of which is assessed with the instruments of logic. Deductions which lead to a norm (or another regulative verbal action) as a conclusion are called *„practical deductions"*. For practical deductions to be correct, the following conditions hold, apart from a correct deduction mode:

- At least one class of premises on which the conclusion essentially depends must contain a prescription demand. If the class of all premises is empty, this is the case of correctness according to an imperative logic.
- Prescriptive conclusions may substantially depend on descriptive premises (the reverse is not true).

Problems arise if there are several options of action, which are different with respect to the probability of an eventuation and the extent of damage. In order to achieve a complete argumentation in those cases, a risk degree has to be determined, i.e., probability and extent of damage have to be quantified or at least comparatively ordered. This shows that, in situations where a decision between different alternatives of action is required, with actions being characterized by a certain probability of consequences occurring and a certain extent of damage, the amenability to argumentation of a certain norm can at least be ensured in cases when the risk degree as the product of probability of an adverse effect and the extent of damage is successfully determined. In the following, the underlying concept of risk that was outlined above will be called *„rational risk concept"*

This risk concept turns out to be problematic in several respects when it is applied to the issue of environmental standards. Its sufficiently exact determination regarding both preferences and the notion of probability depends on a set of conditions, which cannot right away be assumed to be fulfilled in definite situations (1.2.2). One particularly difficult problem with far-reaching consequences here is the assumption that different damages can be ordered by means of a smaller/bigger-relation, i.e. in one dimension.[23]

1.2.2 Problems of the Notion of Risk[24]

If environmental standards are regarded as justified only if they can be reconstructed in the outlined sense as conclusions from risk deductions, it turns out that a risk comparison is part of the evaluation process. This fact, however, remains implicit in the usual formulation of environmental standards, as premises are only implicitly included in most of our assertions andnorms. Should a justification be required, those premises have to be made explicit. So we have to ask under what conditions risk comparisons of the described kind are possible at all and justified. A more detailed analysis shows that some *rationality conditions* have to be observed for the idea of risk comparison to remain within the range of rational decision strategies. Some of these conditions of rationality will be listed below, in the form of distinctions to be observed.

(a) A distinction has to be made between *facing a risk* and *taking a risk* (Rescher 1983, 5–9). The terms „given risk“ and „chosen risk“ will be used here. Basically, only chosen risks are amenable to risk comparison. There are risks which are not really causally connected to our actions and options of action. They are mostly associated with natural processes, like earthquakes, which is why they are often called „natural risks“. Yet also social processes, like revolutions, are among the given risks. However simple this distinction seems to be, a subsumption under one of these two categories is sometimes nevertheless very difficult. We often think that given risks are causally connected with our actions, because we do not know which events follow from what type of actions. So, for instance, revolutions are regarded as consequences of our clinging to private property, or inundations are regarded as consequences of corrections of river-beds.
This shows that scientific progress continuously extends our possibilities to make rational risk comparisons. Tackling risks by means of risk-risk comparisons is therefore an example of the fact that the knowledge provided by science can contribute to the solution of problems of choice under uncertainty.

(b) A distinction has to be made between *direct* and *inverted* risks (Rescher 1983, 10). A direct risk exists when the non-occurence of the undesired state representing the damage could fairly be expected. For instance, somebody buying a lottery ticket may fairly expect to win. As people mostly lose their stakes, which is an undesired state, it is correct to say: „Buying a lottery ticket is a risk“. In the case of an inverted risk, the undesired state is one the non-occurence of which was fairly not to be expected. The finder of a valid lottery ticket has not performed a purposeful action, hence cannot expect to win. Therefore, it is not correct to say: „Finding a lottery ticket is a risk“. We meet examples of inverted talking about risks every day. An accident in a power plant is a direct risk of power generation. The increased birth rate due to a power failure, however, is not a direct consequence of the action of building power plants. If somebody said power generation was a risk since it could lead to an increased birth rate in case of a power failure, this would be an inverted use of the notion of risk. In general: consequences of action which are regarded as possible damages have to be directly connected to the intentions of the action which, with a certain probability, leads to these damages. Even according to this distinction, problems of subsumption may arise, however.

(c) In *standard* cases of risks, we can identify an agent responsible for the action, and someone affected by the consequences of the action (Rescher uses the term ‘maleficiary’, in analogy to ‘beneficiary’).[25] Agent and maleficiary can be individuals or groups, the agent can himself be maleficiary or member of the group of maleficiaries. In *non-standard* cases of risks, „circumstances“, „powers“, „nature“, „evolution“, „society“ etc. are regarded as agents and maleficiaries. With respect to those entities, only metaphorical language is suitable for talking about preferences and expectations of probabilities for an undesirable event.

(d) A distinction has to be made between *routine risk* and *accidental risk.* While the probability of a routine risk can, in principle, be ascertained as an objective probability value on the basis of a large class of events, the probability of an accident largely relies on a subjective expected value. This distinction is complicated, however, by the fact that for many consequences of accidents it is possible to recur to limits or probabilities which are based on knowledge about routine risks. In many cases, however, (above all in rarely occurring events) a risk assessment for accident scenarios is problematic. In contrast to that, there are no basic difficulties with risks of malfunctions, i.e. risks which are calculable from risks of routine cases. When technical risks are regarded, malfunctions in some facility are events which can be foreseen and can thus be constructively taken into account, whereas accidents are events which cannot be constructively taken into account, since, perhaps, the possibility of the event cannot be foreseen, or since the event occurs so rarely that taking it into account would be beyond the scope of any reasonable balancing of costs and benefits. It has also to be mentioned in this context that, in the case of nuclear accident scenarios, a very small probability would have to be multiplied with a very high extent of damage. As our probabilistic assumptions for accidents are heavily prone to errors, the possible range of errors threatens to become so big that a risk consideration can no longer have any normative validity.

(e) The most difficult distinction as to subsumption is the distinction between *risk perception* and *risk itself* (Rescher 1983, 7). The distinction as such is no difficulty; everybody knows that the fact that somebody does not perceive a risk in a certain situation does not imply that he or she is not exposed to any risk. There are probably no risks which are imperceptible *on principle*, but a risk is not necessarily of the same kind as agents or maleficiaries perceive it. And yet it is also trivial that only the actually perceived risk is one which can be reckoned with.

Another discussion in this context deals with the so-called alternative risk concepts, all of them developed in critical confrontation with the standard concept supported here. These alternative suggestions proceed from the individual[26] or collective[27] perception of danger. A basic distinction has to be made, however, between danger and risk. An individual's danger of life, for instance, varies from situation to situation, whereas the risk of death (which is relevant, e.g., for the calculation of a life-insurance premium) varies only little and only weakly as a function of danger. So, in the terminology used here, we distinguish between danger and risk in a way that „danger“ denotes the possible damage in the specific situation of action, whereas „risk“ refers to a type of situation, the extension of which (e.g. risk of mountaineering versus risk of leisure activities) is determined by the purpose of the risk analysis.

Moreover, we have to distinguish between actual and perceived danger in analogy to actual and perceived risk. In the current discussion of the risk problem, above all in the media and the popular social sciences, a far-reaching confusion of perceived danger with actual risk can be observed. Irrespective of the necessary conceptual differentiation, it cannot be denied, however, that the perception of danger is important for a policy of coping with risks. This cannot merely consist in translating subjective perception of danger into political measures, though, for the reason alone that such perception differs from individual to individual or from group to group.[28]

The rational risk concept used here requires a more precise description of its defining elements, namely the concept of probability and that of (negative) preference.

Probability[29]

Basically, we have to ask whether probability is understood as limes of relative frequency, a concept which is exemplified by a die as random number generator. Opposed to this idea is the concept of probability as *subjectively* expected value, as it is determined, for example, by the readiness of a gambler to bet on something. In principle, it seems reasonable to assume a dualism of both concepts, that is to say, each concept of probability is erroneous in some applications.

The idea of a limes of relative frequency underlies our usual statistics. It is just these statistics, however, which show that the idea of statistical probability has its limitations. Statistical probabilistic statements are always statements about a great number of events. Therefore, such a concept of probability is in many cases insufficient for a reconstruction of problems of human agency, since there are problems which cannot be solved by any, however perfect, knowledge about relative frequencies. This is manifest whenever probabilistic statements are to be made about a singular event (i.e. about one particular instance). There is no approximation value of relative frequencies, if such a value is conceived as being objectively linked to a definite event. Everybody knows that the next toss of a die has an uncertain outcome, even if we have perfect statistics at our avail.

The risk of singular actions, which are not to be seen as elements of a class of actions, can thus be expressed only by means of the concept of subjective (personal) probability. The formulation of a subjective expected value may then very well include a class of information, e.g. about previous events. Yet in addition, the expected value depends on our optimistic or pessimistic attitude, for instance.

In many cases it will also be plausible to require that a subjective expected value should replace relative frequency. This is often required implicitly, e.g. when patients are informed by means of mortality and morbidity figures or when air passengers are reassured by means of accident probabilities. It must not be overlooked, however, that here a special claim of rationality is applied, according to which a subjective expected value *should* replace objective probability. This claim of rationality implies that the concept of objective probability has to be preferred whenever a respective statement is available. Otherwise, a subjective expected value has to be used.

Undesirable Outcomes (Extent of Damage)[30]

Actions under risk are those for which, with a certain probability of an event, we run the risk of experiencing, apart from the end of action, consequences which we consider to be undesirable. This „risk" is a side aspect, we never choose the risk (except when the risk is seen as the benefit such as engaging in a high-risk sports activity), we choose a certain action which may have undesirable consequences. That is what risk is all about. For a rational risk comparison we have to refer to negative preferences. In doing so, we have to assume preference orders for individuals, which have to meet a set of formal conditions. In particular, it is obviously insufficient to take only individual preferences into account for actions with far-reaching consequences. And yet we have to start from individual preferences, since there is no possibility to dictate „objective" preference orders with respect to all individuals.

This is not to deny that general preferences can be formulated – starting from basic needs which can be assumed for all people – with the help of which even concrete preferences can be criticized. These general preferences then serve as general criteria for concrete preferences, yet they do not render superfluous a subjective establishment of concrete preferences and of their evaluation interdependence.

If both probability of an eventuation and concrete preference, which underlie the rational risk concept, are based on subjective estimates, the question arises how a risk comparison claiming a generally binding character is possible at all. This question shows that there is a not at all trivial normative element in any risk comparison. After all, the risk comparison is carried out with the (in most cases implicit) aim that, following an assessment, everybody should be ready to accept certain resulting damages, or that you rightly may expect others to accept certain resulting damages, namely those with lower risk degrees. In the present political debate, we experience that this is by no means thought to be self-evident. So the discussion about the rationality of the risk concept finally leads to the question which risks can people be reasonably expected to take at all, and how this can be justified.

1.2.3 The Multiattribute Nature of Undesirability

Environmental standards as rules and regulations for society, as cultural phenomena established by society, arise as the result of an evaluation process and not as a consequence of some scientifically determinable limit. Evaluation processes are multiattribute (multidimensional) decisions. In a multiattribute decision process, heterogeneous, not directly comparable attributes have to be coupled in an overall decision. With respect to each individual attribute, it is usually possible here to arrive at a clear statement. This is due to the fact that a smaller/bigger-relation can be established among quantities of one dimension, whereas this is not directly possible among quantities of different dimensions.

This non-comparability of different dimensions, i.e. of different target concepts, has often been described metaphorically in public debate by saying that apples cannot be compared with oranges. To illustrate it by an example: in one activity I may run the risk of losing my eyesight, another activity may lead to a financial loss. Here is a more realistic example: Power plants can be ordered according to their output of sulfur dioxide, or according to the area they cover, or according to their economic efficiency, or according to their radiological damaging potential, or according to the availability of fuel. These are different attributes (dimensions) which allow of unambiguous evaluations (hence orderings) based on technical know-how and natural laws. The decision process now requires that a correlation among these different dimensions is reached, i.e. that the SO_2 output is weighed against radiological harmfulness or economic efficiency.

Even environmental standards, though limited to curb environmental danage, have to be attained as results of such a multiattribute decision process, with science and technology providing the important data for some of the respective dimensions: e.g., how many tons of SO_2 will still be released per delivered kilowatt-hour of electricity if modern technology is used; or what is the radiation risk per kilowatt-hour of nuclear energy.

In the evaluation process itself, these dimensions are then linked, or, mathematically speaking, a function is defined with as many variables as there are independent dimensions underlying the evaluation process. This link can be reconstructed in a way as if a person individually assessed risks and opportunities and then combined these assessments according to the end of his or her action to an overall decision. It is the subjectively expected utility which is decisive for individual decisions, with the concept of utility being understood here in a broad sense including also non-material ends, such as an increase in perceived fairness or an enhancement of an aesthetic quality.

This overall decision of linking several independent dimensions often have to be made under conditions of uncertainty. In that situation, caution plays a major role, and it causes people to try to postpone decisions until they have acquired more certainty. In practice, in those frequent situations where decisions have to be made under conditions of uncertainty, this leads to the attitude of „risk spreading", i.e. to that mediating choice among conflicting targets and risks which attempts not to endanger the whole but to improve the odds of surviving in the case of the most severe damage. This is a fundamental practice of may insurance companies in their dealing with low-probability, high-consequence risks.[31]

The requirements for risk regulations and the conditional requirement to set quantitative environmental standards cannot be defined irrespective of the order of magnitude of probability and extent of damage. Thus, there is a pragmatical aspect in the question of how to cope with the remaining uncertainties. Characteristic spatial and temporal dimensions of natural systems and processes are typically of little importance for human beings if they are either much smaller or much bigger than measures used for a wide range of human activities. So it is usually irrelevant for people whether a process takes place in one thousandth or one hundredth of a second, although the latter process is ten times slower than the former. If the question is, however, whether an event takes the time of ten or a hundred days, or even one or ten years, using again the ratio of ten, the difference is probably of special relevance. If still longer time-scales are regarded, for instance the difference between ten thousand and one hundred thousand years, this difference begins to fade away and becomes irrelevant. Similar considerations hold for spatial differences, which are found relevant only if they are comparable to those quantities familiar and accessible to normal human activities.

These deliberations are of importance in establishing environmental standards. If potential damage is successfully kept so low that only very little influence on human beings and their life expectancies has to be taken into account, a considerable fuzziness in artiiculatng these environmental standards and a luke-warm commitment to controlling their observance is acceptable. This is, for instance, the case with extremely low level exposure to radiation. Only if the damage is suspected to reach substantial impacts and to be comparable with negative impacts ot other normal human activities, has society a need for a rigid standard which is then to be controlled with high institutional commitment.

1.2.4 Problems of Generalizability

In the political process, there is an interplay of all those social subsystems which develop their own ideas about the correct values of environmental standards according to their own interpretations, and these ideas conflict with each other. The political process may perhaps not succeed in resolving these conflicts, because contradicting opinions are advocated in too uncompromising a way, or because the normative beliefs of one social group seem so „irrational“ to others that they are no longer intelligible to them. There are many historical examples of societies which could not cope with that task and perished. It is just for that reason that a more profound discussion about the concept of rationality of environmental standards is important, since only such a discussion can eventually lead to socially acceptable procedures of handling conflicts.

Yet it must be reckoned with the fact that rationality has to be fought for, first of all because rationality rather complicates the business of those debating in a political arena. But there are another two phenomena making rationality harder to achieve. First, there is „resonance“ in the process of social communication (Luhmann 1986). Here, „resonance“ describes the – unexpected – overdriving of certain effects through social feedback or the – unexpected – amplification or attenuation of side effects (Kasperson et al. 1988). Second, there is the already mentioned discrepancy between individual or collective risk perception and actual risk. This phenomenon may shift individual attention and evaluation patterns to highly amplified risks so that it becomes harder or even impossible to rationally link the same evaluation standards to other forms of damage. After all, complex evaluation and decision procedures are more prone to a distorted perception of reality, due to interests and prejudices. This will require a critical look at the social construction of risks based on cultural myths, political ideologies, or social convictions (Thompson et al. 1992). The mere existence of different cultural and social rationales for perceiving and evaluating risks in society is a compelling reason to develop an overarching framework of rationality that can be justified on normatve grounds and legitimized in a pluralistic society. A detailed suggestion for such a procedure will be desribed in chapters 5 and 6, this section here is meant to lay the philosophical foundations for such an attenpt and to provide the justification for the need of such an attempt.

These few lines of arguments may already suffice to show that the formulation of a rational risk concept may indeed provide an instrument for problem-solving, but the mere existence of this instrument does not ensure the resolution of actual conflicts. This is of particular relevance when certain risk concepts are legally codified by means of environmental standards and will hence be enforced by sanctions, if need be, on persons who do not agree with the premises of this rational risk management strategy.

At first sight, the situation is structurally similar to that of compulsory military service, the necessity of which is justified on the basis of a risk comparison and certain norms. In some democratic legal systems, the problem has been solved by the introduction of an alternative service. This strategy of problem solving is obviously faced with quantitative limitations (the number of men engaged in military service must not fall below a certain number held necessary for defense), and moreover it is often not applicable at all. This is the case when there is no choice involved for the individual recruit.

The example of military draft systems served only as an illustration of the general problem that environmental standards may face. If, for instance, certain limits are defined for pollution of air and drinking water, they may result from a risk comparison between the risk of damage due to an avoidance of air pollution (for instance less electricity available) and the emission in the case of a restriction of pollution which is held economically sound. Yet, those members of society who do not agree with the substantial premises of this risk comparison cannot resort to alternative air or alternative water. And it should be added that possibilities, for example for minorities, to resort to different strategies that help them shape their desirable environment decrease with increasing global universalism in the development of technological change, lifestyles and economic structures. This universalism holds also true for environmental standards, which are increasingly established at the international level.

1.3 NORMATIVE IMPLICATIONS AND ENVIRONMENTAL POLICY

1.3.1 Factual and Normative Validity in a Democracy

Now we have to ask about the quality of environmental standards that would be sufficient to claim general acceptance, i.e., we have to investigate their *acceptability*. Only if acceptability can be inferred, are deliberations possible about the conditions of enforcing their observance by means of sanctions. What are the necessary conditions for morally justifying environmental standards so that they can become legal norms? In Germany, several private organizations have the legal right, sometimes even the obligation, to prepare or suggest standards such as industrial norms or pollution control standards. From a juristic point of view, these sugestions have merely the status of private recommendations (another example are the so called DIN-regulations), yet they also claim normative validity. These recommendations, which do not entail sanctions when they are disregarded, need to be „correct“ or „legitimate“ in the eyes of the affected persons. Only then can they possibly be accepted as precursors of legal regulations.

These considerations are based on the fundamental distinction of *factual* validity (acceptance) and *normative* validity (acceptability). Incidentally, this distinction holds both for descriptive and for prescriptive forms of rationality. If it is disregarded, this will eventually always result in a relativism of the whole validity concept. As to the issue of „environmental standards“, this relativism is especially relevant in its form of *sociologism*. By sociologism we mean all those approaches according to which the validity of a regulative statement follows from factual consents on the part of individuals (which might be specified with different precision). The political process which corresponds with sociologism is the introduction of laws via plebiscite. It should be kept in mind that all classic democratic structures rely on a more or less pronounced distrust of purely plebiscitary legislation. This distrust can be institutionalized by separation of powers, representative system, two-chamber system, relative autonomy of regional authorities etc. Whatever the details of the formulation of such constitutional constructions, they all implicitly distinguish between factual and normative validity by not leaving decisions to plebiscite alone. With reference to the environmental debate, this means that actual acceptance is an insufficient judgment criterion.

With regard to environmental standards, all concepts recurring to the actual acceptance behavior of individuals face an additional problem. Sociological studies about attitudes of factual risk acceptance show that people and social institutions are rather selective in allocating attention to different risks and use „rules of thumb“ to infer the severity of risk.[1] Planning one's own actions according to a combination of probability and preference obviously leads very quickly to overcomplex demands on the human capacity to discern the various consequences of each decision option and to evaluate their probabilities and desirabilities., In most cases, strategies of reducing the complexity are chosen, which are not in accordance with a rational risk framework.

Establishing the *acceptability* of a risky action means ascertaining the required acceptance on the basis of investigating and testing the action under risk according to rational criteria. In short: acceptability is a normative concept. But this normativity does not yet imply that all individuals ceteris paribus have to behave equally. one needs only to think of the subjectivity of expected values and preference orders.

The issue of acceptability raises the question about the justification of norms, hence of universal demands (demands on everybody). Such problems of justification are the subject of the philosophical discipline of *„ethics"*, whereas the actually propagated or executed imperatives form a *„system of morality"*. Description and explanation of morals are the tasks of sociological research. Development and justifcation of ethical norms are domains of philosophical reasoning.

To clarify the specific tasks of ethics, it has to be pointed out that the validity of norms is often self-evident and by no means always of ethical relevance. This holds true for the wide range of legal norms, but also for technical norms or norms of professional and trade associations. Ethical norms are at issue only in the case of demands involving conflicts. By conflict, we mean a situation in which two agents (individuals or groups) strive for incompatible objectives. Two ends are incompatible if it is impossible to realize them simultaneously.. Resolutions of conflicts are possible in various ways, for example (to name two extremes) by liquidation of opponents or by applying the means of a reasoning discourse (justification discourse). Ethics is the art of (and the education in) justification discourse.

Acting under uncertainty does not imply permanent conflict. Risks often concern only the acting individual itself, or what amounts to the same thing, the social consequences of an individual action are not evident, for lack of knowledge or ability. In those cases, however, where others are expected to bear the potential consequences of one's own action, acceptability and hence an ethical problem are at issue.

Only if acceptance and acceptability are distinguished can the state be conceded the right to legally impose risks on its citizens. The state has to be controlled then with respect to the legitimacy of its procedures, not with respect to a consensus about the outcomes that will or can be reached by applying these procedures.

1.3.2 Individual Reasonableness of Environmental Law

Before turning to the ethical problems of generalizability, we must first tackle the question about how to justify the binding character of legally codified environmental standards. Any existing legislation is already proof that the ethical question of generalizability has been answered, though without removing the difficulties connected with the problem of the multiattribute character of these decisions. This situation, however, is not unusual as such, since in many cases legislation cannot wait until every theoretical problem has been solved. Nevertheless, the problem of the legally binding character of environmental standards is particularly dramatic, because there is a greater probability of subjective „deviation" from a majority opinion and far fewer possibilities of extended minority protection compared to other policy arenas. It is therefore not surprising that within the field of environmental problems some groups very quickly tend to claim a kind of right to environmental resistance. As to such a right to resist, two fundamentally different levels have strictly to be distinguished here: the so-called classic right to resist, founded in political philosophy and ethics as a right above positive law, a „suprapositive right", and the right to resist laid down in Art. 20 Abs. 4 GG of the German constitution.

The so-called classic right to resist may, if need be, even be directed against the constitutional order, if, *in the view of the resistant person,* this order is incompatible with the 'suprapositive' right. As a 'suprapositive' right, this so-called classic right to resist is out of reach of any legal formulation and recognition. In an extreme case, when there is no other possibility to prevent the destruction of a polity which ensures human dignity, the individual can only seize those rights, „which are for ever his, as permanent and incorruptible as are the stars upon the crystal round" (Schiller[2]; and Locke: „And where the body of the people, or a single man, is deprived of their right, or is under the exercise of a power without right, and have no appeal on earth, there they have a liberty to appeal to heaven, whenever they judge the cause of sufficient moment"[3]). Along with the dignity of this 'suprapositive' right to resist, and partly entailing this dignity, goes the risk of failure, since this right opposes the legal order. In Germany today, a person who opposes peaceful use of nuclear energy, relying on this 'suprapositive' right to resist, runs a risk of failure, since peaceful use of nuclear energy is constitutionally approved, and the approval is confirmed by the courts.

It is absurd to justify such resistance with reference to the German constitution (Art. 20 Abs. 4 GG).[4] The right to resist formulated in the respective article expresses a right to defense of the state in case of national emergency, meant to defend the constitutional order and not to oppose it on the basis of any 'suprapositive' right whatsoever. The article formulates a right to resist a removal of the social order which is described in its preceding paragraphs 1 - 3, if there is no other remedy. This is equivalent to the set of norms called the free democratic basis order, which is not even amenable to constitutional amendments (Art. 79 Abs. 3 GG).

Actions against a peaceful use of nuclear energy, e.g., are beyond the range of protection of Art. 20 Abs. 4 GG, even if the agents understand nuclear energy as an infringement on their constitutional right to protection of life and health (Art. 2 Abs. 2 GG). The use of nuclear energy, no matter if it should interfere with constitutional rights or even violate them, means in no way a removal of the constitutional order. An active fighting against the use of nuclear energy, whose constitutionality is indeed confirmed by the courts, in particular by the German constitutional court, can in no way be founded on the mentioned constitutional article (Art. 20 Abs. 4 GG), instead it is itself to some extent reason enough for exercising the right to resist.

Mayer-Tasch's observations confound in an unacceptable way the so-called classic 'suprapositive' right to resist with the constitutionally guaranteed right to resist (Art. 20 Abs.4 GG). He does not start from the *objective* removal of the constitutional order as a precondition for the right to resist (as is laid down in Art. 20 Abs. 4 GG), but proceeds from a „scenario looked at from the angle of the disobedient citizen's sense of law" (Mayer-Tasch 1979, 44), in a way that this disobedient citizen sees the use of nuclear energy as an infringement on the constitutional right to life and freedom from injuries (Art. 2 Abs. 2 GG). Such a subjective assessment, however, could at best be basis for the 'suprapositive' right to resist, which is beyond positive law. A constitution acknowledging this right to resist would abolish itself.

Moreover, it is rather doubtful whether and in how far Mayer-Tasch is serious about his founding a right to resist in environmental matters on Art. 20 Abs. 4 GG. For, after focusing on the sense of law of the disobedient citizen he writes: „Whether he was actually right will be finally decided by the criminal judge to whom he explains his reliance on state of emergency or on right to resist. I would like to refrain from prophecies here, but point to the simple fact instead that the right has always to be fought for, and that it is continuously changing" (Mayer-Tasch 1979, 44).

Mayer-Tasch's way of dealing with Art. 20 Abs. 4 GG illustrates how justified most teachers of constitutional law are to be skeptical about this provision, which was inserted into the constitution in 1968, as a kind of tranquilizer, together with the so-called emergency constitution, and which, after all, makes the impossible attempt to constitutionally domesticate revolution.[5] The provision invites misunderstanding and abuse. „There is danger that the right to resist encourages the wrong and paralyzes the legitimate. It could once turn out as a tragic irony that the right to resist was established at the very moment when, for the first time in the history of the German Federal Republic, efficacious groups are trying to break the state monopoly of power by taking matters in their own hands - among other things in reliance on a right to resist“ (Isensee 1969, 103).

1.3.3 Pragmatic Consistency

The reflections on the legally binding character of environmental standards show that the latter cannot be considered irrespective of the social acceptance of the underlying risk assessments. Nevertheless, one has to ask how far the legislator must take social acceptance into account (as has been said, many people cannot be expected to have a reflected and consistent attitude towards risks), and based on what insights social acceptance can be demanded.

These reflections on a risk concept, which often draw on subjective probability and on subjective preference, unfortunately seem to lead to a normative skepticism which holds that nothing generalizable can be said about risk acceptability. This conclusion, however, ensues only if too far-reaching requirements are made on the general validity of ethical considerations. For ethical reflections cannot lead to insights like: 'everybody has to take the risk of flying by airplanes'. To put it in general terms: due to the subjectivity of expectations and preferences, no *categorical* imperatives can be established. It is possible, though, to establish *hypothetical* imperatives; one could, for instance, demand that everybody who takes the risk of going by car has also to take the risk of going by plane.

To illustrate the above, it should be mentioned that ethical imperatives are hypothetical imperatives (almost without exception). So, to be precise, we do not demand that nobody should kill someone else, but that nobody should kill someone else unless in extreme situations of self-defense. Kant's categorical imperative is merely a criterion for testing hypothetical imperatives, and thus cannot be hypothetical itself for reasons of methodology. For instance, the categorical imperative does not justify the categorical imperative that everybody must keep his or her promises, but merely the hypothetical imperative that someone who wants to sustain the institution of promising and to profit from it should keep his or her promises.

Against this background of fundamental considerations, we suggest here the following hypothetical principle of acceptability, which may be called principle of *pragmatic consistency* (comparable to the mentioned imperative of promising) for a solution of the normative problem of acting under risk: *if someone, by choosing a certain way of life has chosen to take certain risks, an option of action bearing the same or less risk may be assumed as acceptable to this person as long as the overall risk burden is not increased.*[6]

In order to justify this principle, traditional stategies of dealing with ethical problems would have to be considered in great detail. Instead, we will merely point at the fact that the principle can both be deduced from Kant's categorical imperative (Kant 1785) and from Hare's principle of generalizability (Hare 1952). According to Kant, maxims are justified if at any time they may serve as (generally binding) norms. Here, „may serve as norms" means „may be realizable norms which do not entail pragmatic contradictions". Hare's principle of generalizability is weaker, since it requires the consistency demanded by Kant only for the individual's „proper rationality". According to that principle, someone who considers something to be good in a certain situation shall consider it to be good in any situation which is equivalent in the relevant aspects. The above-mentioned principle of pragmatic consistency can also be deduced from this weaker idea of generalizability, since the latter implies that an action risk which one is prepared to take may be expected to be taken by others as well.

The principle of pragmatic consistency opposes the intuition of someone who presupposes a complete *incommensurability* of risks for any field. According to the thesis of incommensurability, two different actions under risk, or different types of actions under risk are incomparable. One could say, for instance, that the preparedness to take a risk in skiing had „nothing to do" with a respective preparedness required for living next to a nuclear power plant.[7]

The discussion of this subject is complicated by the fact that reasons for both the thesis of commensurability and the thesis of incommensurability can only be given in a circular way. Asserting that actions under risk are incomparable implicitly means that there can be no claim for consistency. On the other hand, maintaining that there is a principle of pragmatic consistency implicitly means that actions under risk are comparable. It cannot be argued, however, that the *unequal* is incomparable as such, since comparability depends exactly on the existence of a principle of comparison. For people are once again forced to compare the unequal, due to individual and social needs. Whether this is possible depends on whether a principle of comparison can be established. The best-known such principle is that of relating the unequal to a respective monetary value and to compare monetary values. Many taxation schemes, for example, compare in such a manner incomes achieved in very different ways. A person might very well argue that his or her different kinds of income were achieved in unequal ways (e.g. by wage labor, rental earnings and free coal), that this was subjectively experienced very differently, and that therefore these kinds of income must neither be compared nor be taxed according to uniform standards. This objection is not compelling, though, because inequality does not yet mean incomparability. Comparability is established by the principle of expressing everything in monetary equivalents.

This exemplary consideration shows that the comparability of events, in particular of actions, is no genuine attribute of such events. To make the unequal comparable is rather a matter of forming conventions according to an expedient rationality. Whether the unequal is made comparable depends on whether accomplishing a comparison is a means to a generally accepted end. If society has decided in favor of comparability, the different subjective perceptions of the compared items are of no importance.

Of course, one can still insist that any two actions are unequal. For no two actions are congruent in their consequences with respect to all their attributes. A general principle of incommensurability, however, can be disproved by the following reductio ad absurdum: a general incommensurability would imply that there are no social rules whatsoever (of a pre-institutional or institutional kind). Legal systems presume, among other things, that unequal actions are commensurable. This is obvious for those legal systems, where arguments of precedents play an important role. In terms of incommensurability, a defendant could, of course, maintain that his or her action was unequal to the precedent. The judge, however, will, if need be, take the legal rule as a basis for comparison and draw a comparison. A general thesis of incommensurability would hence lead to total anarchism.

Seen positively, this consideration shows that people have always relied on a socially established spectrum of comparability of actions (which are unequal „as such"). The rules of comparison are a constitutive part of the shared life-form. Accordingly, the principle of consistency has to be related to these actual foundations. Hence, the individual citizen can be expected to adhere to those presuppositions which are the foundations of the shared way of life until further notice, i.e. until an explicit contradiction occurs which requires substantiation. Among those presuppositions is the basic acknowledgement of rules of comparability.

Part of the foundations of a social way of life are also regional and temporal limitations of the range of commensurability. This may be explained by the fact that the process of differentiation which led to modern societies has split persons into a multitude of functions. According to this differentiation into functions, people experience a relative subdivision of life as a whole into different sectors of life (job, family, leisure etc.). Nevertheless, the unity of a person as a constant point of reference in this change from one sector to another, and in the interaction of sectors among each other can still be experienced. The diversity of functions is represented, for instance, in the different status which certain maxims of action may have in different sectors. A different subjective binding character could, for example, be attached to the maxims of truthfulness or fairness in the different sectors of job, family and leisure. Accordingly, risk acceptance could as well be differently developed in different sectors of life. It might be for this reason that risk comparisons between leisure activities (skiing) and certain basic needs (choice of abode) are seen by many people as transgressions of thresholds of incommensurability. What is expounded here for sectorial limitations will possibly hold for temporal limitations as well. A preparedness to take risks in a certain period of life cannot be taken as point of reference for any later risk assessment (think of youthful follies, e.g.). Risk-risk comparisons of relevance in a rational risk assessment may therefore rely on a principle of consistency restricted by a sectorial principle of commensurability.

The principle of consistency is applicable in particular when *natural risks* are considered. Nature as such does not expect us to take certain risks, so that we do not have to compare our acceptance behavior with that of a pseudoagent called „nature". A risk acceptance is expressed, however, in the respective behavior of social groups taking a natural risk, or is implied by our actions (or by our refraining from action) by which we consider certain natural risks to be acceptable for those groups. With respect to radiation protection, for example, we consider it acceptable that people live in the German regions of Upper Franconia or the Southern Black Forest (regions with a terrestrial radiation above average) and do not demand compulsory evacuation or a removal allowance. Thus, people in Upper Franconia are expected on their own part and on the part of others to take the risk of living there. Hence, people in general must also accept such a risk imposition.

The demand of consistency is sometimes objected to by saying it forced acceptance of risks just for the reason that fairly high risks have already to be accepted, and that this was preposterous. This argumentation assumes an accumulation of risks, as it is carried out in risk balancing (e.g. in security studies). Risk comparison, however, is concerned with a direct comparison of options of action (which is still sensible if one of the options is already realized). We thus do not assume that the risk in question adds to already accepted risks, but that we have an alternative choice. Considering a balance and considering options are sensible operations for their respective purposes, yet they have to be clearly distinguished.

The principle of pragmatic consistency starts from the assumption that the agents' preparedness to take risks can be definitely concluded from their actions. This will be possible in many cases. Yet complications are conceivable, which might very well be of considerable real import. In many cases it cannot be presupposed that the chosen way of life is due to a person's decision. Apart from social constraints, psychological constraints have to be taken into account. Moreover, there is a question as to whether the term „decision“ is appropriate in cases of ways of life completely determined by traditions, and hence, whether it is appropriate to bind someone to that choice. Finally, someone may be at the point of radically changing his or her ways, so that former options of action do not permit any conclusion about his or her present assessments. These objections show that even actual actions do not provide a reliable basis for actual assessments of acceptance.

Nevertheless, it is by no means rare that a situation may occur where we must make our decisions on the basis of consents which have not actually been expressed. Just think of forms of delegation, representation and anticipation of consents, as they are common in the fields of education and of acting as a doctor or a lawyer. In such cases, a *fictitious discourse* stands for an actual discourse with the persons concerned. A fictitious discourse is imaginarily held with the persons concerned, on the assumption that their acting would be that of perfect individuals (i.e., individuals who are always correctly informed and always act correctly). Additional indicators play a role, which allow for taking into account options and preferences of the persons concerned. So, an argument which has not really been uttered but which nevertheless bears „eloquent witness“ can be part of a fictitious discourse.

Forms of fictitious discourse have to be chosen as a kind of discourse in any situation where a real discourse with all persons affected by an action cannot be imagined. In such cases we have to think out the arguments of a certain person concerned who opposes the action. Besides the usual rules of argumentation, the principle of rational risk comparison and the principle of pragmatic consistency are of decisive importance here.

1.3.4 Social Decision Processes[8]

In the course of a search for rational decision procedures, both cognitive-scientific and institutional-social procedures are considered. It must be admitted, however, that the distinction between cognitive and social procedures is imprecise, since cognitive procedures are in a certain sense social procedures as well. Nevertheless, an „ideally" developed procedure has to be distinguished from its social and institutional implementation; the mathematical procedure of root extraction, for example, is something different from the implementation of this procedure in school education. Yet the difference goes still farther: in many cases, societies need rational social decision procedures, because „ideal" procedures are (still) unknown. The attention to social procedures increases to the extent to which cognitive procedures are not available. This attention grows to a conflict-laden interest if the questions at issue are shaped by differing social values, so that the result which the procedure yields is essentially determined by an evaluative attitude. „In cases of cognitive uncertainty and evaluative dissension, decisions are, to say it in a very pointed way, found appropriate rather on the basis of the procedure yielding them than on the basis of their contents".[9]

If we talk about the „appropriateness" of social procedures, we presume that the latter are not seen as equivalent. In particular, not all social decision procedures are equivalent with respect to their „rationality". It is, therefore, an improper simplification to distinguish in the discussion about environmental standards between a „scientific" (and hence „rational") and a „political" (and hence „irrational") decision part. So, science advising politics must not restrict itself to a discussion of facts which are the subject of (natural) science, and to leave further questions „to politics". The evaluation of social decision procedures as being more or less „rational" is instead a scientific issue as well. Hence the question about a rational standardization of environmental standards indispensably requires that insights of the social sciences are included. A minimal condition for social decision procedures to be rational is that they are organized in a way that scientific insights may win recognition. A view on actual procedures of social decision making shows that this is by no means a trivial demand. Environmental standards are conventional products (with a scientific basis); the point is to organize the formation of conventions according to generalizable rules.

(a) Social reality is always characterized by an ensemble of conflicts and consensus. Seen from a methodological point of view, the *conflict situation* is the starting point - which has to be overcome (this does not mean to start from a pessimistically interpreted „natural state of affairs" in the sense of an anthropological description). The conflict is the methodical beginning, for the reason alone that, due to environmental standards, chances and risks are *unequally* distributed among the members of a society in relation to their subjective preferences. The definitions of the *relevant* environment, of the goods to be protected, of the size of costs and benefits etc., already burden or favor the members of a society quite differently. Even if, for some defined social group, upper limits for burdens (in relation to a certain benefit) are agreed upon, this does not mean that that quantity is invariable in the course of time. Actual acceptance is essentially a changeable quantity, in any respect.

The potential for conflict still increases under conditions of cognitive uncertainty, which society will have to deal with for an unforeseeable length of time in many areas, e.g. in radiation protection. In radiation protection, statements about dose response relations again rely on basic assumptions, concerning, e.g., the procedures of extrapolation.[10] Even if serious experts agree on the necessary assumptions in this context (e.g. about the „safe distance" of such assumptions), a fundamental uncertainty is not abolished. A boundary assumption which, due to the social responsibility of the scientist, is very „conservative" in relation to the actual dose-response relationship could, for example, impose unnecessarily high costs on parts of the population.

Cognitive uncertainty and evaluative dissension lead to *two demands* on public debate about environmental standards: first, the integration of scientific insights into that debate must be possible, i.e., the debate must be free of ideologies resisting science. Second, the plurality of evaluative attitudes must have their say (the debate must show a high „potential for value sensitivity" (Mayntz)), i.e., it must be free of dogmatism as much as possible.

(b) In establishing environmental standards, essentially two *strategies of action orientation* are to be distinguished: a *maximization strategy* concerned with *one* target variable (like „life" or „health") and an *optimization strategy* concerned with a *plurality* of target quantities. The maximization strategy is once again supported in the political discussion about environmental standards; its advantage lies above all in the one-dimensionality of its evaluation aspects and hence in its direct operability. In practice, however, a restriction to *one* target variable cannot be maintained. Public discourse shows that the members of society, even the supporters of a maximization strategy, formulate different target variables. Thus, a social reconciliation of these divergences which is more than a mere enforcement of *one* position by means of power has always to consist in weighing the different target variables against each other, hence in an optimization strategy. For instance with respect to immission levels, very different target states have to be distinguished for humans, animals and plants etc., and among these again for delicate and robust species etc. If again one does not want to define a target state dogmatically, target states have to be reconciled according to the optimization strategy.

So, the same problem arises for establishing environmental standards as for any optimization strategy, namely the problem of making the different effects of one cause comparable in order to be able to establish a comparative order of such effects which meets certain formal requirements, above all transitivity. For instance, it has to be determined which importance should be attached to the extinction of a plant species as compared to certain impairments of human health (affecting how many persons and with what severity).

There is no solution to this problem of *multiattributivity* which provides such a determination in terms of an objective description. Instead, a rational discussion must be held about the circumstances under which a society chooses to live with a broad consensus. Some higher-level decision criteria are helpful here, like, e.g., economic justifiability, feasibility, acceptability and flexibility of standards. There may still be weighing problems even among these formal decision aspects, though. In many cases, however, these can be solved by risk-risk comparisons or be reduced to only a few options. A protection measure for one plant species should no longer be acceptable, for example, if it produces risks endangering even more species. Such considerations presuppose, however, that it is possible to project different set targets onto one target dimension. Somebody rejecting such a reduction of dimensions from the outset has to renounce rational decision procedures for optimization.[11]

(c) With respect to the institutionalization of decision processes, we have to distinguish between decision *makers* (parliaments, ministries, expert bodies etc.) and decision *procedures*. While the question about the decision maker is largely predetermined by the legal system, decision procedures are situated somewhere in the continuum between the two extremes of hierarchical-authoritarian processes and market-determined processes. Procedures existing in social reality occupy a place within this continuum as mixed forms. There are *three* ideal types of procedures which play a prominent role in industrialized countries:

Scientific *expert debate* in its pure form consists of a discussion held exclusively among scientists, about the state of information and the conclusions to be drawn. Yet this procedure is not satisfying because of cognitive uncertainty, which can never be excluded with respect to crucial questions, and because of non-scientific assumptions which hence find their way into those deliberations. Since expert debate virtually takes place behind closed doors, due to the hermetic character of scientific communication, even if the debate is public, the procedure is necessarily met with distrust. However developed the „potential of value sensitivity" of experts might be, it cannot be understood by the broad public.

Procedures of negotiation suggest a reconciliation of interests on the basis of diverging interests. A model for this is, for example, provided by collective bargaining. One advantage of this procedure is the fact that it explicitly proceeds on the assumption of diverging interests. Yet, a standing objection against its pure form holds that a compromise may satisfy those representing the interests but is not appropriate to the issue („bad compromise").

In *quasi-judicial* procedures, there is a neutral instance of the kind of a judge above the interest groups, which has to ensure that all arguments and counter-arguments are presented and dealt with. As opposed to negotiation procedures, here the focus is not on compromise but on controversy. So the parties themselves need not care about a possible balanced outcome of the negotiations. It must be admitted, though, that the institution of the judge is itself subject to the suspicion of partiality in extremely conflicting situations, as they increasingly occur with respect to environmental standards. Weighing up argument and counterargument is done according to standards which themselves cannot be formed neutrally. Hence there is the danger of the procedure being transferred to ever higher metalevels.

Against the background of the mentioned fortes and weaknesses, a mixed form will have to be chosen for decision procedures which are actually to be institutionalized, as this is already the case with procedures established in the U.S. These procedures are characterized by their publicity and their antagonistic character and by a legitimation duty for the agency which makes the decision. Decision procedures in Germany, on the other hand, largely take place behind closed doors; often the legitimation duty cannot easily be located, due to the „dual" system of governmental and private agencies. The „semi-public" position of the German Official Standards Committee (DIN-Ausschuß) is exemplary for this dual structure. Here, a multistage procedure of finding standards takes place, ranging from debate in interested expert bodies to final standardization, which, in some cases, gains access to laws or regulations. It is therefore often doubted whether the traditionally established system in Germany provides a sufficient, procedurally guaranteed „potential of value sensitivity".[12]

ANNOTATION

1 v. Uexküll 1909; v. Uexküll & Kriszat 1934. v. Uexküll's notion of environment has only in a few cases been adopted in biology and ecology (see Weber 1937 and 1939a, 1939b, Thienemann 1956, Stugren 1978). The expert council for environmental questions (Rat von Sachverständigen für Umweltfragen), however, follows this concept of „environment" in 1987. From a biological viewpoint, v. Uexküll's biologism is criticized, above all, by A. Portmann (Portmann 1951 and 1970).

2 As to the different „spheres" being functional and systematic, see, in particular, v. Uexküll and Kriszat 1934.

3 In particular, Heidegger's concept of „being in the world" („In-der-Welt-sein", Heidegger 1927) and Husserl's concept of a „lived world" („Lebenswelt")(Husserl 1936).

4 See Neuweiler 1986.

5 This linkage was emphasized above all by H. Plessner in his criticism of Scheler's and Gehlen's concept of „openness to the world"; see Scheler 1928, Plessner 1928, Gehlen 1940, Hengstenberg 1957, and Plessner 1964, 1974 as well as further texts.

6 For a detailed description of the development of standards of radiation protection, see chapter 3.1.

7 „Naturalism" is generally understood here as the position according to which claims of validity, e.g. scientific knowledge, are based only on natural facts or developments; ethical naturalism, for instance, bases practical claims of validity only on natural explications (see Mittelstraß 1984 and Wimmer 1984). – Incidentally, naturalism is by no means linked necessarily to modern scientific thinking; the specifically scientific understanding of nature is rather based on the subjective constitution of nature, as is shown by the fact, for instance, that statements about nature are possible only by means of man-made measurement equipment (see Janich 1981; for the emergence of modern science seen from this point of view, see Mittelstraß 1970).

8 For the distinction between stochastic and non-stochastic effects, see chapter 2.2.

9 The notion of a „naturalistic fallacy" goes back to the philosopher G.E. Moore; see Moore 1903, and the bibliography in Wimmer 1984.

10 See Janich 1987.

11 A radical critique of anthropocentrism is found in particular in Spaemann 1979, and Meyer-Abich 1984.

12 See Meyer-Abich 1979, and Meyer-Abich 1984.

13 See Murswiek 1988.

14 See Deutscher Bundestag 1988.

15 See Hofmann 1988, 277, ann. 156.

16 As is stated by the expert council for environmental issues in its expertise on environment 1987.

17 See Jarass 1987.

18 Hence, the above-mentioned considerations are based on the idea of individual dignity – in accordance with the German constitution. In contrast to that, modern ethical debate often relates the notion of dignity to the species (in the context of justifying the demand for preservation of the species), as in Jonas 1989, 90f; see also Birnbacher 1990, 270ff. Dignity can only be claimed by a being which is able to assert or fulfill claims; see Gethmann 1990a and 1990b.

[19] See chapter 3.

[20] It is often maintained that conventions as such are arbitrary and not amenable to rational control. Taking the conventionality of environmental standards into account, we would thus be urged to ask 'what's in a limit?' (v. Lersner 1990, 194). Yet it will be shown in the following that conventions, and hence environmental standards, may be justified or unjustified. In addition, environmental standards are influenced by knowledge of the (natural) sciences, in a way which has still to be expounded. (see chapter 5.2).

[21] This approach is expounded in detail in Gethmann 1982.

[22] This approach has been developed by Schwemmer 1971 and 1976, following Kant and von Wright.

[23] The problem of multiattributivity based hereupon is discussed in detail in chapter 5.4.

[24] The following discussion is based on the so far most comprehensive discussion of philosophical problems of the risk concept in Rescher 1983. A detailed discussion of the application of the risk concept in the context of environmental problems is found in Gethmann 1990b.

[25] See Rescher 1983, 8.

[26] Individual perception of danger is treated, above all, in psychological studies; see, e.g., Jungermann & Slovic 1991, Tack 1991.

[27] Collective perception of danger is, above all, the subject of the much-discussed sociological works of Beck 1986, Beck 1988, and Luhmann 1986.

[28] A detailed treatment is found in Gethmann 1990b, § 1.

[29] See the discussion in Stegmüller 1973; for the function of the probability concept in connection with risk problems, see Rescher 1983, 33–43.

[30] For the logic of preferences, see v. Kutschera 1973; as to risk problems, see Rescher 1983, 44–63.

[31] An exact decision analytic elaboration of the problem of multiattributivity will be presented in chapter 5.4.

[32] See the survey of acceptance factors in Birnbacher 1983.

[33] „William Tell" (1805), second act, second scene, lines 1278–1280.

[34] Locke 1690, chap. XIV, § 168.

[35] This position is advocated by Mayer-Tasch 1979.

[36] See Isensee's fundamental criticism (Isensee 1969).

[37] For the first time formulated in Gethmann 1987; see further observations in Gethmann 1990b. The problem of collective risk imposition on individuals with different ways of life requires a specific form of conflict solving.

[38] The thesis of incommensurability underlies Meyer-Abich's criticism of the consistency principle (Meyer-Abich 1989, 35).

[39] This section is based mainly on Mayntz 1990.

[40] Mayntz 1990, 137f, with reference to Majone 1982.

[41] For the procedure of extrapolation, see chapter 2.2.7.

[42] These questions of optimization are treated formally in more detail in chapter 5.4.

[43] On the basis of these general conditions, a proposition for an institutionalization of the procedure of establishing standards will be developed in chapter 6.3.

LITERATURE

Beck, U. (1986): Risikogesellschaft. Auf dem Weg in eine andere Moderne, Frankfurt a.M., Suhrkamp.

Beck, U. (1988): Gegengifte. Die organisierte Unverantwortlichkeit, Frankfurt a.M., Suhrkamp.

Birnbacher, D. (1990): Gefährdet die moderne Reproduktionsmedizin die menschliche Würde?, in: Um Leben und Tod. Moralische Probleme bei Abtreibung, künstlicher Befruchtung, Euthanasie und Selbstmord, A. Leist (ed.), Frankfurt a.M., Suhrkamp.

Birnbacher, D., Koch, D. (1983): Zum Problem der Rationalität in der Akzeptanz technologischer Risiken, in: Der Mensch und die Wissenschaften vom Menschen. Beiträge des 12. Deutschen Kongresses für Philosophie in Innsbruck vom 29. September bis 3. Oktober 1981, G. Frey & H. Zelger (eds.) Innsbruck, Solaris, 487–498.

Deutscher Bundestag (ed.) (1988): Verankerung des Umweltschutzes im Grundgesetz, in: Zur Sache – Themen parlamentarischer Beratung 2/88, Speyer, Klambt-Druck GmbH, pp. 19

Gehlen (1940): Der Mensch. Seine Natur und seine Stellung in der Welt, 13. ed. 1986, Wiesbaden, AULA.

Gethmann, C.F. (1982): Proto-Ethik. Zur formalen Pragmatik von Rechtfertigungsdiskursen, in: Bedürfnisse, Werte und Normen im Wandel, vol. 1, Grundlagen, Modelle, Perspektiven, H. Stachowiak & Th. Ellwein (eds.), Paderborn, Schöningh, 113–144.

Gethmann, C.F. (1987): Ethische Aspekte des Handelns unter Risiko, VGB, Kraftwerkstechnik 67, 1130–1135.

Gethmann, C.F. (1990a): Lebensweltliche Präsuppositionen praktischer Subjektivität. Zu einem Grundproblem der ‚angewandten Ethik', in: Akten des 1. Internationalen Schellingkongresses, W. Jacobs (ed.), Stuttgart, Frommann-Holzboog.

Gethmann, C.F. (1990b): Zur Ethik des Handelns unter Risiko im Umweltstaat, in: Handeln unter Risiko im Umweltstaat, C.F. Gethmann & M. Kloepfer (eds.), Berlin, Springer.

Hare, R.M. (1952): The Language of Morals, The Clarendon Press, Oxford, deutsch: (1971): Die Sprache der Moral, Frankfurt a.M., Suhrkamp.

Heidegger, M. (1927): Sein und Zeit, 16. ed. 1986, Tübingen, Max Niemeyer. Hengstenberg, H.E. (1957): Philosophische Anthropologie, 2. ed. 1960, Stuttgart, Kohlhammer.

Hofmann, H. (1988): Natur und Naturschutz im Spiegel des Verfassungsrechts, Juristenzeitung 43/1, 265–278.

Husserl, E. (1936): Die Krisis der europäischen Wissenschaften und die transzendentale Phänomenologie. Eine Einleitung in die phänomenologische Philosophie, 2. ed. 1962, Husserliana, vol. VI, Den Haag, Nijhoff.

Isensee, J. (1969): Das legalisierte Widerstandsrecht. Eine staatsrechtliche Analyse des Art. 20 Abs. 4 Grundgesetz, Bad Homburg, Gehlen.

Janich, P. (1981): Natur und Handlung. Über die methodischen Grundlagen naturwissenschaftlicher Erfahrung, in: Vernunft, Handlung und Erfahrung. Über Grundlagen und Ziele der Wissenschaften, O. Schwemmer (ed.), München, C.H. Beck, 69–84.

Janich, P. (1987): Naturgeschichte und Naturgesetz, in: Über Natur. Philosophische Beiträge zum Naturverständnis, O. Schwemmer (ed.), Frankfurt a.M., Vittorio Klostermann, 105-122.

Jarass, H.D. (1987): Der rechtliche Stellenwert technischer und wissenschaftlicher Standards, Neue Juristische Wochenschrift 40/1.2, 1225–1231.

Jonas, H. (1989): Das Prinzip Verantwortung. Versuch einer Ethik für die technologische Zivilisation, Frankfurt a.M., Suhrkamp.

Jungermann, H., Slovic, P. (1991): Die Psychologie der Kognition und Evaluation von Risiko, in: Risiko und Gesellschaft, G. Bechmann (ed.), Opladen, Westdeutscher Verlag.

Kant, I. (1785): Grundlegung zur Metaphysik der Sitten, in: Kants Werke. Akademie Textausgabe, 1907, unveränderter Nachdruck 1968, Berlin, Walter de Gruyter und Co., vol. IV, 385–464.

Kutschera, F. von (1973): Einführung in die Logik der Normen, Werte und Entscheidungen, Freiburg i.Br., Alber.

Lersner, H. von (1990): Verfahrensvorschläge für umweltrechtliche Grenzwerte, Natur und Recht 12, 193–197.

Locke, J. (1690): The Second Treatise of Government, in: Two Treatises of Government, 2nd ed. 1970, Peter Laslett (ed.), London, Cambridge University Press.

Luhmann, N. (1986): Ökologische Kommunikation. Kann die moderne Gesellschaft sich auf ökologische Gefährdungen einstellen?, Opladen, Westdeutscher Verlag.

Majone, G. (1982): The Uncertain Logic of Standard Setting, Zeitschrift für Umweltpolitik 4, 305–323.

Markl, H. (1984): Dasein in Grenzen. Die Herausforderung der Ressourcenknappheit für die Evolution des Lebens, Konstanz, Universitätsverlag.

Mayer-Tasch, P.C. (1979): Recht auf bürgerlichen Ungehorsam?, in: Energiepolitik ohne Basis, C. Amery, P.C. Mayer-Tasch & K.-M. Meyer-Abich (ed.), Frankfurt a.M., Fischer Taschenbuch, 40–45.

Mayntz, R. (1990): Entscheidungsprozesse bei der Entwicklung von Umweltstandards, Die Verwaltung 23, 137–151.

Meyer-Abich, K.-M. (1979) (ed.): Frieden mit der Natur, Freiburg i.Br., Herder.

Meyer-Abich, K.-M. (1984): Wege zum Frieden mit der Natur. Naturphilosophie für die Umweltpolitik, Hamburg, Hanser.

Meyer-Abich, K.-M. (1989): Von der Wohlstandsgesellschaft zur Risikogesellschaft. Die gesellschaftliche Bewertung industriewirtschaftlicher Risiken. Aus Politik und Zeitgeschichte. Beilage zur Wochenzeitung Das Parlament 1989/36, 31–42.

Mittelstraß, J. (1970): Neuzeit und Aufklärung. Studien zur Entstehung der neuzeitlichen Wissenschaft und Philosophie, Berlin, Walter de Gruyter und Co..

Mittelstraß, J. (1984): Naturalismus, in: Enzyklopädie Philosophie und Wissenschaftstheorie, vol. II, Mannheim, Bibliographisches Institut – Wissenschaftsverlag.

Moore, G. E. (1903): Principia Ethica, 1959, London, Cambridge University Press; deutsch: (1970): Principia Ethica, B. Wisser (ed.), Stuttgart, Reclam.

Murswiek, D. (1988): Umweltschutz – Staatszielbestimmung oder Grundnorm?, Zeitschrift für Rechtspolitik, 21, 14–20.

Neuweiler, G. (1986): Evolution und Verantwortung, in: Sitzungsberichte der mathematisch-naturwissenschaftlichen Klasse der Bayrischen Akademie der Wissenschaften, München, Verlag der Bayr. Akad. d. Wiss., 12–27.

Portmann, A. (1944): Biologische Fragmente zu einer Lehre vom Menschen, 3. Edition 1969, Basel, Schwabe.

Portmann, A. (1970): Entläßt die Natur den Menschen? Gesammelte Aufsätze zur Biologie und Anthropologie, München, Piper.

Plessner, H. (1928): Die Stufen des Organischen und der Mensch, 3. ed. 1975, Berlin, Walter de Gruyter & Co..

Plessner, H. (1964): Conditio Humana, Pfullingen, Neske Verlag.

Rat von Sachverständigen für Umweltfragen: Umweltgutachten 1987.

Rescher, N. (1983): Risk. A Philosophical Introduction to the Theory of Risk Evaluation and Management, Washington, University Press of America.

Scheler, M. (1928): Die Stellung des Menschen im Kosmos, 11. ed. 1988, Bonn, Universitätsbuchhandlung Bouvier.

Schiller, F. (1805): Wilhelm Tell, in: Sämtliche Werke, aufgrund der Originaldrucke, G. Fricke & H.G. Goeppert, in Verbindung mit H. Stubenrauch (ed.), 4. ed. 1965, vol. 1, München, Hanser.

Schwemmer, O. (1971): Philosophie der Praxis. Versuch zur Grundlegung einer Lehre vom moralischen Argumentieren, Frankfurt a.M., Suhrkamp.

Schwemmer, O. (1976): Theorie der rationalen Erklärung. Zu den methodischen Grundlagen der Kulturwissenschaften, München, Beck.

Spaemann, R. (1979): Technische Eingriffe in die Natur als Problem der politischen Ethik, Scheidewege 9, 476–497.

Stegmüller, W. (1973): Probleme und Resultate der Wissenschaftstheorie und Analytischen Philosophie, vol. IV, Personelle und statistische Wahrscheinlichkeit, erster Halbband, Personelle Wahrscheinlichkeit und Rationale Entscheidung, zweiter Halbband, Statistisches Schließen, Statistische Begründung, Statistische Analyse, Berlin, Springer.

Stone, C. (1972): Should Trees have Standing? Toward Legal Rights for Natural Objects, Southern California Law Review, 45, 450–501.

Stugren, B. (1978): Grundlagen der Allgemeinen Ökologie, 3. ed., Stuttgart, G. Fischer.

Tack, W.H. (1991, im Druck): Risikowahrnehmung und Beurteilung. Die Psychophysik des Risikos, in: Risiko und Gesellschaft, G. Bechmann (ed.), Opladen, Westdeutscher Verlag.

Thienemann, A.E (1956): Leben und Umwelt, Hamburg, Rowohlt.

Uexküll, J. von (1909): Umwelt und Innenwelt der Tiere, 2. ed. 1921, Berlin, Springer.

Uexküll, J. von, Kriszat, G. (1934): Streifzüge durch die Umwelt von Tieren und Menschen, Neuausgabe 1970, Frankfurt a.M., S. Fischer.

Weber, H. (1937): Zur neueren Entwicklung der Umweltlehre J.v. Uexkülls, Die Naturwissenschaften 25, 97–104.

Weber, H. (1939a): Der Umweltbegriff der Biologie und seine Anwendung, Der Biologe 8, 245–261.

Weber, H. (1939b): Zur Fassung und Gliederung eines allgemeinen biologischen Umweltbegriffes, Die Naturwissenschaften 27, 633–644.

Wimmer, R. (1984): Naturalismus (ethisch), in: Enzyklopädie Philosophie und Wissenschaftstheorie, vol. II, Mannheim, Bibliographisches Institut – Wissenschaftsverlag, 965–966.

2

SCIENTIFIC-MEDICAL FOUNDATIONS OF RADIATION PROTECTION

In this chapter, we will present the scientific-medical foundations which are the basis of today's standards for radiation protection. They are the result of decades of intensive research. The case of radiation protection is thus rather outstanding with respect to completeness and reliability of its foundations. Political practice will often require that standards are based on a less solid basis of knowledge. It is exactly for this reason, however, that the standards of radiation protection have been chosen to serve as a model for the establishment of environmental standards in this study.

Of course, even the very comprehensive knowledge of the scientific basis of radiation protection still shows gaps and uncertainties, which have to be filled by assumptions. Such assumptions must not be arbitrary but have to be made plausible by analogous considerations based on experience. Beyond that, however, a scope remains, which may be used more or less pessimistically – or, as it is called, more or less conservatively. This is only partly a matter of science, in part it is already a matter of politics.

In the first place, radiation protection standards, like any other environmental standards, have the purpose of limiting risks. A risk caused by environmental noxae is always determined by two parameters: exposure and effect. If no significant exposure is to be expected, the risk is low even for the most toxic noxa. With regard to chemical substances, the total amount produced plays an important role as the simplest measure for the exposure to be expected. The German "Chemikaliengesetz", for example, requires less strict toxicity tests for substances produced in small quantities than for those produced in higher quantities (ChemG 1982). Another criterion is the application pattern. A by-product of a chemical process, for instance, which is emitted into the atmosphere, may lead, even in case of a smaller production, to a more serious exposure than a substance which is produced exclusively for a well-controllable application.

By emission, we mean the release of a substance at its source. Quantitatively, emission may, for example, be indicated as amount per unit time, or as concentration in the waste stream of some facility. On the other hand, there is immission as a consequence of emission. Immission is the concentration of a substance which reaches the environment or the relevant compartment of the environment. Thus, it is the immission which is relevant for the actual exposure.

If an appreciable exposure is to be expected, quantification of the risk depends on knowledge about the effect as a function of the dose (dose-effect relation). Dose is an expression originating from pharmacology, which describes the applied quantity of a substance. Such a risk quantification may then, together with the determination of the socially acceptable risk, serve as a basis for the determination of an exposure limit (dose limit). More frequently it is a threshold dose, though, which provides this basis. Such limits are typical examples of environmental standards. An exposure limit has also to be technically feasible, however, and one presupposition to this purpose is a sufficiently reliable determinability of exposure, hence dosimetry. For this reason, research on effects could be called the cognitive and dosimetry the operative component of risk management.

In the present political system in Germany, there is no environmental standard which is based on an explicit determination of an acceptable risk for the population at large. Radiation protection limits for the population, which are treated as an example in this study, had not to refer to risks; instead, they could be based on the actual and unavoidable exposure to natural radiation, more exactly: on its natural variation. Since, however, the connected risk can be assessed quite well, such a determination of limits then also implies information about the size of a risk considered to be acceptable (cf. chapter 3.3).

2.1 EXPOSURE TO RADIATION

2.1.1 Survey

A precondition for the observance of an exposure limit is the possibility of measuring the exposure, i.e. the dose, in a sufficiently reliable way and with adequate technical expenditure. Here, adequate technical expenditure means an expenditure adequate to the number of persons concerned. For the measurement of a noxa affecting only astronauts, even a relatively complicated technique would be acceptable. On the other hand, measurement of radioactivity affecting a potentially large number of persons has reached a technological standard today which allows application by trained personnel. After the Chernobyl accident, when parts of the German population were extremely alarmed and many people tried to make measurements on their own, it turned out, however, that there is still a technical border to be crossed before an application by lay people is conceivable.

Furthermore, determinability of exposure is a precondition not only for handling limits but also for establishing them, since quantification of exposure plays a role already in obtaining dose effect relations. So, even the reliability of the latter depends on the quality of dose determination and on a reasonable dose definition.

Exposure of the population at large (this notion serves as differentiation against the smaller group of occupationally exposed persons) is due to very different sources. It is usually subdivided into four components, which differ with respect to their sources and hence with respect to our ability to influence them.

The first component cannot be influenced at all, it is the regionally very different exposure to radiation from natural sources. Natural sources comprise the external sources of cosmic and terrestrial radiation, and the internal sources of radioactivity (above all potassium-40). This component amounts to 30% of the average natural exposure to radiation in Germany. The second component is exposure to radiation from radon in the air of closed rooms. This component is also of natural origin (building ground, construction material), but it becomes manifest only due to accumulation of radon in the air of closed rooms, and is hence dependent on civilization. It accounts for an average total exposure of almost another 30%, with a particularly large variation according to housing conditions. The third component is exposure to artificial radiation for medical purposes, which, in Germany, amounts to just under 40% of the average total exposure. It occurs for the individual benefit of the patient and is not subject to any legal limitation. Here, doctors are bound to weigh benefits against harms according to the rule of economical use.

Now, the fourth component, exposure to artificial radiation due to every other use humans make or made of nuclear fission and radioactivity (with the exception of medical treatment of patients) is the one which is limited by standards of radiation protection. It contributes to the average exposure of the population with an average fraction of just under 2%. In this portion, German nuclear power plants are involved with a few tenths of one per cent. As a consequence of the radioactive fallout resulting from the Chernobyl accident, this fourth component will, in future, amount to about 2.5% instead of 2.0%

(as a mean value for Germany), averaged over the next fifty years. (In the first year after the catastrophe, the fallout contribution to total exposure might have been much higher, though, and might have reached the order of exposure to natural radiation.) All contributions to radiation exposure aggregated in this fourth component stem principally from artificially produced radioactive substances (including the fallout of nuclear test explosions), which are dissipated in the environment. They may lead to external irradiation (e.g. from the soil) and to internal irradiation (e.g. by inhalation or by ingestion from food). Thus, radiation protection standards for the population serve to limit a normally very low share of total exposure to radiation. They are supposed to, above all, prevent a continuous increase of anthropogenic environmental radioactivity.

The following two sections will describe the different types of radiation, together with their different properties, and then explain dose concepts and units of dosimetry which are applicable for each kind radiation. Section 2.1.4 deals with quantitative determination of exposure to radiation, i.e. dosimetry. Section 2.1.5 will demonstrate how a dose is calculated, taking exposure to internal radioactive iodine as an example. This section requires more specific knowledge than the others, but it is not essential for getting a general understanding of the subject. In section 2.1.6, finally, the different components of general exposure to radiation are treated in detail as to quality and quantity. Natural components as basic exposure provide an important point of reference for establishing standards for additional anthropogenic exposure. In this context, their variation is of particular importance.

2.1.2 Types of Radiation

Matter is built up of atoms, the nucleus of which consists of positively charged protons and neutral neutrons, surrounded by a shell of negatively charged electrons. The number of protons, which is equal to the number of electrons, is called the atomic number. It determines the chemical nature of a type of atom, hence its position in the periodic table of chemical elements. There are natural elements with atomic numbers between 1 (hydrogen) and 92 (uranium). Elements with higher atomic numbers, the transuranium elements, one of which is plutonium, are artificial elements.

The sum of protons and neutrons determines the atomic weight. With a definite number of protons, or atomic number, the number of neutrons, and hence the atomic weight, may vary within a certain range. Types of atoms with the same atomic number but with differing atomic weight are called *isotopes* of the respective element. Isotopes of an element are stable only within a narrow range of atomic weight, below or above which they are unstable. Decaying, they emit high-energy radiation, therefore they are called *radioactive*. The rate of decay per unit time is characterized by the *half-life*, i.e. the

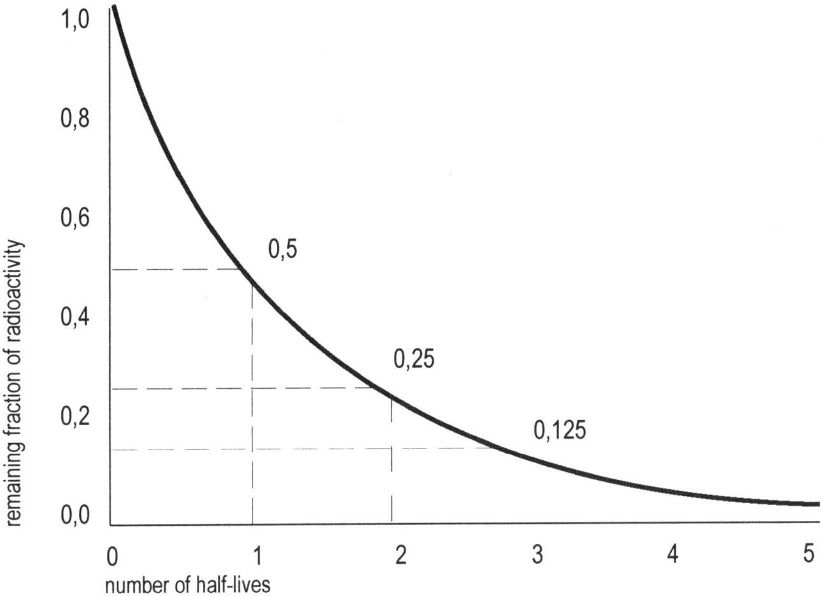

Fig. 2.1 Temporal progress of radioactive decay

period in which half of the currently existing atoms decay. The rest then radiates with half the intensity. Thus, radioactive decay follows an exponential law (cf. fig. 2.1). There are radioactive types of atoms *(radionuclides)* with half-lives between milliseconds and 100 billions of years. The longer the half-life, the longer, but also the less intensive, the radiation emitted by a given mass of a radionuclide.

There is a large number of radionuclides. Only a small portion of them occurs in nature in appreciable quantities, a far larger proportion can only be produced artificially. Most artificial radionuclides are isotopes of chemical elements, the natural isotopes of which are inactive. Only a few elements have natural radioactive isotopes (e.g. potassium-40). Some, however, occur only as radioactive elements (e.g. uranium).

Relevant for radiation protection are three types of possible products of the disintegration of a nucleus: α-, β-, *and* γ-*radiation*. α– and β–radiation consist of particles, γ–radiation consists of photons, i.e. quanta of energy. α–particles are the nuclei of helium, i.e., they are combined of two protons and two neutrons, β–particles are electrons. Emitting one α–particle results in a reduction of the atomic weight of the decaying nucleus by four units, and of the atomic number by two units. The product is therefore an isotope of a lighter element, which may be radioactive as well. Emitting one β–particle increases the atomic number by one unit (the neutron becomes a proton, due to the loss of the emitted electron). The product is therefore a – possibly radioactive – isotope of the

atomic number	symbol	name	half-life	mode of decay	energy of the γ–radiation (MeV)
1	^{3}H	tritium	12,3 a	β(very soft)	–
6	^{14}C	carbon	5736 a	β	–
11	^{24}Na	sodium	1 5,02 h	β, γ	4,122
19	^{40}K	potassium	1,28 x 10^{9} a	β, γ	0,161
27	^{60}Co	cobalt	5,26 a	β, γ	2,505
36	^{85}Kr	krypton	10,73a	β, γ	0,005
38	^{89}Sr ^{90}Sr	strontium	50,5 d 28,6 a	β β	– –
44	^{106}Ru	ruthenium	368,2 d	β	–
53	^{129}I ^{131}I	iodine	1,57 x 10^{7} a 8,04 d	β, γ β, γ	< 0,001 0,383
[55 56	$^{137}Cs^{*)}$ ^{137m}Ba	cesium barium	30,17 a 2,5 m	β γ	 0,662]
58	^{144}Ce	cerium	284,2 d	β, γ	< 0,001
86	^{222}Rn	radon	3,82 d	α, γ	< 0,001
92	^{235}U ^{238}U	uranium	7,04 x 10^{8} a 4,47 x 10^{9} a	α, γ α, γ	0,067 < 0,001
94	^{239}Pu	plutonium	2,41 x 10^{4}a	α, γ	< 0,001

*) in equilibrium with ^{137m}Ba

Table 2.1 Characteristics of some selected radionuclides

chemical element with the next higher atomic number. Emitting a photon only changes the energy state of the nucleus. Another characteristic of any radioactive decay is the energy of the emitted particles or photons, which is mostly of the order of 0.1 to 10 MeV.

There are also types of radiation relevant for radiation protection which do not originate from radioactive decay. These are primarily *X-rays*, which consist of photons, like γ–rays, but which are poorer in energy. They are produced by bombarding matter with electrons which are accelerated in the high-voltage field of the X-ray tube. *Fast neutrons*, which arise primarily from nuclear fission, may be of relevance for radiation protection of occupationally exposed persons, but not for the environment.

α-, β–, and γ– or X-rays have very different properties with respect to radiation protection. α–particles carry the highest energies, and release them during their passage through a medium on a short distance (high Linear Energy Transfer – LET). The consequence is that they are completely absorbed even by very thin layers of matter, e.g. a sheet of paper, and therefore do not reach the inner organs of the body from outside. Hence they hardly lead to external exposure to radiation. If, however, α–emitters are incorporated, they may have a particularly high impact, since α–particles deposit their energy in the tissue in a very concentrated and hence very effective manner. Correspondingly, their biological effectiveness is higher than that of other types of radiation at the same energy dose. This will be explained later in more detail. β–radiation has a lower LET-value than α–radiation. It is shielded completely only by layers of matter of at least a few millimeters thickness. γ–radiation, finally, is never completely absorbed during passage through matter but is only weakened according to the thickness of the medium. For this reason, walls of some meters thickness are possibly required to shield sources of heavy γ–radiation in a way that they have virtually no effect. Another consequence of this high penetration capacity of γ–rays is, of course, the fact that the major part of radiation emitted by an incorporated[1] γ–emitter passes the body without being absorbed, i.e., without having any effect.

2.1.3 Dose Concepts and Dose Units

Radiation always comes from a radiation source. Radiation is emitted by natural and anthropogenic sources. Radiation which is of anthropogenic origin and for which standards can be fixed usually consist of radioactive substances which are dissipated in the environment. Examples are:

- krypton-85, a gaseous radioactive fission product, which is contained in low concentrations in the air near nuclear installations,
- strontium-90, a long-lived fission product, which originates from nuclear weapon tests and has dissipated all over the earth's surface,
- cesium-137, which, as a relatively volatile fission product, accounts for the major part of long-lived released in the Chernobyl reactor accident and which therefore occurs today quite commonly in the top soil layer.

The *source intensity* of a radiation source may be expressed in different quantities. An appropriate measure for radionuclides is the *activity* A_S (radioactivity), that is, the number of radioactive disintegrations per unit time. Activity is indicated in Becquerel today (1 Bq = 1 disintegration per second). The activity concentration in materials from environmental sources of radiation (air, water, soil, biomass, food) can be measured with great sensitivity and accuracy. As a rule, even source intensities which are far below any level of relevance for practical radiation protection can be determined accurately.

The source intensity as such, however, does not tell much about the extent of radiation to which some object is exposed, for instance because the source radiates in all directions and because this radiation (in the case of α– and β–radiation) is partly absorbed already in the radiating material itself. Yet we approach an answer to the question about exposure by looking at the *radiation field* in which the exposed object is situated. The radiation field is characterized by the *density* φ *of particle flux (or photon flux)*, that is, the number N of particles which, in the time t, traverse a certain sectional area A perpendicular to the direction of radiation.

$$\varphi = \frac{N}{At} \tag{2.1}$$

A unit often used to indicate flux density is $cm^{-2}s^{-1}$. Multiplied with the exposure time, this yields the particle fluence Φ, to which an object was exposed per square centimeter of sectional area:

$$\Phi = \frac{N}{A} \tag{2.2}$$

The sum of particle fluence over the sectional area of the exposed object yields the number of particles which hit the object, and is hence a measure for exposure. It corresponds roughly to the pharmacological dose.

Now, one important task of impact research was to refine dose measurement or dose calculation to such an extent that the biologically actually effective portion of an applied dose could be known. This problem has by and large been solved for ionizing radiation today, so that particle flux or photon flux are no longer of relevance for practical radiation protection.

To determine the biologically effective radiation dose, it is essential to understand that effects are caused only by that part of radiation which not only hits the radiosensitive object but which is also absorbed by it (the part which is not absorbed passes the object without any effect and can be measured at its backside). One important process taking place during absorption is the ionization of atoms or molecules of the irradiated matter. The *ion dose* is defined as the number of ion pairs[2] which are thus produced in one

Activity

Becquerel – Bq -
(former unit: Curie – Ci -)
1 Bq = 1 disintegration per second
1 Ci = 3.7 x 10^{10} Bq

Energy

microscopic: electron volt – ev -
(1 ev = energy needed to move an electron against a potential difference of 1 volt)

macroscopic: Joule – J -
(former unit: calorie – cal -)
1 J = 1 Watt-sec
1 cal = 4.184 J

Radiation Dose

energy dose: Gray – Gy -
(former units: rad, Roentgen – R -)
1 Gy = 1 J/kg
1 rad = 0.01 Gy $\approx$ 1 R
dose equivalent = energy dose x quality factor:
Sievert – Sv – (former unit: rem)
1 rem = 0.01 Sv

Prefixes

mega – M – :10^6, kilo – k – :10^3
micro – μ – :10^{-6}, milli – m – :10^{-3}

cubic centimeter of air. It is measured in Coulomb per kilogram (C/kg). The Roentgen (R) is an obsolete historical unit today.[3]

All types of radiation relevant for radiation protection have the capacity to ionize, directly or indirectly. Hence the notion "ionizing radiation" is a collective name for these types of radiation as opposed to, e.g., visible light or radiowaves, which have less

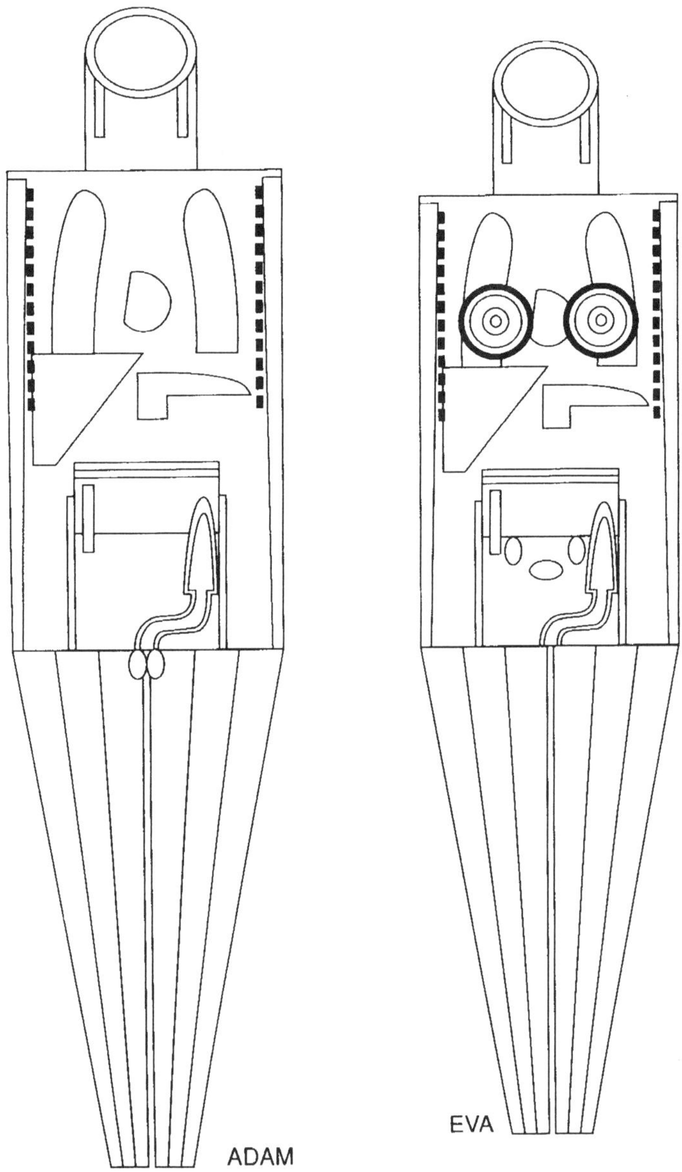

Fig. 2.2 Human phantoms ADAM and EVE (mean areas including breast or testes) (Kramer & Drexler 1982)

energy and therefore normally do not ionize. Since the ion dose is defined for air only, and no analogous quantity can be defined for water or tissue, the degree of an exposure is not characterized by the ionizations which the radiation causes but by the energy it transfers to the respective matter. The energy dose is defined as the energy transferred by radiation, divided by the mass to which it is transferred. Under circumstances which are of relevance for radiation protection, however, there is almost always a fixed ratio of energy dose to ion dose, the latter being more easily measurable. Hence ionization is the phenomenon to which most of the measurement equipment used in radiation protection responds. Yet there are also types of instruments for dose measurement which are based on different physical principles (e.g., thermoluminescence and nuclear-magnetic resonance). There are also sensitive instruments for the measurement of dose rates (dose per unit time), for example instruments with scintillators, which are appropriate even for dose rates as low as usually occur in the environment.

So, the energy dose is the energy absorbed per unit mass, thus the potentially effective dose. It is measured in Gray today (1 Gy = 1 Joule/kg). It can either be measured in the air at some place (local dose) or at the surface of the exposed object, especially of a person (personal dose). Instruments of dosimetry sum up the radiation doses over time. There are, for instance, film dosimeters, usually used for monitoring occupationally exposed persons, and dosimeter clips, which operate according to the principle of the electroscope, and which the persons concerned can read themselves during their occupation in a radiation field.

Local dose and person dose, however, do not yet characterize the actually risk-relevant extent of an exposure. As a next step, we need to know the energy dose in the body or in the organ exposed, that is, the body dose or *organ dose*. For external exposures, this step can be carried out experimentally with the help of human phantoms. These are 1:1 dummy human organisms, which are exposed to external radiation, and at the surface of which the person dose is measured (fig. 2.2). Inside the phantom, the dose, i.e. the organ dose, can be measured at the respective interested organ. So, empirical relations between person dose and organ doses can be ascertained, with the help of which models for these relations and then general conversion algorithms are deducible.

As to internal exposure, we cannot resort to empirically found relations between person dose and organ dose. Yet, in this case as well, it is possible to carry out measurements with phantoms, or calculations with the help of models based on physical considerations and calculations. For any standard case, there are tested algorithms and even tables of conversion factors available. The conversion factors permit a calculation of the organ dose from the amount of an incorporated radionuclide at the moment of its incorporation.

The energy dose D – be it a person dose or an organ dose – is defined as the ratio of energy E to mass M of a volume V in which this energy is absorbed:

$$D = \frac{E}{M} \quad or\, D = \frac{1}{\rho} \cdot \frac{E}{V} \tag{2.3}$$

(ρ is the density of the absorbing material)

The absorbed dose per unit time (dt), that is, the dose rate (Gy x s^{-1}), is:

$$D = \frac{d}{dt} D \tag{2.4}$$

So far, we have discussed energy absorption as a consequence of exposure to radiation, hence the purely physical phenomenon of interaction of radiation with an arbitrary object. For the purpose of establishing standards of radiation protection, however, we are interested in biological impacts on human beings. So we have to ask whether the energy dose is always a measure for the biologically effective dose. The absorption of energy, particularly by ionization, is the primary step for biological effects as well, but the energy dose is nevertheless not a measure of biologically effective and hence risk-relevant exposure, independent of the type of radiation. Apart from the total energy absorbed, the biological effect depends also on the microscopic spatial density of this absorption and therefore both on the type of radiation (α,β,γ) and on the energy of individual particles or photons. For chemical noxae with their very different ways of acting, it would probably be unrealistic to look for a universal biological measure for exposure, independent of the specific nature of the noxa.[4] Since biological impacts of radiation, on the other hand, are always due to the same primary processes of ionization and excitation, and since the yield of these primary processes per unit energy differs little for different types of ionizing rays, the dose equivalent H, which is closely linked to the energy dose, could be introduced as a universal biological measure of exposure here.

To get the dose equivalent, we multiply the energy dose D with a *quality factor Q*, which roughly corresponds with the relative biological effectiveness (RBE) of the respective type of radiation as to radiation impacts relevant at small doses:

$$H = D \times Q \tag{2.5}$$

The dose equivalent is indicated in Sievert (Sv) today.

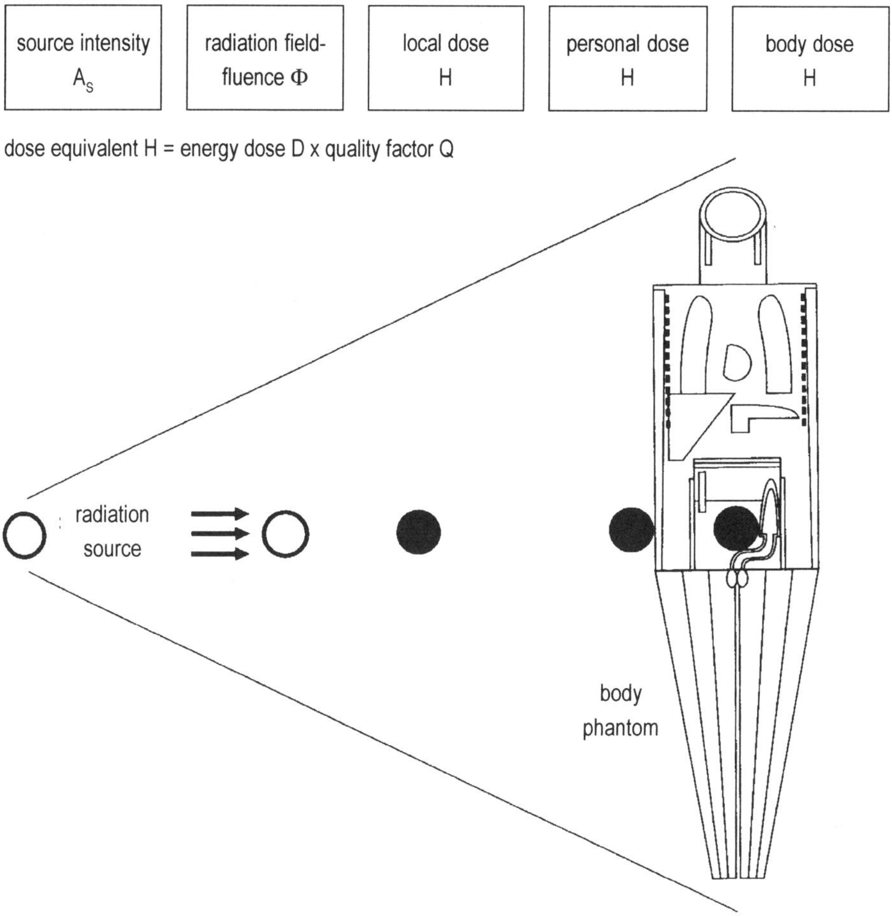

Fig. 2.3 From source intensity to body dose

The RBE and the quality factor derived from it are explained in more detail in section 2.3.2. Here, it should only be mentioned that they both depend particularly on the ionization density, i.e., on the amount of energy absorbed in the tissue per unit path length of the particle or photon. The ionization density itself is characterized by the linear energy transfer (LET). Densely ionizing radiation (e.g. α–radiation) and sparsely ionizing radiation (e.g. γ–radiation) are distinguished. The quality factor is the higher – at least within the LET-range of interest here – the higher the ionization density of the radiation is. Moreover, it depends on the radiation energy of each type of radiation. These estimates are based on comparisons of dose response relations of different types of radiation.

In the German radiation protection regulations, the most important values for the quality factor are:

γ-rays, X-rays, β–rays:	Q = 1
neutrons, protons:	Q = 10
α-particles:	Q = 20

To round off the picture, another measure of exposure should be introduced here, namely the commonly used *effective dose equivalent* H_{eff}, although its definition can only be understood in the context of a more detailed consideration of biological effects which are discussed later in section 2.2. Meanwhile, the shorter term "effective dose" has become common. The effective dose has been defined to be able to easily compare stochastic risks[5] of any exposure – organ or whole body exposure (Jacobi 1975). It is thus a dose measure which refers to risks and which, with the help of weighting factors, takes into account the different dispositions of different organs to radiationinduced tumors and the possibility of genetic mutations:

$$H_{eff} = W_1 H_1 + W_2 H_2 + \ldots + W_T H_T \tag{2.6}$$

H_T = mean dose equivalent in organ or tissue T;
W_T = weighting factor for organ or tissue T with respect to the relative risk of stochastic radiation damages.

The effective dose permits a summation over all exposures, for the purpose of risk assessment. In concrete terms, this may, for example, mean the following: At an irradiation of the whole body the lungs participate with a dose equivalent of H_{lungs}, the bones with H_{bones}, and the thyroid with $H_{thyroid}$. This means, for each organ, a certain probability to be affected by cancer, which leads to the death of the irradiated individual. The risk that we perceive is the total risk of dying from cancer, which is proportional to the effective dose according to equation 2.6. Therefore, the German regulations, with respect to radiation protection, have established dose limits in the form of effective doses. The effective dose is indicated in the same unit as the dose equivalent, namely in Sievert (Sv).

Another quantity of a completely different type, relevant for radiation exposure, is the collective dose. While all dose concepts discussed so far serve to quantify individual exposure to radiation, the collective dose is the sum of the individual doses of a collective. If 100'000 persons were exposed to 0.3 mSv each, the collective dose is 30 personSv.

The relevance of this concept is illustrated by an example from the field of workers' radiation protection. Let us suppose that a ten days' job has to be carried out in a sector of a plant where the local dose rate is so high that a worker is exposed to a dose of 0.01

Sv during a working day. No dose higher than 0.05 Sv per annum is admissible for occupationally exposed persons in Germany. Then those regulations may be met by a reduction of the local dose rate by fifty per cent, due to shieldings or decontaminations, or by substituting another worker for the first one after five days. Both solutions comply with the limits, yet radiologically, they are not equivalent. In the first case, the collective dose is 0.05 personSv, in the second case, it is 0.1 personSv. Although this is of no relevance for the individual risk, it tells something about a collective risk, which is of relevance from the viewpoint of society.

Hence the collective dose can help environmental policy in optimization considerations. The effectiveness of radiation protection measures or measures for an emission reduction may be characterized quite well by comparing different measures with respect to their effect on the collective dose. For instance, we could weigh whether, with the same financial expenditure, additional retention measures in nuclear installations (i.e., measures which are not obligatory to meet the legal limits for individual doses for people living in the neighborhood) or modernization of medical X-ray technology would have a greater impact on the collective dose of the population.

2.1.4 Dosimetry

The following three subsections will discuss the question of how to actually ascertain risk-relevant dose values, i.e. body doses or organ doses. Two types of cases can be distinguished here, which we will call the monitoring case and the planning or prognosis case.

In the monitoring case, there is an existing radiation source. So the dose rate for external exposure can be either directly measured or calculated from a measured source intensity. Here, source intensities are concentrations of radioactivity (e.g. Becquerel krypton-85 per m^3 of air) or radioactive surface density (e.g. Becquerel cesium-137 per m^2 of surface), which can be measured with very high sensitivity and accuracy. In the case of internal exposure, the amounts of radionuclides taken in with food, drinking water or breathing air, and those eliminated are determined, and then the dose or the dose rate in an organ in which this radionuclide is absorbed is calculated. One interim step of this calculation deals with the internal radiation sources and their source intensities which result from the distribution of radionuclides to different organs. Figure 2.4 depicts this diagrammatically.

In the planning or prognosis case, we start from planned or precalculated emissions of nuclear installations in order to calculate resulting external or internal exposures. From these, we calculate, with the help of radioecological models, the external radiation

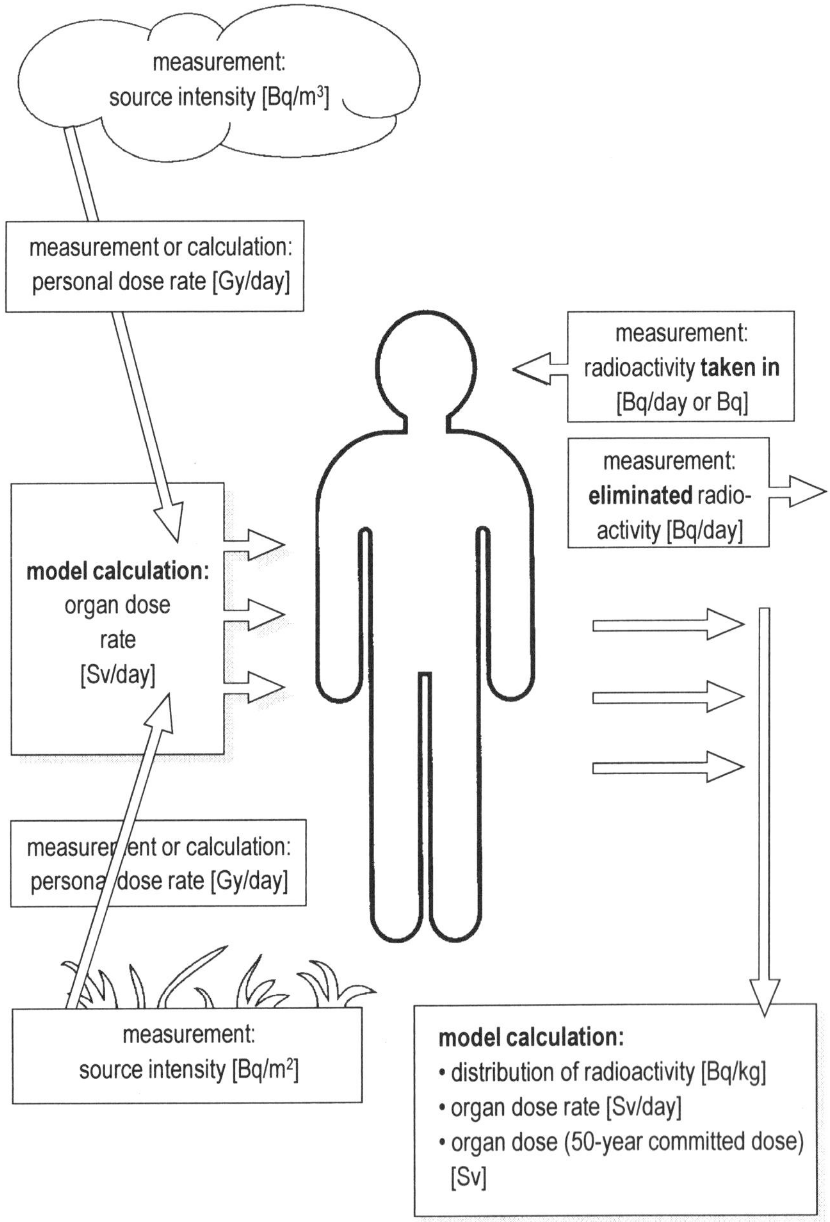

Fig. 2.4 Ways of determining doses (the units were chosen arbitrarily)

sources which will originate from radioactive air plumes and fallouts and the ingestion of radioactivity which is to be expected for the population.

2.1.4.1 Dosimetry of External Radiation Sources

What has to be achieved here is the determination of a dose rate in a person's organ or whole body which results from exposure to an external radiation source the intensity of which is known. The person exposed could, for example, be a farmer who works for some hours every day on a meadow contaminated with cesium-137 due to Chernobyl fallout. If the source intensity, hence in this case the density of radioactive Cs-atoms on the meadow surface (Bq/m^2), and moreover, the type and energy of the radiation and the self-shielding of the source are known, the following equation holds for the dose rate (Gy/s):

$$D(t) = C(t) \times DRC \tag{2.7}$$

with DRC a dose rate constant.

The dose rate refers to a point in time t, since the concentration of radioactivity C(t) is time-dependent. First, radioactivity disappears according to the respective half-lives of the radionuclides involved, second, concentration also changes because of processes of dispersion and washout. Time-dependence has to be taken into consideration in all those cases where a dose is to be integrated over a long period of time – long with regard to these processes (dose is dose rate multiplied with exposure time). Cs-137 has a half-life of about 30 years, so that this time-dependence is virtually of no importance in the case of a one-year dose over a meadow contaminated with Cs-137, yet it is quite relevant if a 50-year dose is determined.

The dose rate constant in equation (2.7) is the conversion factor from source intensity to dose rate. Its calculation is too complicated to be explained here. The main elements involved are:

- those geometrical conditions which essentially determine which part of the emitted radiation reaches the exposed object (the whole body or an organ) and is absorbed in it. The simplest geometry, which, however, does almost not occur in environmental radioactivity, is a point source. In the context discussed here, the most frequent source geometries are the extended (extended as compared with the exposed object) homogeneous area source (contaminated area of soil) and the again extended – at least in upward direction – homogeneous volume source (radioactive cloud), in which the object is immersed completely. According to the source geometry, the different organs are hit by very different radiation flows.

- scattering and absorption processes which photons or β–particles undergo on their way to the exposed object and which lead to a distribution of intensity and energy different from that in the emitted radiation. Among these are scattering or absorption in air, at roughnesses of the surface, and at compact obstacles, which lead to a weakening. Yet there may also be an intensification due to scattering processes, e.g., when radiation emitted by a radioactive cloud is backscattered from the ground.
- absorption of radiation in the exposed object, which leads to different exposure of the individual organs in the human body, in addition to radiation distribution for geometrical reasons.

In principle, the step from source intensity to organ dose requires a few interim steps:

$$\text{source intensity} \rightarrow \text{local dose} \rightarrow \text{personal dose} \rightarrow \text{organ dose.}$$

These interim steps are implicitly included in the tabulated dose rate constants (cf. table 2.2). Yet the scheme also shows that instead of starting from a measured source intensity one can also start from a measured local dose or person dose. This case is rather relevant for radiation protection of occupationally exposed persons.

Dose rate constants are tabulated for many types of exposure situations. Table 2.2 is an example (Eckerman et al. 1980; Kocher 1983). This is an achievement of research into radiation protection, which has been carried out for decades at great expense. Yet on the other hand, it is always possible to simulate exposure situations by means of human phantoms and to either determine body doses or organ doses experimentally or verify calculated values by experiment.

2.1.4.2 Dosimetry of Incorporated Radionuclides

In external dosimetry, which we are primarily interested in with regard to radioactive substances in the environment, source intensity and the respective dose rate constant are the quantities by means of which the *dose equivalent rate* in the irradiated organ can be determined. Then, the total *dose equivalent* which a person receives depends on the duration of exposure. Internal dosimetry deals with exposures due to radioactive substances which are taken in from the environment, e.g. by ingestion from food. Once a certain amount of radionuclides is taken in, total exposure is predetermined by biological and physical parameters, irrespective of the behavior of the exposed person (except for medical measures to accelerate the elimination of radionuclides). The essential quantities here are the amount of radionuclides taken in and the corresponding dose factor, from which we reasonably do not calculate a dose rate but the so-called 50-year committed dose. That is the dose equivalent which accumulates in the fifty years following intake of the radionuclide, hence the lifetime dose, if fifty years are seen as mean value of remaining lifetime for adults after the intake of radionuclides. Lifetime

radio-nuclide	skin (β)	breast (γ)	lungs (γ)	bone marrow (γ)	ovaries (γ)	bones (γ)	testes (γ)	whole body (γ)
^{3}H	0	0	0	0	0	0	0	0
^{14}C	0	0	0	0	0	0	0	0
^{85}Kr	2,6E-4	6,7E-7	5,4E-7	5,6E-7	4,9E-7	6,3E-7	7,3E-7	5,8E-7
^{90}Sr	4,4E-5	0	0	0	0	0	0	0
^{90}Y	3,4E-3	0	0	0	0	0	0	0
^{95}Zr	6,9E-5	2,1E-4	1,8E-4	1,8E-4	1,6E-4	2,0E-4	2,4E-4	1,9E-4
^{99}Tc	0	2,3E-10	1,3E-10	9,8E-11	1,0E-10	2,3E-10	2,0E-10	1,5E-10
^{129}I	0	1,3E-5	2,9E-6	1,1E-6	2,1E-6	5,1E-6	8,4E-6	5,5E-6
^{131}I	5,8E-5	1,2E-4	9,4E-5	9,6E-5	8,2E-5	1,2E-4	1,3E-4	1,0E-4
^{133}Xe	0	2,2E-5	9,6E-6	6,0E-6	7,2E-6	1,7E-5	1,7E-5	1,2E-5
^{135}Xe	5,6E-4	8,6E-5	6,1E-5	6,3E-5	5,1E-5	8,2E-5	8,4E-5	6,6E-5
[^{137}Cs$^{*)}$	7,4E-5	0	0	0	0	0	0	0]
[^{137m}Ba	2,9E-4	1,8E-4	1,5E-4	1,5E-4	1,3E-4	1,6E-4	1,9E-4	1,6E-4]
^{210}Pb	0	2,3E-6	3,9E-7	1,7E-7	2,8E-7	7,1E-7	8,6E-7	6,9E-7
^{210}Bi	1,1E-3	0	0	0	0	0	0	0
^{210}Po	0	2,5E-9	2,1E-9	2,1E-9	1,9E-9	2,3E-9	2,7E-9	2,2E-9
^{222}Rn	0	1,2E-7	9,5E-8	9,7E-8	8,5E-8	1,1E-7	1,3E-7	1,0E-7
^{226}Ra	0	2,6E-6	1,7E-6	1,6E-6	1,3E-6	2,5E-6	2,3E-6	1,8E-6
^{235}U	0	5,9E-5	3,8E-5	13,6E-5	3,0E-5	5,6E-5	5,2E-5	4,1E-5
^{236}U	0	9,1E-7	3,8E-8	1,8E-8	2,4E-8	6,5E-8	1,7E-7	1,7E-7
^{238}U	0	8,0E-7	3,3E-8	1,5E-8	2,1E-8	5,7E-8	1,5E-7	1,5E-7
^{238}Pu	0	1,1E-6	3,3E-8	1,2E-8	1,9E-8	5,5E-8	2,0E-7	2,1E-7
^{240}Pu	0	1,1E-6	3,2E-8	1,2E-8	1,9E-8	5,4E-8	2,0E-7	2,0E-7
^{241}Am	0	1,5E-5	5,4E-6	2,9E-6	4,0E-6	1,0E-5	9,6E-6	7,2E-6

$^{*)}$ in equilibrium with ^{137m}Ba

Table 2.2 Dose rate constants (Sv/annum per Bq/cm^2) for irradiation of an adult person who stands on contaminated ground, in relation to source intensity (radioactive surface density) for different radionuclides and organs (E-8 means x 10^{-8}) (Eckerman et al. 1980; Kocher 1983)

doses for children are often calculated with a 70-year committed dose. For most radionuclides, however, the 50-year dose and the 70-year dose have been virtually completely accumulated after one year already, because of rapid elimination and radioactive decay. Yet most real situations do not consist in one single intake but in a continuous intake of radionuclides. Then we have to deal with several overlapping 50-year or 70-year periods.

Clearly, a direct measurement of source intensity is much more difficult for internal than for external exposure. For sufficiently energetic γ–emitters, the total amount of radioactivity in a body can be determined with *whole-body measurements*. With the

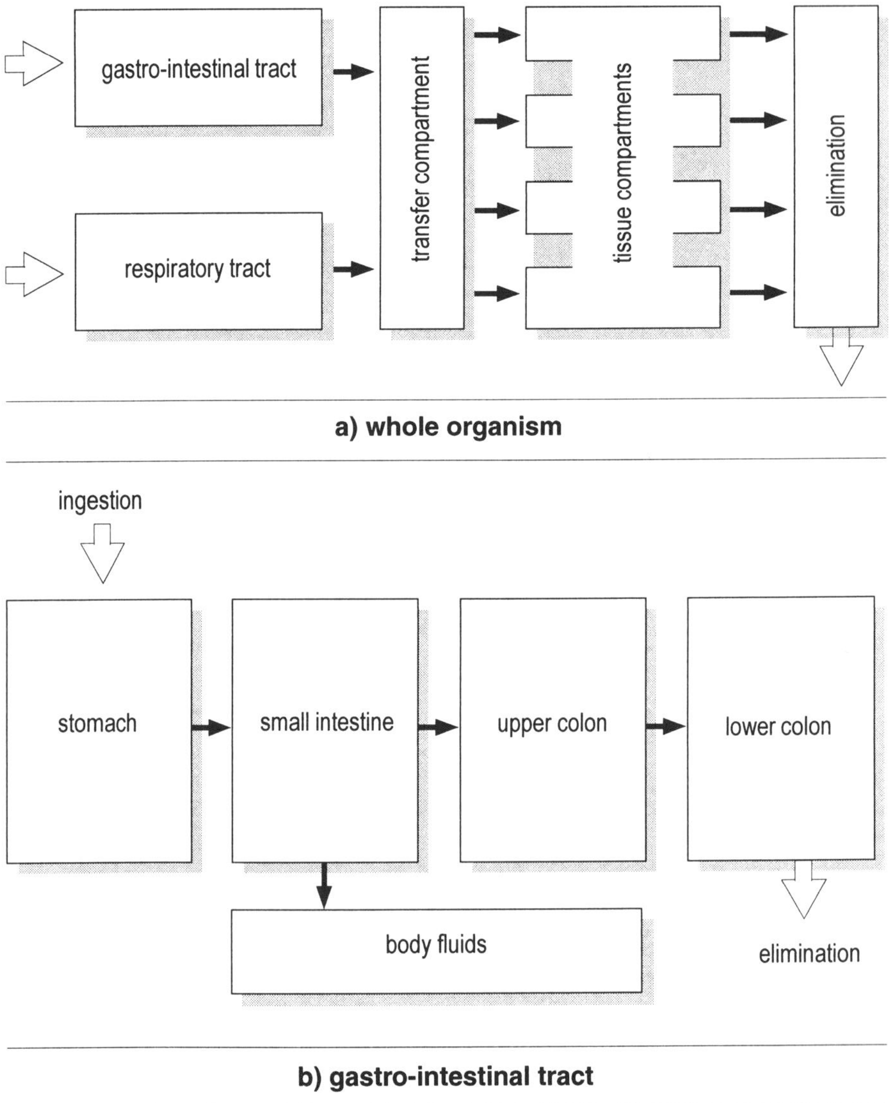

Fig. 2.5 Model scheme of metabolic kinetics of radionuclides

help of the so-called γ–ray spectrometry we may then perhaps also find out what kind of radionuclides are involved. Pure α– and β–emitters cannot be detected by whole-body measurement but can only be determined by elimination measurement. For some organs, e.g. the thyroid, the incorporated radioactivity can even be measured organ-specifically from outside. Organ-specific measurements are difficult, though, and are required only in special cases or for a sample. To determine an internal exposure we have, as a rule, to proceed from radionuclides eliminated by urine and faeces.

Various radionuclides are distributed differently in the human organism. Exemplary for those radionuclides which spread virtually homogeneously over all body tissues is potassium-40, which is always contained in normal potassium and therefore exists as a natural internal radiation source everywhere in the body. Some anthropogenic radionuclides, like the radioactive cesium isotopes and tritium (as tritiated water) behave similarly. Yet there are also many anthropogenic radionuclides in the environment whose distributions in the body show marked heterogeneity, e.g. the iodine and strontium isotopes.

Hence there are two problems to be solved in internal dosimetry. We have to determine:
- with the help of a biological model, the distribution – and its dynamics – of a radionuclide in the organism, in order to determine the organ-specific source intensities and their change in time;
- with the help of a physical model, the portions of radiation emitted from these sources which are absorbed in the different organs, in order to determine lifetime exposure of the person concerned.

Intake of radionuclides occurs via respiratory tract (inhalation) or via gastro-intestinal tract (ingestion), in rare cases even via circulatory system (e.g. injection) or via skin (percutaneously). Biological distribution models subdivide the human organism into compartment systems. Respiratory tract and gastro-intestinal tract are such compartments, as well as the transfer compartment blood together with some connected organs. A radionuclide reaches the transfer compartment blood via lungs or via intestines (unless it is injected directly). Some portion remains within the compartment system by which it was taken in, and disperses into the different subcompartments, as for instance, the different parts of the digestive tract. Elimination occurs via kidneys, intestines and lungs. Figure 2.5 shows a diagram of the metabolism model used for internal dosimetry by the ICRP.[6]

The mechanisms underlying transitions between different compartments or subcompartments are understood fairly well. The heavy concentration of radioactive iodine in the thyroid follows the normal iodine metabolism of this organ, which leads to the formation of the hormones thyroxine and triiodothyronine. Hence iodine is an essential trace element. Lack of iodine may lead to a growth of gland tissue (goiter). Additionally, lack of iodine favors intake of radioactive iodine by the thyroid. Yet there are many radionuclides the distribution of which in the organism is less understood and known. The majority of them, however, are such radionuclides which do not significantly contribute to radiation exposure of the population.

Distribution of radionuclides in the body is a dynamic and not a static process, since a certain portion of the amount taken in is continuously eliminated at the end of the metabolic chain. Hence it is determined on the one hand by the shares of different compartments in the respective total body amount and on the other hand by rate constants for transitions. Rate constants can often also be indicated in terms of biological half-lives, which are analogous to half-lives of radioactive decay. Of course, radioactive decay also leads to a reduction of the concentration of radionuclides in the body, which becomes significant when physical half-life is of the same order or shorter than biological half-life. The total rate of both processes can then be characterized for radionuclides with simple biokinetics by an effective half-life:

$$T_{eff} = \frac{T_{phys.} \times T_{biol.}}{T_{phys.} + T_{biol.}} \tag{2.8}$$

If there is a continuous input of a more or less constant amount of radionuclides into the body, a stationary concentration is reached as soon as the amount disappearing per unit time is equal to the amount added.

It has to be taken into account that radioactive decay may lead to a radioactive daughter nuclide with different radiation characteristics and, as an isotope of a different chemical element, with different chemical characteristics as well. Such a parent-daughter situation occurs with radon, which, as a radioactive noble gas (resulting from natural decay of uranium), decays with a half-life of about four days, producing several radioactive daughters. While radon itself, as a chemically inert gas not involved in metabolism, is exhaled, its solid decay products, which are mostly bound to aerosols, are deposited in a finely dispersed form in the respiratory tract. Thus, the problem known as "radon problem", of a sometimes alarmingly high indoors exposure to radioactive radon emitted from the building ground or from building material, is rather a radon-daughter problem.

The biological behavior of a radionuclide is essentially characterized by its chemical properties. Hence, the chemical form in which it is taken in is of considerable importance for distribution and duration of incorporation (cf. table 2.3). Of equal importance may be the path along which it enters the body. One example of this is plutonium. It makes a crucial difference for its duration of incorporation whether plutonium is ingested via the gastro-intestinal tract, from where it is eliminated rather quickly, or whether it is inhaled as particles of aerosols (above all PuO_2). Insoluble particles may be retained for relatively long period in the respiratory tract, so that the tolerable concentration of plutonium is higher in the case of ingestion than in the case of inhalation. With regard to a common error, it should be noted that the chemical toxicity of plutonium is negligible as compared with its radiotoxicity.

element	class	compounds
cobalt (Co)	Y W	oxides, hydroxides, halides, nitrates all other compounds
strontium (Sr)	Y	titanates, soluble compounds
zirconium (Zr)	Y W D	carbides, oxides, hydroxides,halides,nitrates all other compounds
molybdenum (Mo)	Y D	oxides, hydroxides, sulfides, all other compounds
ruthenium (Ru)	Y W D	oxides, hydroxides halides all other compounds
cerium (Ce)	Y W	oxides, hydroxides, fluorides all other compounds
technetium (Tc)	W D	oxides, hydroxides, halides,nitrates all other compounds
bismuth (Bi)	W D	all compounds except nitrate nitrate
uranium (U)	Y W D	insoluble oxides sparingly soluble compounds readily soluble compounds
plutonium (Pu)	Y W	dioxide all other compounds
iodine (I)	D	all compounds
cesium (Cs)	D	all compounds

Table 2.3 Retention classes for inhaled compounds (Killough & Eckerman 1983)

Thus, shares of total activity in different compartments, and time constants of biological transitions and of radioactive decay are the parameters which determine the time-dependent distribution pattern of a radionuclide in an organism. Time constants of radioactive decay have fixed values, which are exactly known. As to biological parameters, individual dispositions may be of great importance. Of utmost relevance is size variability of the important organs, due to individual differences, sex and age, and particularities of metabolism. So, for the distribution model, a reference person has been defined. (Age is often separately taken into consideration, though, at least with respect to the dimension of the organs.) This situation, which requires neglecting individual particularities and calculation of mean values, is reflected in a correspondingly rough classification of radionuclides in their respective chemical form into three retention classes, which comprise substances the effective half-life of which is of the order of days (class D), weeks (class W), and years (class Y). Table 2.3 gives examples of classifications of radionuclides according to this scheme (see also ICRP 1979).

Notwithstanding the imperfect approximation of reality through model description of the distribution of radioactive substances in the organism, the important fact is that some radionuclides by and large disperse homogeneously in the body, while others heavily concentrate in specific organs.

The radionuclides distributed in the organism turn some organs or tissues into radiation sources, the respective source intensity of which can be derived from the distribution model. As far as particle rays with low penetrating power are concerned (α-rays and β-rays), the radioactive material contained in an organ more or less irradiates only this organ itself. With γ-rays, on the other hand, other organs are affected as well. In principle, the instruments for considering scattering and absorption of radiation are the same as for external dosimetry. Mathematical treatment, however, becomes more complicated, since not only the geometry of the irradiated organs but also the geometry of the radiation sources, which are completely or partially identical with these organs, is complicated.

The dose rate in an organ T irradiated by different organs S at the time t is described by the following expression (one equation for each radionuclide):

$$\dot{D}(t) = \sum_S A_S(t) \times \Delta E \times AF_{(T \leftarrow S)} \tag{2.9}$$

$A_s(t)$ = activity in the organ S at the time t (according to the biological distribution model)

ΔE = energy per radioactive disintegration

$AF_{(T \leftarrow S)}$ = fraction of the energy emitted in organ S which is absorbed in organ T per unit mass (S may be the same as T)

From $\Delta E \times AF_{(T \leftarrow S)}$, an organ-specific value SEE for the absorbed radiation energy can be derived. SEE is the specific effective energy, i.e., the energy which is absorbed per radioactive disintegration in one gram of the irradiated organ. The SEE-value is the

radionuclide $T_{1/2}$=half-life	retention class	gonads	female breast	bone marrow	lungs	thyroid	remaining parts of the body	effective
Sr - 90 ($T_{1/2}$ 29 a)	Y	2,96E - 10	2,69E - 10	3,28E - 08	2,86E - 06	2,69E - 10	5,73E - 09	3,51E- 07
Sr -89 ($T_{1/2}$ 50 d)	Y	7,96E - 12	7,69E - 12	1,07E - 10	8,35E - 08	7,96E - 12	3,79E - 09	1,12E - 08
I - 131 ($T_{1/2}$ 8 d)	D	2,53E - 11	7,88E - 11	6,26E - 11	6,57E - 10	2,92E - 07	8,31E - 11	8,89E - 09
Cs - 137 ($T_{1/2}$ 30 d)	D	8,76E - 09	7,84E - 09	8,30E - 09	8,82E - 09	7,93E - 09	9,12E - 09	8,63E - 09
Pu- 239 ($T_{1/2}$ 24.000 a)	W Y	3,17E - 05 1,20E - 05	2,98E - 11 2,79E - 11	1,97E - 04 7,60E - 05	1,73E - 05 3,23E - 04	4,36E - 12 2,34E - 12	1,06E - 04 4,16E - 05	9,19E - 05 1,19E - 05

Table 2.4 Examples of dose factors (Sv/Bq), related to the inhaled radionuclides (ICRP 1979)

result of the second step of internal dosimetry, which aimed at determining the distribution of radiation energy in the body for a given distribution of radionuclides.

In a third step, we combine the values of $A_s(t)$ and SEE, take the dose equivalent, and sum up the dose rates over 50 years, if the persons concerned are adults. So we obtain dose factors for the 50-year committed dose equivalent per amount of radioactivity taken in, usually in Sievert per Becquerel (Sv/Bq). Such dose factors have been calculated and tabulated for many radionuclides on the basis of a series of simplifying assumptions, for different ways of intake and different ages. Table 2.4 shows some examples (ICRP 1979). Those values draw attention to some radiological characteristics of different radionuclides: heavy retention in the respiratory tract increases exposure of the lungs and decreases exposure of other organs (Pu-239 W/Y); longer half-life leads to higher exposure (Sr-89 Y/Sr-90 Y); uniformly distributed radionuclides mean uniform exposure and vice versa (Cs-137/I-131, Pu-239).

These dosimetric calculations also serve as the basis for determining a value of maximum admissible annual intake of a radionuclide, by recalculation from a given dose limit – from an effective dose of 0.3 mSv per annum in the given case. These derived values are tabulated in the radiation protection regulations, for air on the one hand and for drinking water and food on the other hand. Allowing for a mean annual consumption of 7'200m³ of air, this leads to maximal admissible concentrations (MAC) for air.

radionuclide	breathing air (Bq/a)	drinking water and food (Bq/a)
strontium-90 (insoluble compounds)	5×10^4	3×10^5
strontium-89 (insoluble compounds)	2×10^6	5×10^6
iodine - 131	6×10^5	3×10^5
caesium - 137	6×10^6	4×10^6
plutonium - 239 *	1×10^2	2×10^4

* for some Pu-compounds, higher limits are established

Table 2.5 Selected values for maximum annual intake of radionuclides by inhalation or ingestion, related to the annual limit for occupationally exposed persons, allowing for the 50-year committed dose and the lifetime limit. (As to the release of radioactive substances into the environment, the German radiation protection regulations require that 1m³ waste air must not contain more than a fraction of 10^{-6} of the values listed in the column for breathing air (Bq/m³), and that 1m³ sewage water must not contain more than an annual average fraction of 10^{-2} of the values listed in the column for drinking water and food (Bq/m³) – if no other values are prescribed by the responsible authorities.) (StrlSchV 1989)

In section 2.1.5, we will outline the calculation of internal dose equivalents from radioactivity taken in, using the example of radioactive iodine isotopes.

2.1.4.3 Ways of Exposure

External radiation sources can be measured, incorporated activity can be ascertained by determination of intake or elimination. An important type of situation where both these procedures are not applicable are situations of planning. Here, only projected emission values are available, from which the intensities of expected internal and external radiation sources, i.e., immission values, have to be calculated. Situations of prognosis are somewhat similar, since expected intensities of internal and external sources are to be calculated from actual emissions, e.g., resulting from accidents. While the preceding two subsections described the way from source intensity to external dose rate or from intake of radionuclide via source intensity to internal dose, this subsection will outline calculation procedures which cover the way from emission to intensity of external source or intake of radionuclides.
In principle, such calculation procedures would also permit the calculation of emission limits as secondary standards, starting from admissible dose values. Because of the large variation range of boundary conditions it is, however, not expedient to establish universal limits for emissions. Instead, calculation procedures were standardized, for instance as "Allgemeine Berechnungsgrundlagen" for normal operation of nuclear power plants. This exemplifies that methods as well may have a standard-like character.

The basis of all calculations which go from emissions to source intensities in the environment are radioecological models for a description of the manifold transfer processes undergone by emitted radioactive substances. Emission itself may either take the form of gaseous substances or aerosols being emitted into the atmosphere or of liquids being emitted into surface waters or into ground water. Radionuclides emitted into the air may become internal radiation sources

- by inhalation,
- by dry or wet fallout on plants and subsequent direct or indirect ingestion from food (vegetables, or meat and dairy products from cattle which was fed with the contaminated plants),
- by dry or wet fallout on the soil and subsequent indirect ingestion (in the following years, radionuclides are taken in by plants through their roots and hence enter the human food chain again directly or via meat and dairy products). By resuspension of soil particles, radioactive substances deposited on the soil may also contaminate the surface of sprouted plants.

They may become external radiation sources

- as a radioactive cloud,
- by fallout on the soil and resulting ground radiation.

Radionuclides emitted with water may become internal radiation sources
- by getting into the drinking water,
- by getting into the ground water and from there into plants through their roots and then into the human food chain,
- by contaminating plants and/or soils due to agricultural irrigation measures,
- by getting into the human food chain via marine or limnic organisms.

There are, above all, two types of transportation processes which have to be taken into consideration in modeling, namely physical and biological ones.

Among physical transportation processes are, for example, atmospheric distribution, distribution in ground water, vertical transportation in upper soil layers, and distribution in surface waters, including the oceans. Such processes are only roughly understood in spite of their physical nature, since they are so complex. Furthermore, modeling and mathematical formulation of such processes are faced with limits when representative data are difficult to gather, since local values vary greatly (e.g. sorption coefficients in soils and rocks).

Biological transportation processes are transitions between environmental compartments, involving biological steps. Examples are the transfer from soil to plant or the transfer from water to aquatic organisms. One characteristic biological quantity here is the transfer coefficient, which is, e.g., defined as relation of the activity concentration in a plant to that in the soil. One particular group of biological transportation processes are those within an organism, which involve metabolic processes. Examples are the transitions from feed to milk and from feed to meat. They are of analogous nature and are described by similar models as the distribution of radionuclides in a human body. Experimental values for biological parameters are, of course, more easily attainable here than for human beings.

Today, efficient models are available to describe paths of exposure from source of emission to a human being. The model system ECOSYS for terrestrial ecosystems, developed in Germany (cf. fig. 2.6), stood the test in prognosticating the behavior of fission products which have been deposited in Southern Bavaria after the Chernobyl accident (Matthies et al. 1982; Pröhl et al. 1986). Measured values of exposure have largely confirmed the prognosticated values (Jacobi 1987). Therefore, considerable confidence is placed in this model, also with respect to exposures to radioactive emissions which do not result from accidents.

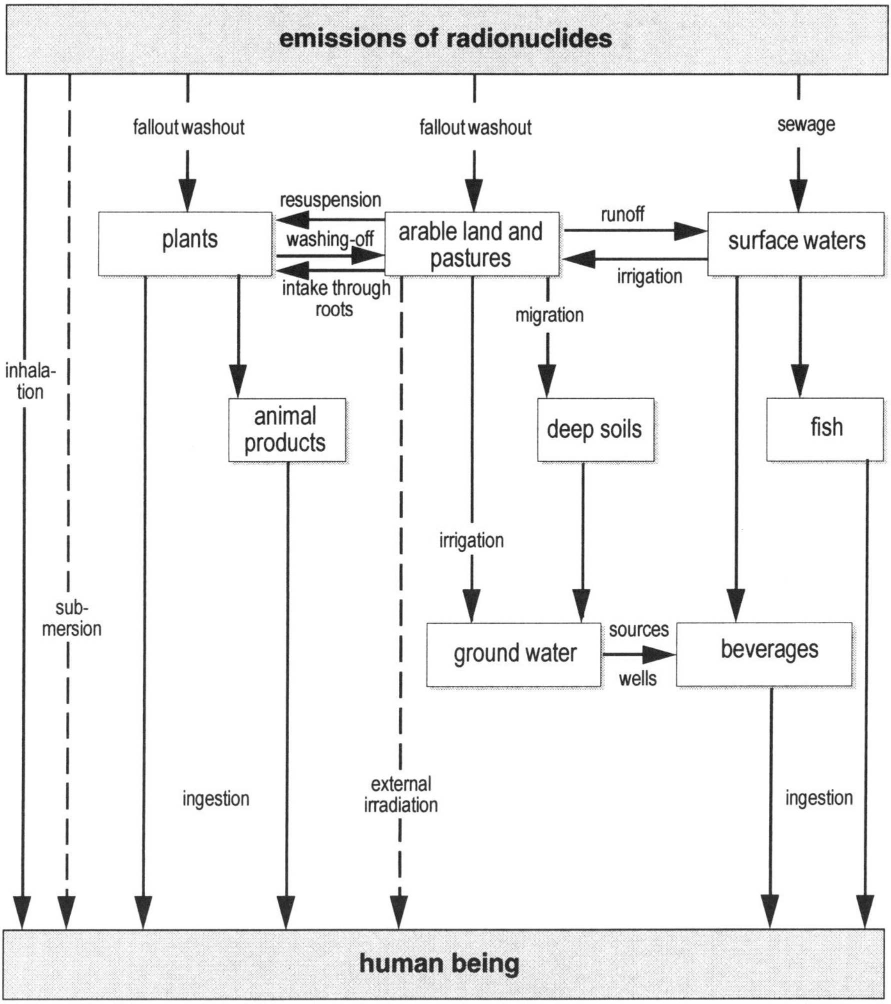

Fig. 2.6 Compartment scheme of ECOSYS – (ECOSYS is a radioecological simulation model describing the transfer of radionuclides through terrestrial food chains up to human beings.)

2.1.5 *Calculation Example:* Thyroid Dose Factors for Incorporated Radioactive Iodine Isotopes[7]

The following text is meant to illustrate a large part of the principles described in the preceding sections by applying them to a concrete example, namely the recalculation of thyroid dose factors. In addition, the reliability of this calculation is discussed, again to demonstrate by an example how carefully this is done. Both tasks require a more detailed and more professional presentation than other parts of this chapter. The reader may omit this section without a loss in information necessary for an understanding of the remaining text.

2.1.5.1 Biokinetics of Incorporated Iodine Isotopes

Iodine is taken into the organism through breathing air and ingesting food. Transition from the lungs and the gastro-intestinal tract (from the stomach and the small intestine) into the blood happens very quickly (within a few hours) and completely – almost independent of the chemical compound in which the iodine occurs. A high percentage of the iodine in blood is taken in by the thyroid, the rest is eliminated with the urine.

The ICRP metabolism models assume the following biokinetic data (fig. 2.7) for incorporated radioactive iodine isotopes in a *reference person:*

- All iodine compounds are classified in the *retention class D* (cf. table 2.3).
- *Resorption* occurs *by 100%* in the small intestine.
- *Intake* of iodine from the blood *in the thyroid* amounts to *30%*.
- *Biological half-life* of iodine in the thyroid is *80 days*.

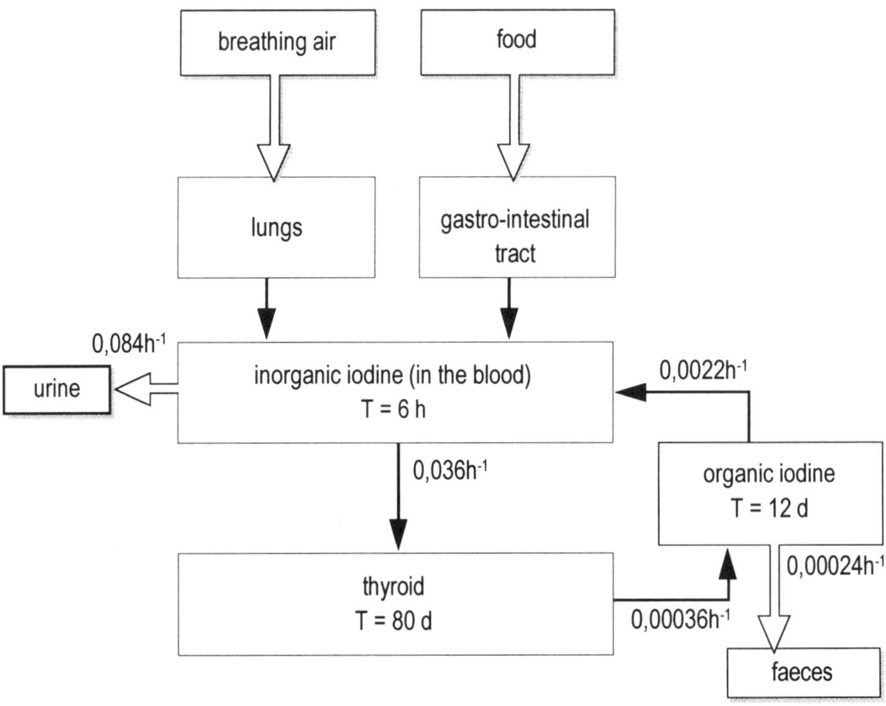

Fig. 2.7 Diagram representing the ICRP-model for the description of the iodine kinetics in human beings (indicated are transition rates and half-lives of residence in individual compartments)

- Iodine leaving the thyroid in the form of an organic compound is uniformly distributed in the whole body and remains there with a biological half-life of 12 days. One part of it is excreted, another part again gets into the thyroid. This recycling leads to an *effective* increase in *biological half-life* to about *120 days.*

Since metabolic data are strongly dependent on total intake of iodine with food, on age, and on the general functional state of the organism (thyroid, kidneys), the applicability of the ICRP-values is restricted to a healthy adult, "sufficiently" (100–200 µg/d) supplied with iodine (reference person).

Figure 2.8 shows the experimentally ascertained uptake of iodine by the thyroid as a function of daily iodine supply from food (Glöbel 1978). Due to the insufficient iodine supply in Germany (an average of about 60–80 µg iodine per day), an intake of about 40%–60% into the thyroid is more likely to be correct for a German adult than the ICRP-value of 30%. Age-dependence of this parameter cannot significantly be asserted because of big individual differences (Karhausen et al. 1973). This calculation, in order to be on the safe side, therefore assumes a value of 50% for all ages.

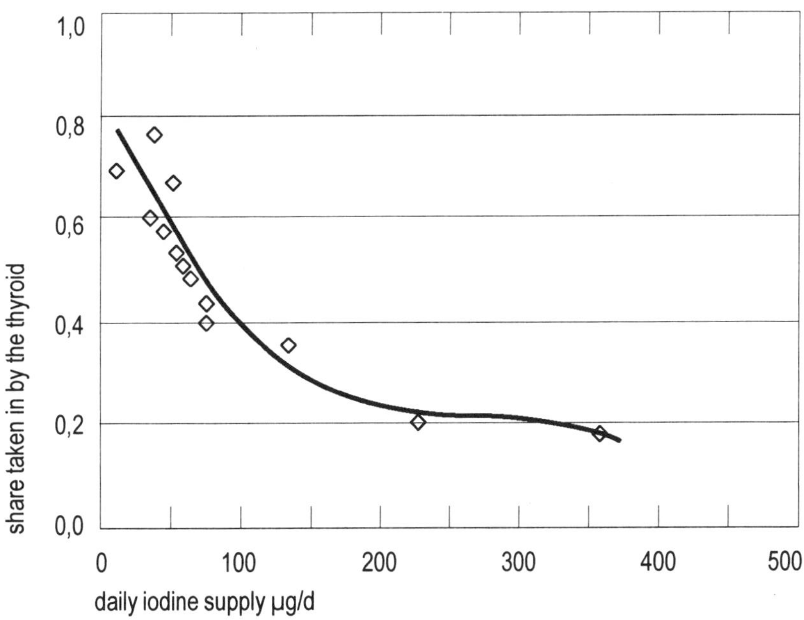

Fig. 2.8 Dependence of intake of iodine into the thyroid on daily iodine supply (Glöbel 1978)

Under balanced conditions, the total elimination of iodine (with urine, feces, and perspiration) approximately matches the iodine intake (Snyder et al. 1969). According to that study, elimination of iodine through urine amounts to about 50% to 75% of the daily ingestion of iodine from food, for persons with and without goiter equally. Hence the remaining elimination of iodine with feces and sweat may amount up to 50%.

Biological half-life, as opposed to iodine intake into the thyroid, shows a strong age-dependence. Table 2.6 lists typical values (allowing for recycling after deiodization of the thyroid hormones), which underlie the subsequent calculations (Hofmann 1973; Kaul & Roedler 1980). For the case of poor iodine supply, these half-lives are identical with the respective data of a recent study (Stather & Greenhalgh 1983).

age (a)	biological half-life (d)
0	14
1	25
5	30
10	44
15	50
> 20	120

Table 2.6 Biological half-lives of iodine in the thyroid as a function of age

age (a)	thyroid mass (g)
0	3,0
1	3,5
5	7,0
10	15,0
15	24,0
> 20	45,0

Table 2.7 Thyroid mass as a function of age

2.1.5.2 Specific Effective Energy

With the help of data for radioactive decay of iodine isotopes, SEE-values for the reference person can be calculated on the basis of values of Absorbed Fractions (AF).

Of additional relevance for the calculation of the SEE, however, are the masses of those organs for which doses are to be determined. ICRP calculations for the reference person assume a thyroid mass of 20g for purposes of radiation protection of occupationally exposed adults. Apart from its age-dependence, however, this value varies strongly due to iodine supply from food; hence in Germany we have to proceed from a higher value than that for the ICRP reference person. The age-dependent thyroid masses listed in table 2.7 have been derived from autopsy data collected in Saarland, Germany (Fischer 1981).

Starting from AF-values (cf. equation 2.9) of the reference person, AF-values were calculated for the ages: newborn, one year, five years, ten years, and fifteen years (Henrichs & Kaul 1982). From these values, age-dependent SEE-values for the mentioned ages were derived from the values in table 2.7, and were listed for the thyroid in table 2.8.

nuclide	age (years)					
	0	1	5	10	15	>20
I - 123	1,2E - 2	1,1E - 2	5,7E - 3	2,9E - 3	1,9E - 3	8,6E - 4
I - 125	1,0E - 5	8,9E - 6	4,8E - 6	2,5E - 6	1,6E - 6	7,3E - 7
I - 129	2,1E - 2	1,9E - 2	1,0E - 2	5,2E - 3	3,4E - 3	1,5E - 3
I - 131	6,3E - 2	5,6E - 2	3,0E - 2	1,5E - 2	1,OE - 2	4,6E - 3
I - 132	1,8E - 1	1,6E - 1	8,4E - 2	4,3E - 2	2,8E - 2	1,3E - 2
I - 133	1,4E - 1	1,2E - 1	6,6E - 2	3,4E - 2	2,2E - 2	1,0E - 2
I - 134	2,2E - 1	1,9E - 1	1,0E - 1	5,4E - 2	3,5E - 2	1,6E - 2
I - 135	1,3E - 1	1,2E - 1	6,3E - 2	3,2E - 2	2,1E - 2	9,5E - 3

Table 2.8 Values of the Specific Effective Energy SEE (MeV/g) for different iodine isotopes in the thyroid

2.1.5.3 Dose Factors for Reference Persons of the Population

Dose factors (in Sv/Bq) indicate the dose (in Sv) in a part of the body per delivered activity (in Bq). Values of dose factors for the incorporation of iodine isotopes result from equation (2.9) and the respective age-dependent figures in tables 2.6, 2.7, and 2.8.

Table 2.9 lists the results of these calculations as a function of age, where a value of 50% was assumed for intake of iodine into the thyroid, as was mentioned before. Differing intake values have an only approximately proportional effect on calculated dose factors, due to the complex correlations among total iodine supply, thyroid mass, biological half-life and intake factor.

Table 2.9 permits a comparison of the newly calculated dose factors for adults with those of the ICRP (ICRP 1979). The values obtained here are about 25% below the ICRP-values; this is mainly due to the higher thyroid mass and the higher intake into the thyroid assumed here.

Age-dependence of dose values is essentially influenced by the physical half-life of the respective nuclide; for short-lived iodine isotopes it is almost exclusively based on the different organ mass, while for long-lived iodine isotopes this influence is compensated by a reduced biological half-life in children and adolescents as compared to adults.

nuclide	supply	age (years) 0	1	5	10	15	adults	ICRP adults	StSchV adults*)
I-123	inhal.	2,1E-8	1,8E-8	1,OE-8	5,4E-9	3,6E-9	1,7E-9	2,2E-9	-
	ingest.	3,8E-8	3,6E-8	2,0E-8	1,0E-8	7,0E-9	3,3E-9	4,4E-9	-
	inject.	4,2E-8	3,7E-8	2,1E-8	1,1E-8	7,0E-9	3,4E-9	-	-
I-125	inhal.	5,3E-7	7,5E-7	4,8E-7	3,5E-7	2,3E-7	1,6E-7	2,2E-7	-
	ingest.	9,2E-7	1,3E-6	7,8E-7	5,7E-7	3,8E-7	2,6E-7	3,4E-7	-
I-129	inhal.	1,9E-6	2,8E-6	1,9E-6	1,4E-6	1,1E-6	1,2E-6	1,6E-6	2,0E-6
	ingest.	2,9E-6	4,6E-6	3,0E-6	2,3E-6	1,7E-6	1,9E-6	2,5E-6	2,7E-6
I-131	inhal.	1,9E-6	2,1 E-6	1,2E-6	6,4E-7	4,4E-7	2,1 E-7	2,9E-7	3,8E-7
	ingest.	3,2E-6	3,3E-6	1,9E-6	1,OE-6	7,0E-7	3,5E-7	4,8E-7	5,1 E-7
	inject.	3,2E-6	3,4E-6	1,9E-6	1,1 E-6	7,0E-7	3,6E-7	-	-
I-132	inhal.	1,6E-8	1,4E-8	7,8E-9	4,1E-9	2,7E-9	1,2E-9	1,7E-9	1,4E-8
	ingest.	3,6E-8	3,2E-8	1,8E-8	9,3E-9	5,9E-9	2,8E-9	3,9E-9	1,7E-8
I-133	inhal.	4,6E-7	4,2E-7	2,3E-7	1,2E-7	7,6E-8	3,5E-8	4,9E-8	1,0E-7
	ingest.	8,8E-7	8,0E-7	4,2E-7	2,2E-7	1,4E-7	6,7E-8	9,1E-8	1,4E-7
I-134	inhal.	2,7E-9	2,4E-9	1,3E-9	6,9E-10	4,4E-10	2,0E-10	2,9E-10	6,2E-9
	ingest.	5,7E-9	5,2E-9	2,8E-9	1,4E-9	9,9E-10	4,5E-10	6,2E-10	8,4E-9
I-135	inhal.	8,0E-8	7,0E-8	3,9E-8	2,1E-8	1,3E-8	6,3E-9	8,5E-9	-
	ingest.	1,7E-7	1,5E-7	8,3E-8	4,3E-8	2,9E-8	1,3E-8	1,8E-8	-

*) German radiation protection decree adults

Table 2.9 Values of the thyroid dose per absorbed activity (Sv/Bq) in case of incorporation of iodine isotopes. Values are calculated for an intake factor of 50% and – in case of inhalation – for a mean aerodynamic cross-section of 1μm. (For a comparison: values from the ICRP publication 30 and from the calculation basis for the German radiation protection regulations StrlSchV)

2.1.5.4 Reliability of Dose Calculations

As the preceding description of the calculation procedure shows, the calculation of dose factors involves various biological and physical parameters, the reliability of which determines that of the dose factors.

Our previous calculations of dose factors started from parameters which were assumed as "best estimates"; deviations from "reality" are, on the one hand, dependent on individual variations of parameters (e.g. iodine intake into the thyroid), on the other hand, they are due to the lack of quantitatively and qualitatively sufficient information (e.g., about dependence of thyroid mass on iodine ingestion from food).

Parameters which essentially determine the reliability of the calculated dose factors are:
- Absorbed Fractions (of energy),
- thyroid mass,
- the portion of supplied iodine absorbed in the thyroid,
- biological half-life of iodine in the thyroid.

Now, we will discuss in detail the reliability of these parameters and their influence on the reliability of calculated doses.

Specific Effective Energy (SEE)

Two causes contribute to the variability of values of the Absorbed Fractions (AF):
- actual calculation of AF-values involves statistical errors due to application of the Monte-Carlo procedure, or systematical errors due to the use of simplifying procedures (calculation for isotropically emitting point sources in infinitely extended medium);
- the phantom used for the calculation of AF-values is a mathematical representation of human anatomy, which, apart from connected simplifications, describes a "representative" person, not a specific individual.

The mentioned statistical errors depend on the distances between source and target organs, and on the energy of photons. For photons of 200 keV, e.g., they range from 1% to 30% (Roedler 1977). These errors increase with increasing distances and decreasing energy. If the statistical error reaches a value of 50%, a different calculation procedure is applied: we proceed from isotropically radiating point sources, embedded in an infinite homogeneous medium. This simplification leads to an overestimation of AF-values, which, in extreme cases, may reach the factor 2 (very low energies and large distances), but which is of a mean size of 5 to 10%.

The uncertainties introduced into calculations by applying a phantom can be quantified by means of the organ masses. Since the influence of these uncertainties on the reliability of the SEEvalues depends on the energy and the respective dose contribution, the easiest way to quantify them is based on a Specific Absorbed Fraction SAF:

$$SAF(T \leftarrow S) = \frac{AF(T \leftarrow S)}{m_T} \tag{2.10}$$

If we proceed from standard deviations of organ masses, as they are listed for adults by the ICRP (ICRP 1975), there is a mean relative error of 15% for SAF in the case of non-penetrating radiation, of 10% (S = T) or 5% (S ≠T) in the case of penetrating radiation.

These errors are, of course, much bigger when the adult reference person phantom is used for children and adolescents (up to a factor of 20); if there are age-dependent SAF-values available, though, (Henrichs & Kaul 1982), they show only slightly bigger errors than the above-mentioned values.

Biokinetic Parameters

Biokinetic parameters essential for the accuracy of dose calculation are
- the portion of activity which reaches the thyroid,
- biological half-life.

Both parameters show great regional and individual variations, among other things, due to different iodine supply and corresponding reactions of the organism.

Values for iodine intake given in the literature extend over too broad a range to indicate age-dependence. For information about standard deviation of measured values of intake, this study relies on a survey of respective publications (Stather & Greenhalgh 1983), according to which the relative standard deviation amounts to about 25%. With a mean intake factor of 0.5, this yields a standard deviation of 0.13.

According to a study of Dunning and Schwarz (1981), the intake factor may, as a first approximation, also be assumed to follow a logarithmic normal distribution. In 95% of the cases, the intake factor is below 0.75.

As to the variation range of biological half-lives, only very few comparable studies are available. Standard deviations for conversion rates and hence for biological half-lives given by Stather & Greenhalgh amounted to an average of 25% for persons with normal thyroid function; for an unselected sample of persons, these authors' survey of literature yields standard deviations of about 70%.

In the following, we will assume a logarithmic normal distribution of biological half-life, the mean values of which are listed in table 2.6 and the standard deviation of which is 70%, independent of age; the 95%-percentile then is the double of the mean value.

The influence of this parameter on the effective half-life T_{eff} depends on the relation of biological half-life $T_{biol.}$ to physical half-life $T_{phys.}$ (and hence on age):

$$\frac{\Delta T_{eff}}{T_{eff}} = \frac{\Delta T_{biol.}}{T_{biol.}} - \frac{\Delta T_{biol.}}{T_{biol.} + T_{phys.}} \tag{2.11}$$

With a short physical half-life ($T_{phys.} < T_{biol.}$), the contribution of biological half-life to the error of the effective half-life is negligible, in the reverse case, the relative errors become more and more identical.

The correlation between the relative error of the effective half-life and the relation of biological half-life to physical half-life is represented in figure 2.9.

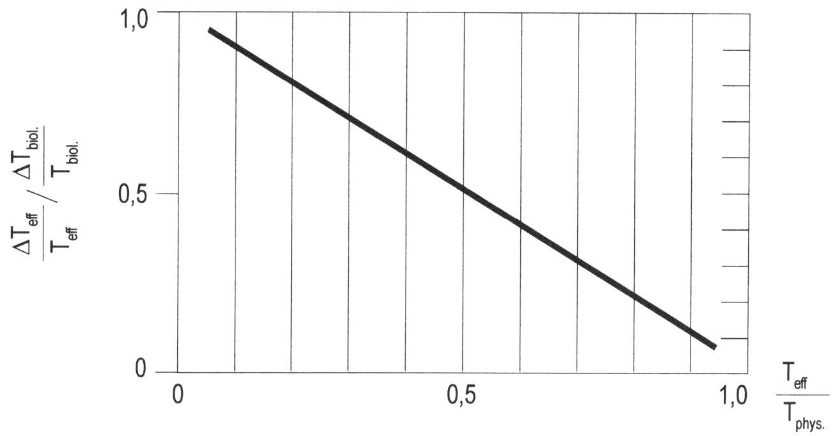

Fig. 2.9 Dependence of the relative error of the effective half-life on the relation between effective and physical half-life

Influence of Individual Errors on the Error of Dose Factors

The individual parameters determining the thyroid dose show great variation ranges, due to individual differences, different iodine supply and individually different reactions to lack of iodine.

There is one common cause for a presumably crucial share of these uncertainties (organ mass, intake factor, and biological half-life), namely lack of iodine, which suggests a correlation among the individual parameters or their uncertainties. If such a correlation existed, it would reduce the uncertainty of resulting doses as compared to the uncertainty in the case of uncorrelated parameters. Nevertheless, the available data do not distinctly indicate such a correlation or do not allow us to statistically prove its existence.

For the case of one single incorporation of iodine, we will therefore, in the following, proceed from uncorrelated parameters in the sense of a "best estimate".

To determine the uncertainty of the resulting dose, the underlying distributions of individual parameters were folded; the resulting values of the 95%-percentiles of the thyroid dose after one single incorporation are 3.5 times the respective mean value (table 2.9) for the long-lived iodine isotope I-129, 2.2 times the mean value for the short-lived iodine isotopes (I-123, I-131, I-132, I-133, I-134, I-135) and 2.3 times the mean value for I-125.

In the case of a continuous intake of iodine isotopes, we may assume a stationary state. The uncertainty of the thyroid dose can then be derived from the variation in iodine concentration, since this dose is proportional to the then constant activity concentration in the thyroid. Autopsy data gathered in Saarland, Germany (Hellstern et al. 1978) again yield a logarithmic normal distribution of this quantity, with a (not age-dependent) standard deviation of 30%. Therefore, in that case the 95%-percentile of the dose is 50% above the mean value, that is, by a factor of 1.5.

2.1.6 Radiation Exposure of the Population

The contributions of different sources to the average radiation exposure of the population have already been mentioned and classified in section 2.1.1. Among them, about 60% are of natural origin, and about 40% are artificially produced radiations. Exposure to artificial radiation is for the most part due to medical applications of radiation and radioactive substances. Normal operation of nuclear installations accounts for only a few tenths of one per cent. Releases of radioactivity because of malfunctions, which would noticeably increase this value, have not yet occurred in Germany so far. German territory has been affected by the fallout of nuclear tests and of the Chernobyl accident. Only in a few places did the Chernobyl fallout reach a similarly high activity concentration per unit area of soil as did fallout due to nuclear tests. In the following, we will study radiation exposure of the population in more detail.

2.1.6.1 Exposure to Natural Radiation

Human beings, like all other beings, have always been exposed to natural ionizing radiation. This radiation may be classified into three categories according to its origin, though only two of which are of importance:

- cosmic radiation
  (often very energetic, hence penetrating, rays of particles and photons, partly from the sun, and partly from galactic and extragalactic sources which are not yet fully understood)
- primordial radionuclides on earth
  (radionuclides, which originate from the formation of terrestrial matter, and the half-lives of which are so long that there are still existing amounts of them today; they include, primarily, uranium-238, uranium-235, thorium-232 and potassium-40)
- cosmogenic radionuclides
  (radioactive types of atoms which originate from interaction of cosmic radiation

with terrestrial matter, e.g. tritium, C-14; their share in exposure to natural radiation is only very low)

Cosmic radiation, which has been known under this name since about 1912, and which was discovered because it presented an insuppressible disturbance factor in measurements of low radioactivity, consists of particles and photons of very different energies. The hardest component are the μ-mesons, which originate as secondary particles in nuclear reactions, with energies of the order of 1'000 MeV. They penetrate lead walls of one meter thickness and more. The soft component, which is already largely absorbed by lead of 10 cm thickness, consists of electrons, positrons (positively charged electrons) and photons. At sea level, this component contributes only 10% to cosmic radiation, at an altitude of 3'500 m, however, about 75%. It is therefore this soft component of cosmic radiation which is essentially responsible for the strong dependence of exposure to natural radiation on altitude. It reaches a maximum in periods of heavy solar flares. Increased activity of sunspots reduces its intensity, probably due to magnetic phenomena.

At sea level, cosmic radiation causes an effective dose of about 0.3 mSv per annum in Germany. Figure 2.10 shows the dependence of dose on altitude, with the dose

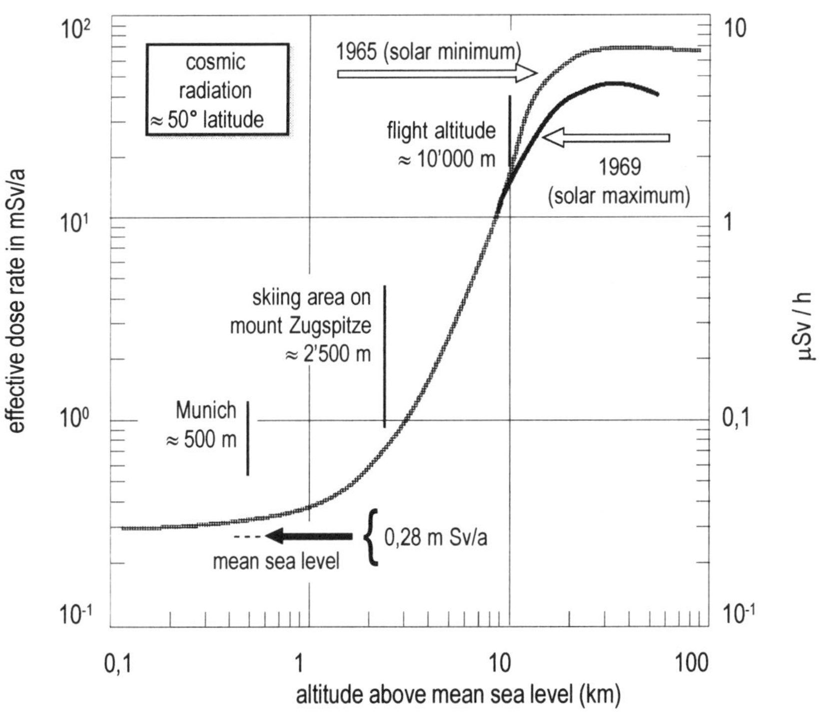

Fig. 2.10 Dependence of cosmic radiation on altitude

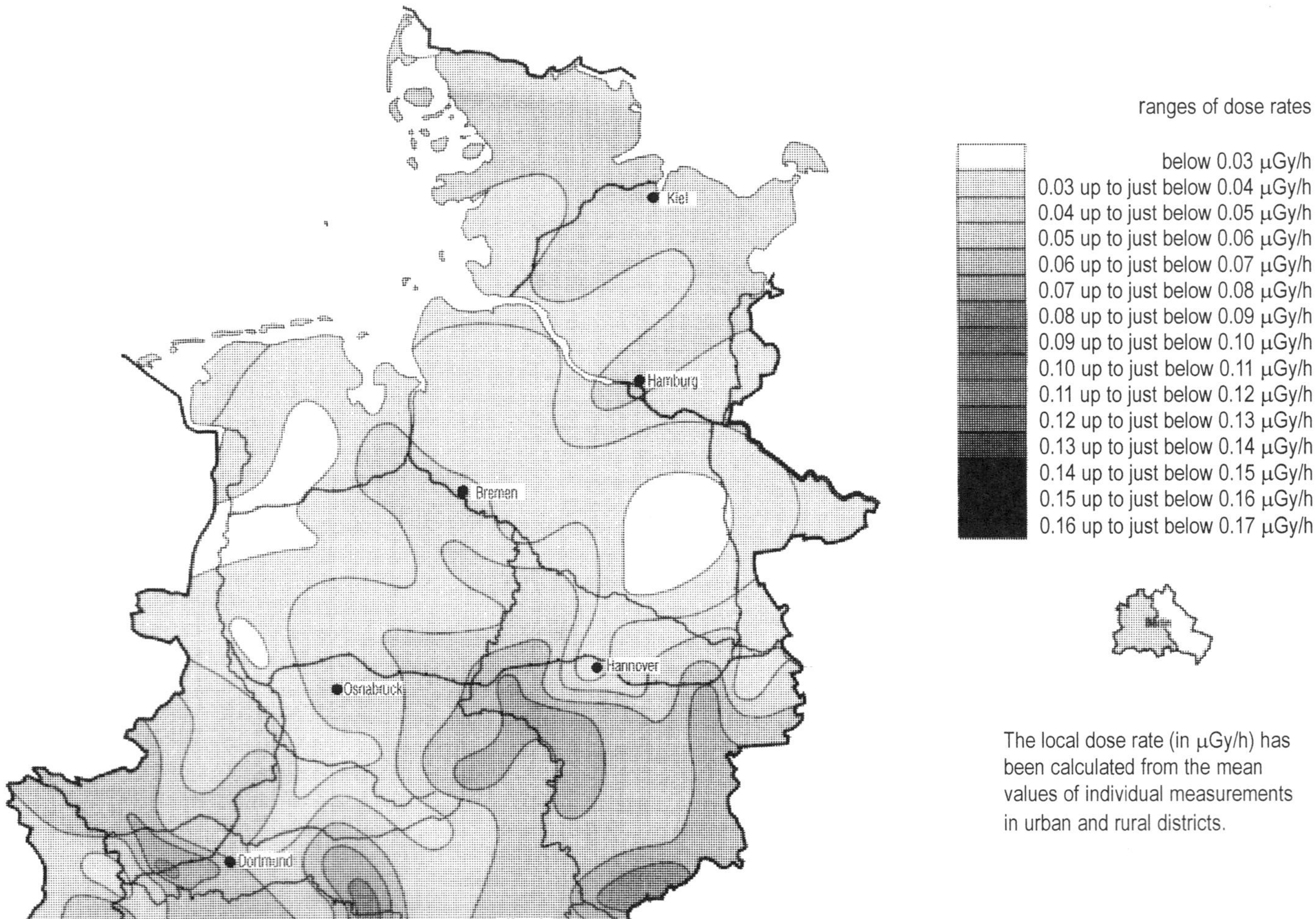

The local dose rate (in µGy/h) has been calculated from the mean values of individual measurements in urban and rural districts.

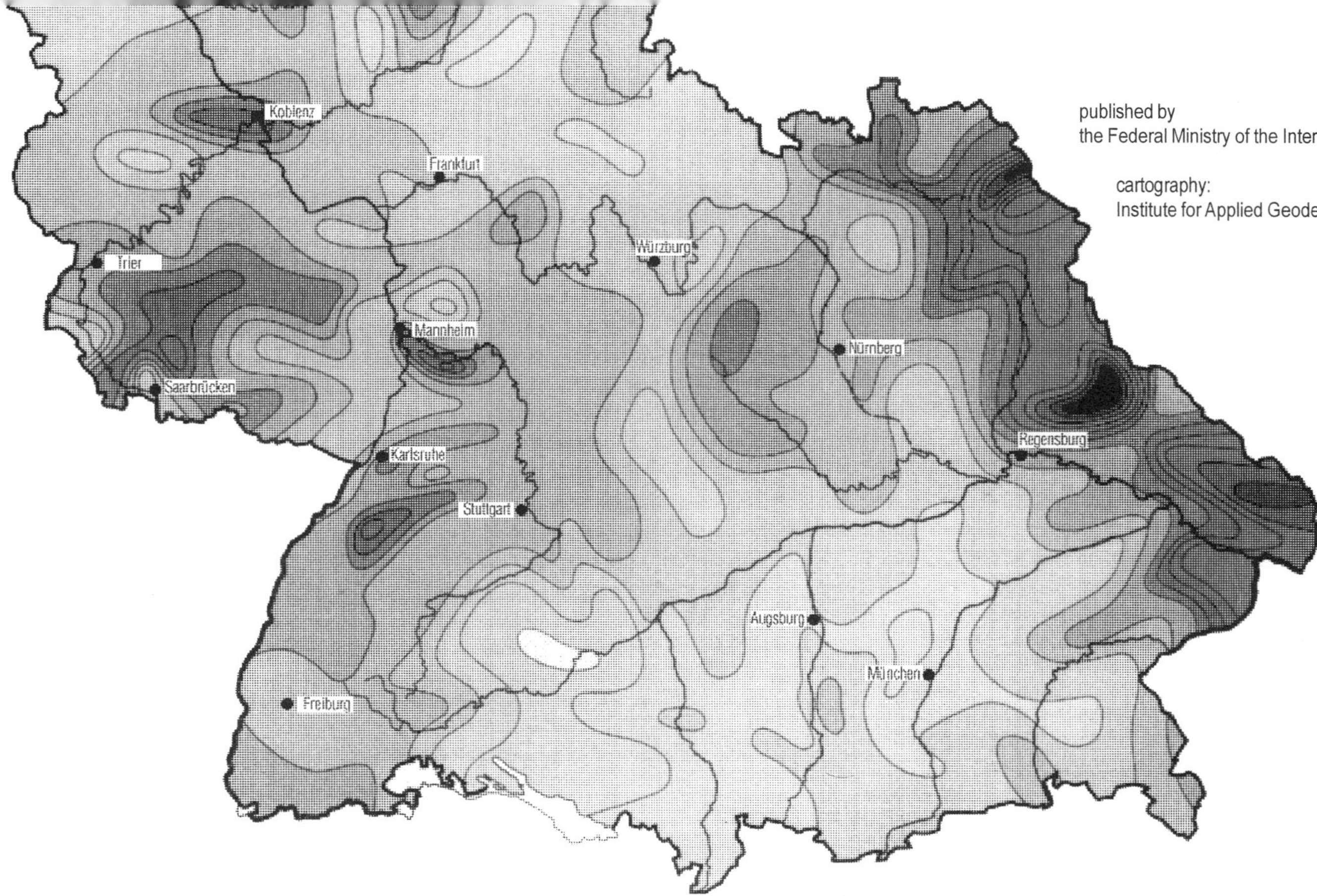

Fig. 2.11 Map of the former Federal Republic of Germany (Western states) showing the distribution of exposure to terrestrial radiation

approximately doubling at an altitude of 1'500–2'000 m (Jacobi et al. 1981). It should be noted that one single transatlantic flight (Frankfurt–New York, at 12 km altitude) entails an additional dose of 0.03–0.04 mSv, hence increases the annual dose by 10%.

Primordial radionuclides contribute in different ways to exposure to natural radiation. As to external exposure, mainly terrestrial radiation is of relevance. It is due to the content of uranium, thorium and potassium in the earth's crust and varies according to local concentrations. Kerala in India is notorious for its extreme values, since the thorium mineral monazite is found there in the form of sand on the earth's surface. In Germany as well, there are distinct sites with top levels of terrestrial radiation. Figure 2.11 shows the geographical distribution of the terrestrial local dose rate (measured outdoors) in the eleven states of former West Germany. Values for sites with minimal dose rate (north of Hanover) and for such with maximal dose rate (e.g., the Fichtelgebirge mountains) differ by about the factor five. In most parts of the country, however, values range between 0.03 μGy/h and 0.07 μGy/h.

Staying indoors means on the one hand being shielded against terrestrial radiation, on the other hand, inhabitants of houses are in most cases exposed to an additional external irradiation, since construction materials are more or less rich in natural radioactive substances of primordial origin. In a house made of slag bricks or clinker bricks, this leads to a total increase of external exposure by about one half, in a wooden house it leads to a slight reduction by a few per cent.

Figure 2.12 shows the frequency distribution of measured values of the annual dose resulting from exposure of the German population to external terrestrial radiation, both for measurements outdoors and indoors. The mean values of the annual effective dose amount to 0.35 mSv or 0.45 mSv, respectively. Based on these values, a mean dose of 0.4 mSv/a can be assumed for the German population. But figure 2.12 also shows that for 10% of the population this value is higher by at least 50% and amounts to ≥ 0.6 mSv/a, and for another 10% it is lower by at least 50% and amounts to ≤ 0.2 mSv/a.

But exposure to natural radiation is not only due to external irradiation but has also an *internal component*, caused again by primordial radionuclides. It amounts to 0.3–0.4 mSv/a – radon and its descendants excluded – and its distribution has a total variation range of more than 0.1 mSv. Thus, the variation range here is lower than that of the external terrestrial component. The reason for this is that more than half the internal exposure to natural radiation is caused by potassium-40. Potassium, which is an essential constituent of any animal or plant cell, always contains 0.012% of the radioactive isotope K-40. The share of internal exposure to natural radiation caused by this radionuclide is therefore constant to ± 10%. The share caused by uranium and thorium, on the other hand, is subject to similar variations as exposure to external terrestrial radiation. Different

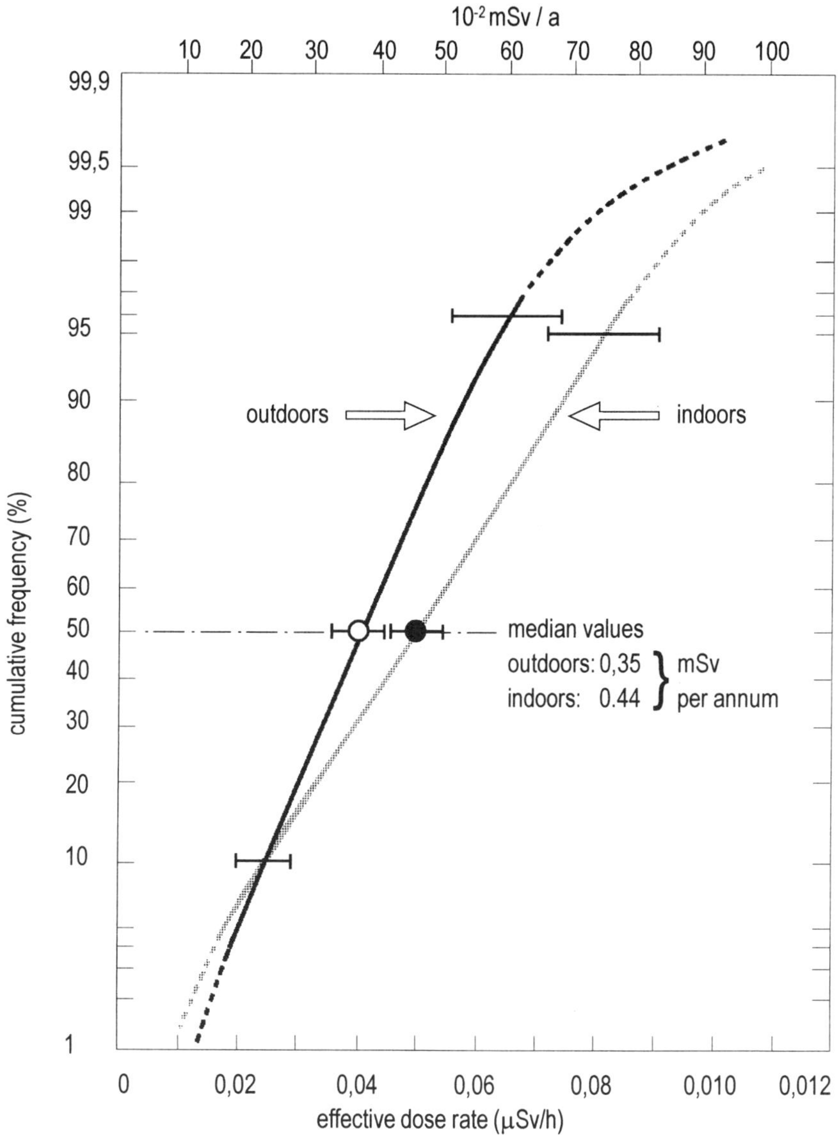

Fig. 2.12 Frequency distribution of the effective dose due to terrestrial γ-radiation (Jacobi et al. 1981)

amounts of these elements are ingested from agricultural products and drinking water, according to their respective concentration in different soils. Local extreme concentrations may, here again, have a strong effect on exposure in individual cases.

Another cause of internal exposure to natural radiation is the content of radon and its daughter nuclides in the air in closed rooms. Radon is a noble gas, while the daughter nuclides are isotopes of heavy metals (Po, Pb, Bi) which are fixed by aerosol particles and are deposited in the lungs. It is only in the past three decades that we have understood that this source of radiation is a problem. The different isotopes of radon (Rn-220 and Rn-222) originate in radioactive decay of uranium and thorium, which are both contained in construction materials and in the building ground of houses. Radon from the ground diffuses into houses through cracks and leaks in the floor. As to construction material, apart from its content of uranium and thorium, its porosity is also of importance. Generally speaking, it is primarily the building ground which is responsible for high radon concentrations in many houses. Construction material contributes up to one third. Measurements in more than 6'000 houses in Germany yielded radon concentrations above 50 Bq/m^3 in half of the houses and below 50 Bq/m^3 in the other half. Figure 2.13 shows the frequency distribution of measured radon concentrations in German houses. The mean effective dose caused by radon, a quantity which does not tell too much, since variations of concentrations in closed rooms are extreme, amounts to 1 mSv.

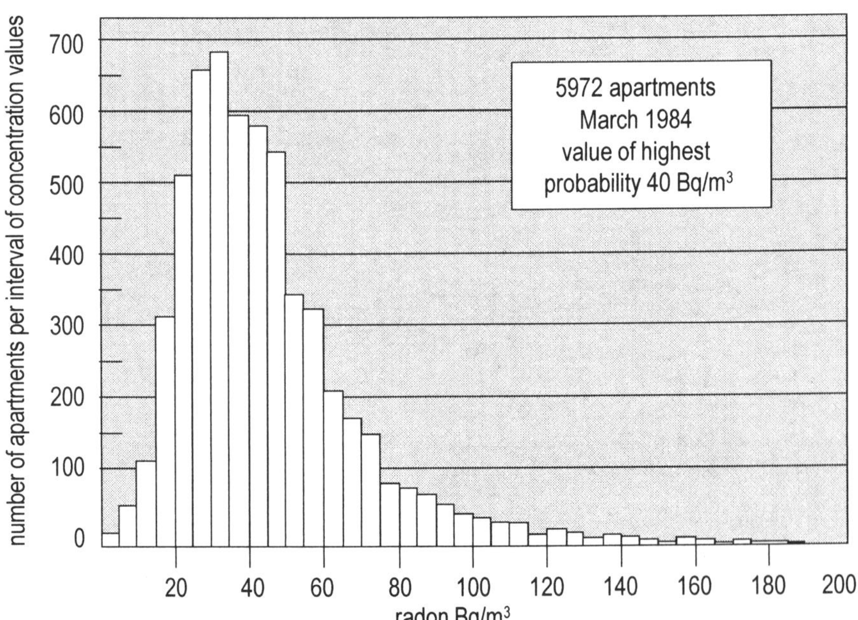

Fig. 2.13 Frequency distribution of all measurements of radon concentrations in the air in apartments

In its recommendation of 30 June 1988, the German commission on radiation protection established 250 Bq/m^3 as a reference value for an upper boundary of the normal range. If the long-term mean value is higher, rehabilitation measures should be proposed to the inhabitants.

Table 2.10 recapitulates mean effective annual dose equivalents caused by all sources of natural radiation, and their scattering around the mean value. Variations of exposure due to radon are too little known to be indicated quantitatively, nevertheless we can be sure that the sum of all contributions is scattered around the mean value by at least ± 0.3 mSv/a.

	mSv per annum	variation range (mSv)
cosmic radiation	0,3	0,15
terrestrial radiation (mainly from potassium-40 and the decay products of the uranium-radium family as well as from the thorium family)	0,4	0,35
incorporation from food and water (decay products of uranium and thorium, as well as potassium-40)	0,3	at least 0,1
inhalation of daughter nuclides of radon, outdoors and indoors (this value may vary extremely, according to subsoil and habits of ventilation)	1	cannot be specified but is very high
sum of exposure to natural radiation	2,0	at least 0,6 mSv

Table 2.10 Average values of exposure to natural radiation

2.1.6.2 Exposure to Artificial Radiation

By far the highest share of exposure to artificial radiation is due to medical applications of radiation and radioactive substances, primarily for diagnostic purposes. There is a very broad application range, and virtually all people are subject to some application in the course of their lives, therein protected only by a rule of minimization without established limits. The mean annual medically caused Effective Dose ranges from 1 to 2 mSv. Mean Effective Doses for some typical applications are:

radiography of the lungs	0.1 mSv
mammography (organ dose: about 5 mSv)	1.0 mSv
computer tomography of the pelvis	20–30 mSv

In former medical examinations of the thyroid function with the help of iodine-131, adults received thyroid doses of 500–1'000 mSv, hence up to 1'000 times the limit of the German radiation protection regulations for the population at large. Although examinations entailing such exposures occur only rarely today, there is still a vast technological potential for further dose reductions, in particular by elimination of obsolete equipment and avoidance of suboptimal application. And in this sector financial expenditures for a significant reduction of exposure can be expected to be more moderate than with regard to many other efforts of risk reduction.

Mean exposure of the population to all other artificial radiation sources lies below 0.05 mSv per annum. About one half of this is due to the fallout of nuclear tests, and less than 20% is due to use of nuclear energy. There may be another 0.05 mSv per annum in regions highly affected by the Chernobyl fallout, the average value for Germany, however, is much lower.

The radiation protection limit for the German population at large of 0.3 mSv per annum as effective dose due to artificial, non-medical radiation sources corresponds with the inevitable variation range of exposure to natural radiation. This is discussed in detail in section 3.1. In Germany, both the actually measured exposure to radiation emitted by nuclear installations, which is much lower than the dose limit, and the dose limit itself, which is restrictive copared to international regulations, are often critically commented upon by saying that facts as well as mandatory regulations only hold for normal situations and not for accidents. Environmental standards, as far as they refer to emissions by technical installations, can only hold for proper operation of those installations. Under conditions of malfunctions or accidents, higher emissions are usually unavoidable and evidently controllable only to a certain extent or not at all. Therefore, environmental standards have to be supplemented by technical standards, which make malfunctions sufficiently improbable, or which sufficiently guarantee that malfunctions will have no heavy impacts on the environment. One crucial element of those technical standards is a licensing procedure which has to ensure that these objectives are reached.

Nuclear technology has probably been the first technology which has been subjected to this principle very rigorously. Here, the basis of the licensing procedure is a safety report. It has to prove that the installation to be licensed is laid out in a way that no serious impacts on the environment must be expected in case of a design basis accident to be defined (formerly GAU = severest credible accident). Of course, the exposure limit for the population of 0.3 mSv/a cannot be maintained in that case. For such situations of malfunction, a total dose of 50 mSv, which should not occur more than once in a human life, is defined to be tolerable. This value is identical with the annual limit for occupationally exposed persons.

The licensing procedure is supplemented by technical controls on the part of the licensing authority. A combination of these two institutions guarantees that no nuclear installation violates radiation protection standards under normal operation conditions. Experience shows that, in practice, nuclear power plants operate far below those standards. Normal operation of a reprocessing plant would entail emissions closer to those standards, but would, of course, be bound not to exceed them either. A final disposal of radioactive waste in geological formations of deep bedrock normally does not entail any radioactive emissions into the environment, neither during storage nor during the post-operation phase.

It can never be excluded completely that an accident of greater proportions than a design basis accident occurs. Such accidents are called hypothetical accidents, because they are of extremely low probability. Since up to now no nuclear facility has been designed so as to completely control such an hypothetical accident, its occurence could possibly lead to severe environmental impacts. It would then no longer be possible to meet radiation protection standards.

Accidents entailing catastrophic impacts on the environment can hardly be conceived for reprocessing plants and not at all for final waste repositories in deep geological formations. If emissions occurred in such a repository, due to some malfunction, it would take them an extremely long time to reach the biosphere. The existing safety analyses show that this would, at the earliest, happen after some period of the order of 1'000 years. For nuclear power plants, where catastrophic accidents are conceivable, enormous technological efforts are made to virtually exclude them. The question of whether we are willing to take the risk of these low-probability accidents means to weigh these risks against the advantages of nuclear energy, in particular the advantage of its marked environmental compatibility in normal operation. But here we leave the field of environmental standards and turn to questions of technical risks and their acceptability. There are no standards for these questions; respective decisions are made rather in licensing and legislation procedures.

If an accident with environmental impacts occurs, ad-hoc standards must be set, if need be, which enable us to handle the situation. One relevant example is the Chernobyl accident. There was, e.g., the necessity to establish ad-hoc standards for admissible concentrations of iodine-131 in milk. Such limits, which have the character of intervention standards, were established in many European countries, but at different levels. In most German states, a limit of 500 Bq/l recommended by the German commission on radiation protection was adopted. Finland and Sweden established a limit of 2'000 Bq/l. Such intervention values for concentrations of radioactivity in food are calculated from an exposure limit and an average value of consumption. In Germany, exposure limits were based on dose values between the general limit for the population of 0.3 mSv/a and the malfunction reference values of a single effective dose of 50 mSv or a thyroid dose of 150 mSv. Different interpretations of this scope led to different MAC-values for iodine-131 in milk.

When maximum admissible concentrations of radioactivity in food are established in accident situations, there will always be a question of weighing different protection objectives against each other. As a rule, one objective competing here with the objective of radiation protection will be the ensuring of food supply; that is, enough food meeting the ad-hoc standards will have to be made available. Another objective, if of lower priority, is that of securing the economic situation of food producers. Because of the necessity of such balancing considerations, establishing standards as a precaution for accident situations is not a sensible thing to do. Such standards will always have to orient by the actual circumstances.

2.2 BIOLOGICAL IMPACTS OF IONIZING RAYS

2.2.1 Survey

The following four problem fields must be considered in a discussion of biological impacts and risks after exposure to ionizing rays:
- death of cells, which may lead to a damage of tissues and eventually to the death of the organism, if cells of particular importance are concerned (2.2.3)
- induction of development anomalies after prenatal exposure to radiation (2.2.4)
- induction of hereditary defects (2.2.5.1)
- induction of malignant diseases (leukemia and solid tumors) (2.2.5.2).

Exposure to radiation may affect cells of all kinds of tissues, i.e. somatic cells and gametes. In the first case, damage to a cell leads to somatic[8] radiation effects, which affect the exposed individual. An exposure of gametes may lead to changes of its genetic material. If the general operability of the cell is unharmed, a fertilization involving the mutated gamete is possible. In that case, hereditary defects may occur in offspring of the exposed individual.

For a quantitative assessment of radiation risks, knowledge of the dose effect relation is necessary. Two basically different types of dose effect relations are distinguished, which are represented in figure 2.14. As to the first type, a threshold dose has to be exceeded before the described effects may be induced. Once this threshold is crossed, the number of persons affected, but primarily the severity of the effect, increases with increasing dose (ICRP 1977). This type of radiation effect is called *non-stochastic*[9] radiation effect. Here, a multicellular mechanism operates, that is, many cells have to be damaged for effects to become manifest. Acute and late radiation damages of tissues and organs (leukemia and solid tumors excluded) are assigned to this category.

Developmental anomalies belong to this class as well. Threshold doses are higher, often much higher, than the exposures to be expected due to environmental radioactivity.

As to the second type of radiation impacts, the so-called *stochastic effects*, it is not the severity of such effects but the probability of their incidence which increases with increasing radiation dose. Their occurrence follows a stochastic distribution. Experimental observations and reflections about the mechanism make us assume that induction of these effects is a unicellular process, that is, a process which starts from one single damaged cell. As we lack experimental or epidemiological data in the lower dose range, we assume that there is no threshold dose and that the dose effect relation is a linear one for low doses. Induction of hereditary defects and of malignant diseases (leukemia and solid tumors) is assigned to this category of radiation effects.

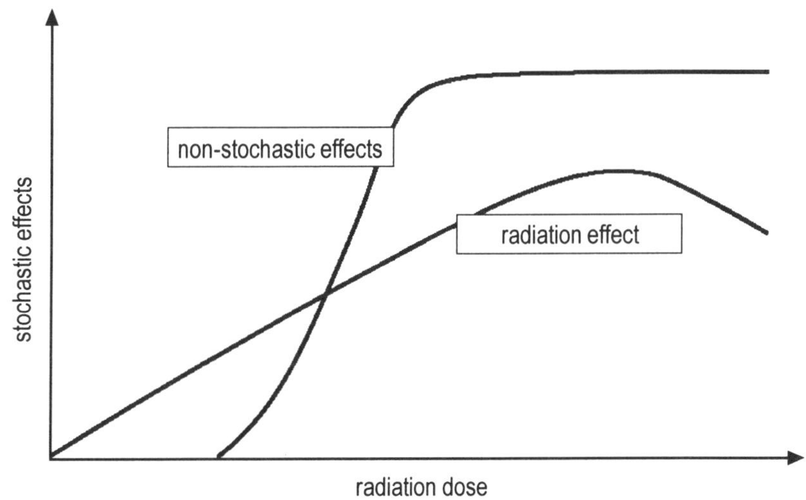

Fig. 2.14 Basic types of dose effect relations of exposure to radiation

As to hereditary defects, there is no doubt that damage of one single germ cell suffices to possibly lead to a mutation in the next generation after participation of the cell in a successful fertilization. With respect to induction of leukemia and solid tumors (malignant diseases) we may also reasonably assume that malignant transformation of one single cell is sufficient to cause, with a certain probability, a malignant disease, since a monoclonal tumor growth has been observed (UNSCEAR 1986).

Yet, epidemiological studies found a significantly increased occurrence of malignant diseases in human beings only at radiation doses of at least some tenths to one Sievert. A statistically significant increase of genetic defects through radiation has not yet been observed epidemiologically so far, it has only been found experimentally in animals. Since an increase of those damages at low doses may, however, be veiled by the statistical variation ranges of "spontaneous"[10] rates, we nevertheless assume that there is no threshold dose for induction of stochastic effects. Therefore, these radiation effects are of crucial importance for an assessment of the radiation risk at low doses, and hence for environmental standards, which are to be discussed here.

Killing of mammalian cells to an extent that leads to functional impairments of tissues and organs occurs only after exposure to radiation doses of at least 0.5 Sv to 1.0 Sv. Irradiations of the whole human body with several Sievert may reduce crucial cell populations so much that acute clinical symptoms and eventual death are the consequence. Developmental anomalies, on the other hand, are induced at lower doses. Animal experiments show that radiation doses of about 0.1 Sv may induce such effects.

2.2.2 Basic Radiobiological Phenomena

2.2.2.1 Processes of Formation of Radicals and Ionization

Exposure of biological material leads to energy absorption (as was already discussed in section 2.1), which results in the formation of radicals and in ionization events in the irradiated matter. These processes may directly (direct radiation effect) affect biological macromolecules (nucleic acids, proteins). But energy absorption may also lead to formation of radicals and ionizations in water molecules in the cell, the products of which may then perhaps react via interim steps with biomolecules and induce radiochemical changes (indirect radiation effect). The ionization density is determined by the type of radiation and its energy. Type and energy of radiation are subsumed under the term radiation quality (Kiefer 1981).

Microscopic spatial distribution of energy absorption is very different for different radiation qualities. The energy dose only indicates the spatial mean value of energy deposit related to mass. If mammalian tissues are exposed to sparsely ionizing radiation (radiation with low LET, e.g. γ–radiation of cobalt-60) at a dose of 10 mGy, this leads to a mean absorption of the energy of three ionizing particles or γ–quanta in every cell nucleus; at a dose of about 3 mGy the energy absorbed is only that of one γ–quantum or ionizing particle. So far, dose distribution over all cells is thus relatively homogeneous. If the radiation dose decreases further, the absorbed amount of energy in each hit cell nucleus remains constant, but the number of hit cell nuclei decreases in proportion to the radiation dose. For radiation with high ionization density, however, the border dose, below which not all cell nuclei are exposed, is higher. So, as to an irradiation with neutrons of 14 MeV, not every cell nucleus is hit at doses below about 50 mGy. At a dose of 5 mGy, energy is deposited, on average, in every tenth cell nucleus only, but the dose in each of these nuclei is the same. Hence small radiation doses lead to an inhomogeneous distribution in the tissue. The degree of inhomogeneity depends on the radiation quality.

2.2.2.2 *Molecular Changes of the DNA and their Repair*

With respect to radiation damages considered here, molecular changes of the deoxyribonuleic acid (DNA) due to radiation are essentially involved in the development of the damage chain. Participation of proteins and membrane lipids in damage development has not been proved, yet cannot be excluded either. Molecular-biological studies have shown that after an irradiation of DNA, ruptures of polynucleotide chains – breaks of single strands and of double strands – (von Sonntag 1987) and damages or loss of the nucleotide bases occur. These damages are depicted in figure 2.15. A double-strand break occurs when both polynucleotide strands are ruptured due to an ionizing particle traversing the DNA helix or due to several independent radiochemical events which happen so closely that a dissociation of the ruptured DNA parts occurs. A considerable number of such changes of the DNA, particularly changes of the bases, has been characterized on the molecular level (von Sonntag 1987). It is not clear, however, which of these damages lead to the biologically relevant effects (e.g. cell death, malignant transformation of cells).

Measurements of damages to the DNA showed that irradiation of mammalian cells with sparsely ionizing radiation at a dose of 10 mSv damages the DNA in the genome of a cell in about 20 places (Altmann et al. 1970; Feinendegen 1977). Under these circumstances, single-strand breaks and base damages prevail, whereas the fraction of double-strand breaks following irradiation with sparsely ionizing rays is small. These impairments of the DNA may lead to structural modifications of chromosomes – chromosome aberration (fig. 2.16). Such chromosome aberrations in mammalian cells occur with a frequency of 10^{-2} to 10^{-3} per cell at a radiation dose of 10 mSv.

Most of these radiation damages are removed by highly efficient repair systems. Quickness and completeness of repair depends on the molecular kind of modifications of the DNA (Hanawalt et al. 1979; Generoso et al. 1980). Kinetic studies have shown that molecular differences may lead to a different rate of repair of these DNA impairments, even in the case of single-strand breaks. During these experiments, three groups of single-strand breaks and their respective repair mechanisms were observed (Dikomey & Franzke 1986).

The break of a DNA polynucleotide chain may occur at the bond between a phosphate group and the neighboring deoxyribose in a way that one single enzymatic step (by the enzyme ligase) leads to a complete repair of the radiation damage.

The DNA polynucleotide chain may be modified by rupture of bonds in a deoxyribose molecule. In that case, several enzymatic steps are necessary for the "healing" of the damage: first, the damaged parts of the DNA have to be removed from the polynucleotide

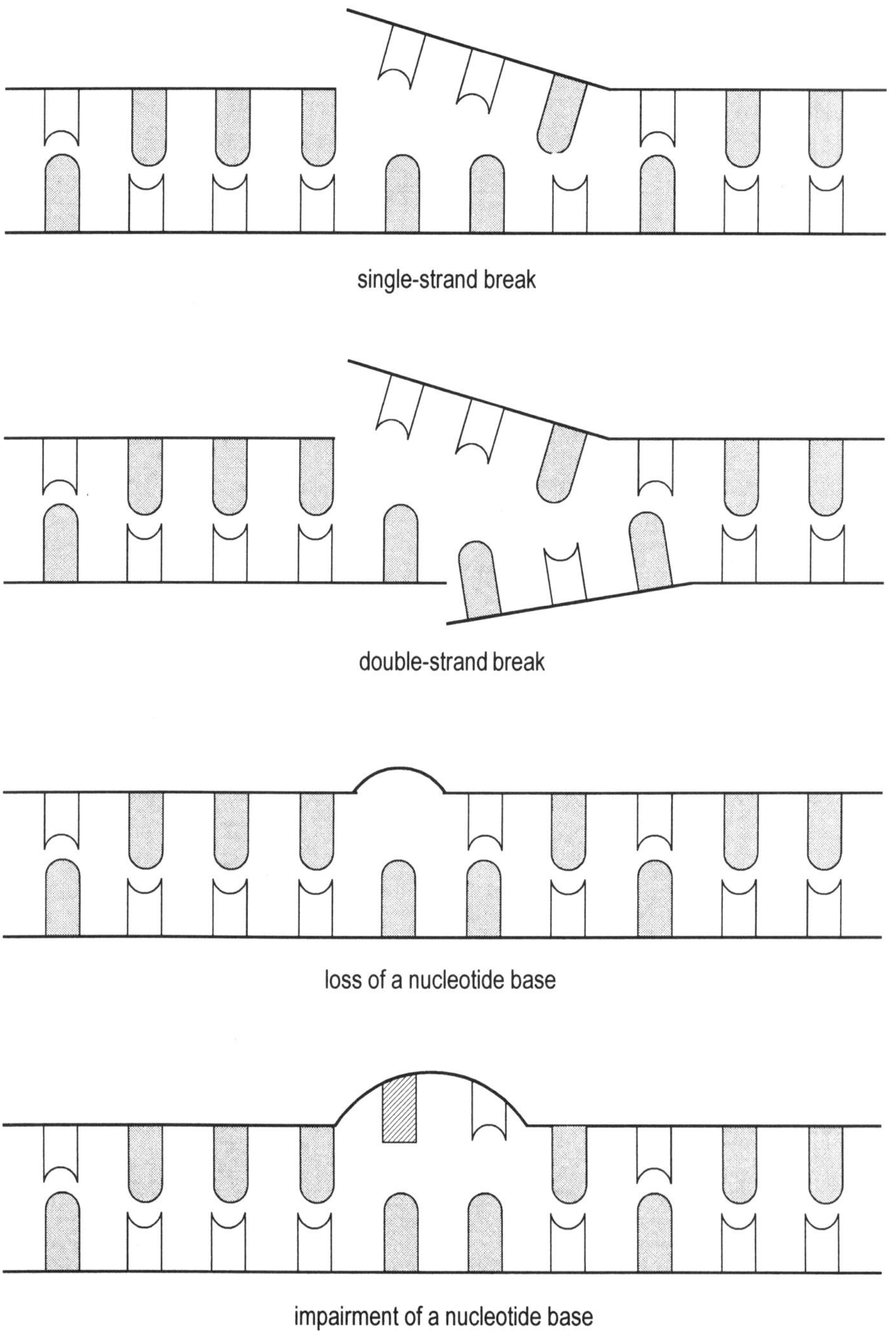

Fig 2.15 Radiation damages in the DNA

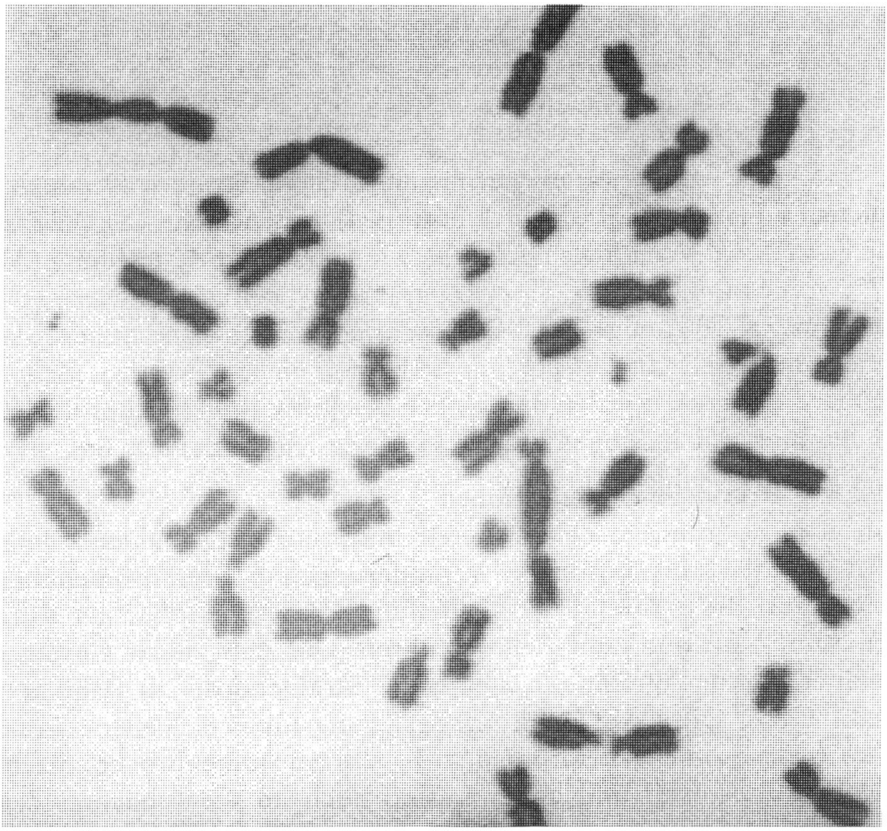

Fig. 2.16 Mitosis (metaphase) of an irradiated human lymphocyte with chromosome aberrations (2 dicentric chromosomes and chromosome ruptures)

chain; then a reconstruction of the damaged DNA polynucleotide chain with the original base sequence occurs, based on the information stored in the complementary, undamaged DNA strand. Finally, the DNA polynucleotide chain is closed again by the enzyme ligase.

In a similar way, impairments of the DNA bases due to radiation are repaired. In that case, however, once the radiation damage has been detected, the polynucleotide chain has to be ruptured near the damage location, so that the modified parts of the DNA can be removed.

Yet errors may happen during these highly complex repair processes, "misrepair" occurs. Moreover, damages may remain undetected by the repair systems or may not be repaired in time. The rate of repair depends on the kind of damage. Repair processes and their efficiency are genetically controlled (Hanawalt et al. 1979; Generoso et al. 1980).

Detailed mechanisms of this regulation will need further investigation, mainly in mammalian cells.

More double-strand breaks occur after exposure to densely ionizing rays (rays with high LET; e.g., α–rays, neutrons) than after sparsely ionizing irradiation (Alper 1979; von Sonntag 1978). Their repair is far more difficult and requires more time than that of single-strand breaks, in particular when they are due to radiation of high ionization density. Hence cellular repair processes take place only to a limited extent after effects of densely ionizing radiation.

2.2.2.3 Chromosome Aberrations

Impairments of the DNA due to irradiation leads to chromosomal modifications (chromosome aberrations) in irradiated cells, which can be observed during cell division (mitosis) (see fig. 2.16) and which lead to cell death. The number of these chromosome aberrations increases with increasing radiation dose (Ishihara & Sasaki 1983). Dose effect relations have been determined for different structural chromosome aberrations, e.g., chromosome ruptures, dicentric chromosomes and others, even for human cells. Hence, quantitative determination of aberrations in lymphocytes may be used for estimating doses. This cytogenetic method has proved to be useful in cases of radiation accidents, to obtain information about the level of exposure of the persons concerned. By means of this procedure, whole-body doses can be estimated down to the range of 50–100 mGy (mSv). Cells affected by such chromosome damages will die. Damages to health do not occur at the death of single cells.

2.2.2.4 Cell Death

Of crucial importance to the topic of damage to mammalian cells after irradiation is cell proliferation (propagation or regeneration of cells). The extent of cell proliferation is very different for different tissues and organs. So, on the one hand, there are tissues and organs with very low cell proliferation (e.g. nerves and muscles) and on the other hand, there are organs with continuous high cell proliferation (e.g. bone marrow and small intestine). Highly proliferating cell systems have shown to be generally more sensitive to radiation than less proliferating ones. On the other hand, cell proliferation is low in differentiated (specialized) cells. Radiosensitivity therefore decreases with increasing differentiation of cells during development, e.g., with the formation of new blood cells in the bone marrow or with maturation of spermatozoa (Alper 1979; Streffer & van Beuningen 1985).

When a cell is irradiated, it usually proceeds through its present cell generation cycle, including the next cell division. Even further cell cycles and cell divisions may follow before the radiation damage is completely expressed on the chromosomal level and eventually the death of the cell occurs. Hence, this kind of cell death is called reproductive cell death (Hall 1978; Alper 1979). In the dose range of up to some Sievert, this is usually the only mechanism for cell killing to be reckoned with. For non-proliferating lymphocytes and a few other cell types (e.g., oocytes) there is, however, a different mechanism of damage development after exposure to radiation in the range of 1 Sv, where cells die during the same cell generation cycle in which irradiation occurs, before further cell divisions can take place. For other types of cells, far higher radiation doses are necessary to produce that kind of cell death. This mechanism is called interphase death (Streffer & van Beuningen 1985).

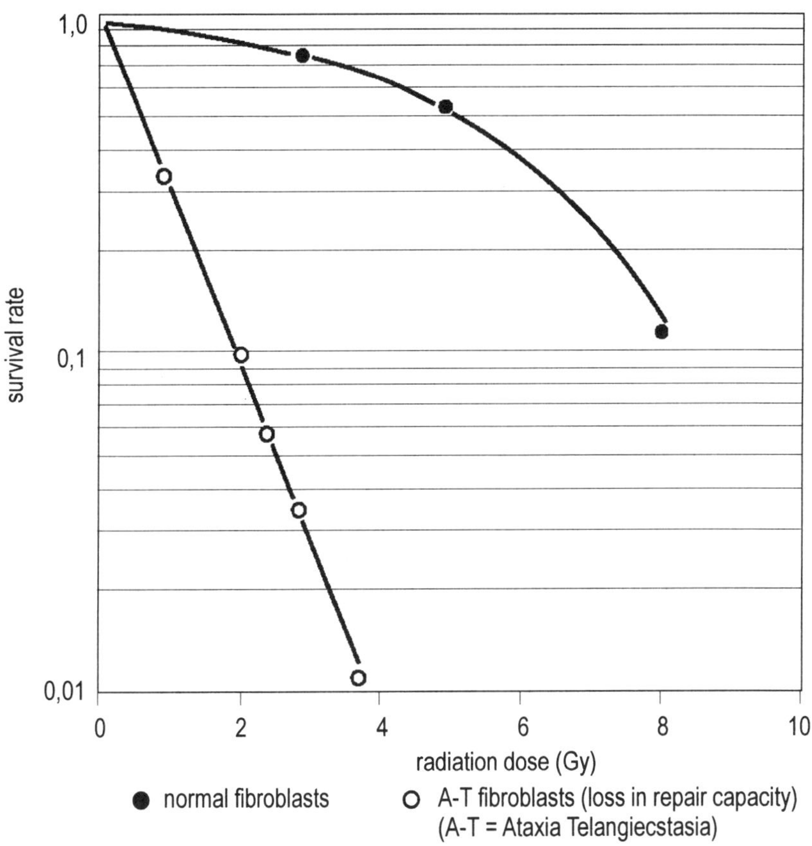

Fig. 2.17 Survival rates of cells after g-irradiation at low dose rates (Patterson et al. 1984)

Radiation-induced cell killing can be measured for proliferating cells in vitro, with the help of the colony-formation test. If the survival rates of the cells ascertained in that way are plotted in a diagram on a logarithmic scale against the radiation dose on a linear scale, we usually obtain a dose response relation for sparsely ionizing radiation which is curved for low dose ranges (fig. 2.17, normal fibroblasts). For higher dose ranges, the dose response relation approximately becomes a line with this plotting mode. This kind of dose response relation is usually called a shoulder curve. Survival of cells (S) is indicated in relative values referring to the number of colony-founding cells before irradiation. This curve of cell survival (S) may be described by a linear/ quadratic function of the dose (D) (Chadwick & Leenhouts1981).

$$S = \exp - \alpha D + \beta D^2 \tag{2.12}$$

2.2.2.5 Intracellular Recovery

It has been discovered that the curvature of the survival curve for normal fibroblasts in the lower dose range is due to intracellular recovery from radiation damages (Elkind & Sutton 1960). This means that in the lower dose range the radiation effect and hence the killing of cells is reduced by processes of recovery. Although it has not yet been proved, there are many hints which suggest that these intracellular processes of recovery are closely connected to repair of radiation damages in the DNA (UNSCEAR 1982; Streffer & van Beuningen 1985). Yet it is also known that there are molecular radiation-induced damages of the DNA, which are not repaired completely. Many experimental data suggest that it is these damages (e.g., breaks of both polynucleotide chains) which lead to chromosome aberration and eventually to the death of the cell.

Yet there are persons in whom, due to a genetic defect, certain DNA repair systems either no longer operate or do so only to a reduced extent. Among these genetically caused defects are diseases like Xeroderma pigmentosum and Ataxia telangiecstasia. In the first case, damages to the DNA in cells exposed to *ultraviolet light* (UV) cannot or can only partly be repaired (Hanawalt et al. 1979). Such changes due to UV are provoked in skin cells by sunlight and cause heavy ulcerations. The second disease means a major reduction of repair of damages to the DNA which occur after a *γ-ray or X-ray* irradiation. So the persons concerned show a significantly higher radiosensitivity than normal persons. The dose response curve loses its shoulder and is linear in the low dose range as well, with a semi-logarithmic plotting mode (fig. 2.17, A-T fibroblasts) (Patterson et al. 1984; UNSCEAR 1986).

Ataxia telangiecstasia is a disease which only affects homozygous carriers of heredity, which are rare in the German population (1 : 40'000 to 1 : 100'000). In every cell there are two places on homologous chromosomes for genetically determined characteristics. If these genes are identical, the person is called homozygous. If they are different, the person is called heterozygous. Recent studies show that heterozygous carriers of this hereditary disease also show higher sensitivity to radiation than normal persons – if to

Severe acute damages to different organs, primarily to the skin, have been observed after radiotherapeutic treatment of homozygous persons, which demonstrates their increased radiosensitivity. Moreover, persons affected with Ataxia telangiecstasia show an increased rate of malignant diseases, suffer from immunodeficiency and often show failures in the development of the central nervous system.

Due to intracellular recovery processes, fractionated irradiation leads to lower radiation impacts than short-term irradiation, if equal total doses are compared. This fractionation effect is of considerable importance for radiotherapy of malignant diseases. But similar recovery effects are also found if an acute irradiation at high dose rates is replaced by chronic irradiation at very low dose rates. Then, radiation impacts with respect to cell death but also to other biological effects like malformations after prenatal irradiation, and genetic defects are significantly lower (Hall 1978). So, obviously, the temporal distribution of doses is of importance for the magnitude of the radiation risk.

Although we may generally assume that prolongation of the irradiation period with the total dose remaining the same leads to a reduction in radiation effects, reverse effects have also been observed in some experiments on cells in vitro. Furthermore, it has been found that, with regard to the induction of malignant diseases, radiation effects are only slightly influenced by the temporal dose distribution (UNSCEAR 1986). Moreover, marked phenomena of recovery and hence less severe radiation effects after chronic exposures are found only for sparsely ionizing radiation.

2.2.2.6 Relative Biological Effectiveness (RBE)

For densely ionizing radiation, like neutrons and alpha-rays, recovery effects such as DNA repair occur much more rarely or cannot be observed at all. If we therefore compare dose response relations for the killing of cells after irradiation with X-rays and with neutrons, we find that the shoulder of the dose response curve is significantly reduced and the exponential part is steeper after irradiation with neutrons (fig. 2.18). Thus, radiation of high ionization density is much more efficient than that of low ionization density at the same dose (energy dose) (Alper 1979; UNSCEAR 1982; Streffer & van Beuningen 1985). It is therefore necessary for an assessment of radiation risks to know not only the level of the energy dose but also the radiation quality. For a quantitative evaluation of this difference, we determine the so-called *R*elative *B*iological *E*ffectiveness (RBE). For this purpose, we set the RBE of a sparsely ionizing radiation type, e.g. of X-rays, to 1 as a reference radiation and then determine the respective ratios of radiation doses which lead to the same biological effects.

$$RBE = \frac{D}{D_{Ref}} \quad \text{(for the same biological effect)} \tag{2.13}$$

These RBE-value have turned out to depend on many parameters, such as the biological effect considered (e.g. killing of cells, induction of cancer etc.), the examined type of cell or tissue, the dose level, and others (UNSCEAR 1982; UNSCEAR 1986). Therefore, no generally valid RBE-value can be specified for a certain radiation quality. For pragmatic reasons, quality factors have been established for an assessment of the radiation risk and for radiation protection, which have been recommended by the International Commission on Radiological Protection (ICRP 1977). These quality factors are determined by experimental studies and clinical experiences about the level of RBE-values, and usually correspond with the upper range of the ascertained values. According to definition, they are 1 for X-rays, gamma-rays and beta-rays, and have been set to 10 or 20 for fast neutrons or alpha-rays, respectively. Data from animal experiments, primarily concerned with the induction of mammary neoplasms in rats by X-rays and neutrons (Shellabarger et al. 1980), led to a discussion about whether quality factors for neutrons have to be increased with respect to induction of leukemia and cancer (UNSCEAR 1986).

Multiplication of the energy dose with these quality factors yields the dose equivalent. The dose equivalent is generally used in radiation protection to include an assessment of the radiation quality into the radiation dose (cf. 2.1.3).

2.2.3 Non-Stochastic Effects in Tissues and Organs

2.2.3.1 Acute Radiation Effects

Acute effects only occur at radiation doses of 0.5 Sv and higher (fig. 2.19). Exposures to radiation emitted during proper operation of nuclear facilities remain far below these threshold doses. Hence acute radiation effects are of no importance for an assessment of risks which have to be reckoned with if the environmental standards discussed here are observed (dose limit for people living in the vicinity of nuclear installations: effective dose equivalent or whole-body dose of 0.3 mSv per annum). Yet such effects might occur due to exposures after serious accidents.

As has been shown in section 2.2.2, tissues with high cell proliferation are particularly sensitive to radiation. In such tissues, radiation damages are expressed rather quickly. Typical examples of the occurrence of high proliferation rates are the regeneration of cells in spermatogenesis, the formation of blood cells in the bone marrow, and cell regeneration in the skin, in the mucosae, and in the small intestine. In all these systems of cell regeneration, cell division is rapidly reduced after irradiation. For radiation

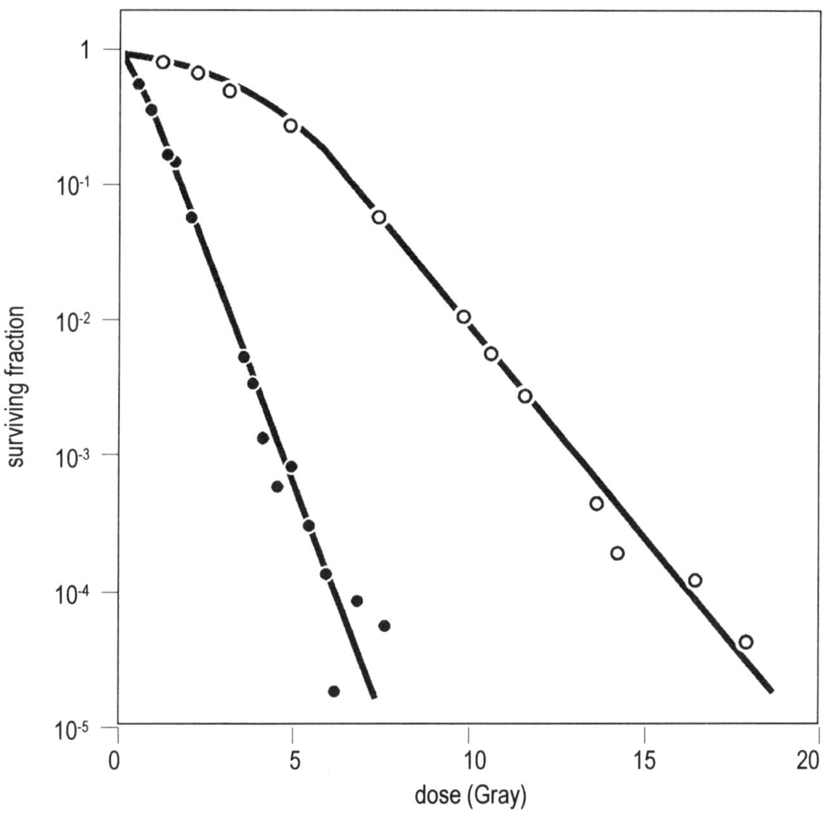

Fig. 2.18 Dose effect relations for the survival of fibroblasts of hamsters after irradiation with 280 kV X-rays (O) and fast neutrons (●)

doses below 2 Sv, the reduction of the rate of cell division is usually reversible. Some hours or days after exposure to radiation at such doses, a recovery takes place.

For higher radiation doses, these processes are irreversible, though. Above all, a considerable decrease of parent cells in cell regeneration systems occurs. As a result, regeneration of cells is reduced, while natural physiological loss of cells continues, which leads to a major decrease in the number of cells in these tissues or organ systems. After an irradiation of the gonads at radiation doses of about 4 Sv, a man has to reckon with irreversible impairments of his fertility, or with sterility. Radiation doses below 1 Sv may already cause fertility impairments, which are, however, reversible. Irradiation of female gonads at doses above about 4 Sv also leads to sterility, although the respective cells are not proliferating ones (UNSCEAR 1982).

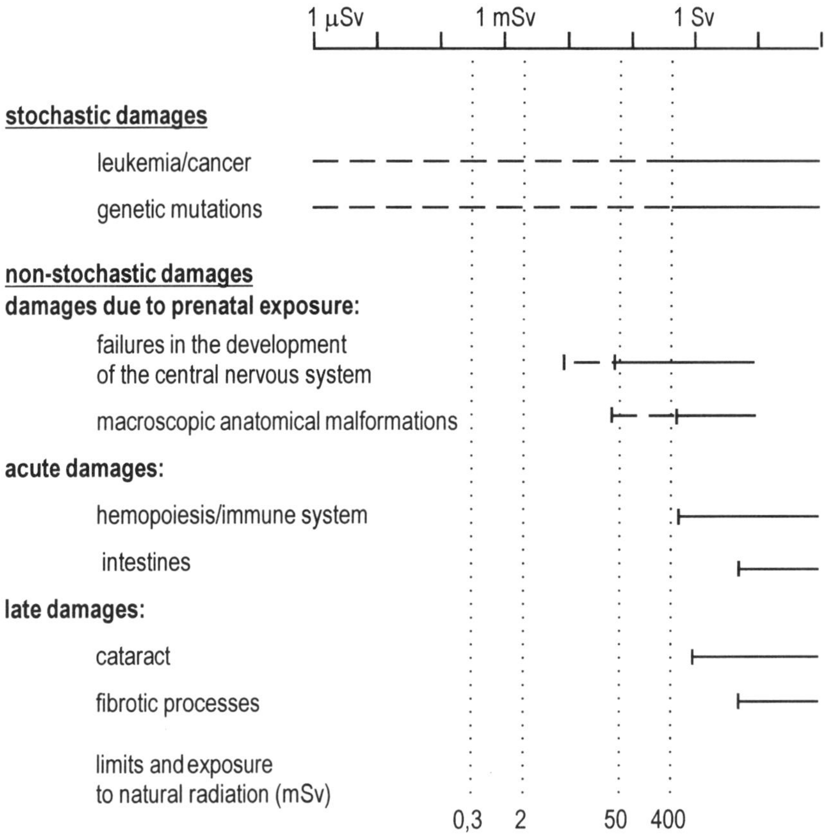

Fig. 2.19 Dose ranges for radiation effects
Straight horizontal lines: dose ranges for which measurement data are available; dashed horizontal lines: dose ranges for which available data are uncertain or derived from extrapolation.
Left hand bounds in horizontal lines indicate threshold doses for the respective effects.
Vertical dotted lines indicate dose limits (effective dose equivalent) for the population at large (0.3 mSv per annum) and for occupationally exposed persons (50 mSv per annum, 400 mSv for a whole professional lifetime). In addition, the mean effective dose equivalent of exposure to natural radiation is indicated (about 2 mSv per annum).

Irradiations of the whole body at doses of 2 Sv and higher lead to manifest disturbances of cell regeneration in the bone marrow. Hence, these irradiations result in a decrease in peripheral blood cells. For radiation doses up to the range of about 4 Sv, recovery of the bone marrow is possible, so that the blood picture can regenerate if the critical phase is bridged by transfusions of blood cells and if infections are avoided. Due to the relatively short lives of granulocytes, the number of these blood cells decreases within one to three days after irradiation of the whole body, whereas a decrease in thrombocytes and, particularly, in erythrocytes occurs only later. Thus, reduction of these blood cells is not due to radiation-induced destruction of already circulating mature cells, but to decreased cell regeneration (Bond et al. 1965; UNSCEAR 1982; Nothdurft 1985).

On the other hand, a decrease in the number of lymphocytes is due to their high radiosensitivity, which leads to autolytic processes in the differentiated cells themselves after irradiation at doses above 1 Sv. Since lymphocytes and lymphatic tissue in general are highly radiosensitive, irradiation of the whole body at doses above 1 Sv results in hypoimmunity (UNSCEAR 1982; UNSCEAR 1988), so that the danger of infections is very high for the persons concerned. Therefore, they need adequate treatment and aseptic isolation (UNSCEAR 1982).

If regeneration of the bone marrow is no longer possible, the person concerned dies. Acute clinical symptoms due to a deficiency of the bone marrow appear about 14 to 30 days after irradiation (Bond et al. 1965). Among them are petechiae (point hemorrhages due to rupture of the capillaries in the tissues) and anemic states. This is called hematopoietic radiation syndrome. In the case of whole-body irradiation ranging from 4 to 5 Gy(Sv), or at a bone marrow dose of about 3 Gy(Sv), we expect a lethality of 50%. Whole-body irradiations in the range of about 7 to 9 Gy(Sv) are lethal. These findings all refer to sparsely ionizing radiation. Therapy may be possible, if suitable immunocompatible bone marrow is available for a transplantation. So far, however, the only suitable donors of bone marrow are close relatives of the irradiated patients (parents, brothers and sisters) (UNSCEAR 1988).

The consequence of still higher whole-body doses are severe damages to the gastro-intestinal tract (Trott 1985). Above all, a regeneration of epithelium cells of the small intestine, which are the functioning cells for formation of the epithelium of the villi, is seriously reduced. It is here that the metabolic processes between the interior of the intestines and the tissues take place. The epithelium cells live for about five days and are then shed into the interior of the intestines. After this period, severe clinical symptoms occur, like diarrhea, loss in electrolytes, and impaired absorption of nutrients from the intestines, since no new epithelium cells can be generated as a consequence of irradiation. Therapy of this so-called gastro-intestinal radiation syndrome is not possible. Yet the gastro-intestinal system has proved to have a high capacity of recovery. With respect to a fractionated irradiation with sparsely ionizing rays, far higher radiation doses are tolerable.

2.2.3.2 Non-stochastic late effects in normal tissues

Apart from acute changes, there are effects which only appear some months or even years after exposure to radiation. Among them are cataracts[11] (opacity of the eye lens) and fibroses in different tissues. The eye lens is a particularly sensitive system, so that opacity may be expected even at radiation doses of 2 Sv. Usually, it shows only years after irradiation and increases in frequency and severity with increasing doses. This was experienced after therapeutic irradiation. In particular, cataracts appear with a considerable probability (about 30%) after therapy of leukemia with high-dose whole-body irradiations (about 10 Sv) and subsequent transplantation of bone marrow (UNSCEAR 1982; UNSCEAR 1988).

Other late effects are skin deformations after chronic exposure to radiation. Necrotic modifications and eventually skin carcinoma are found, in particular, in physicians whose hands were heavily exposed to X-rays during their professional activities. Yet these effects result from radiation doses of some Sievert or even some ten Sievert. In radiation accidents, heavy exposures of the skin to β–radiation may occur, which induce severe burn-like symptoms within a period of one or several weeks.

Furthermore, fibrotic processes are observed in tissues after local irradiation at doses of the level of some ten Sievert, as it happens primarily in therapeutical situations. Then, a decrease in cellular elements and an increased deposit of extracellular protein fibers occurs, which means reduced function. For these fibrotic processes there is also a long induction phase, so that changes normally become manifest only several years after exposure to radiation (UNSCEAR 1982).

2.2.4 Induction of Developmental Anomalies

Prenatal development of the mammalian organism is subdivided into three main phases (the periods of which are indicated for human beings):
- preimplantation phase (from beginning of pregnancy to about the tenth day post conceptionem; p.c.),
- main period of organ formation (from about two weeks p.c. to the end of the second month of pregnancy),
- fetal period (from then on to birth).

Due to different development-physiological phenomena, the radiosensitivity of an embryo or a fetus differs drastically during these periods, with respect to both extent and quality of radiation damages (SSK 1985). During the preimplantation phase, radiosensitivity is high. Recent studies in the context of animal experiments showed that even malformations may be caused during this period (Pampfer & Streffer 1988). Yet, during the preimplantation phase, an embryo is usually able to repair radiation damage and develop into a normal organism, or it dies, so that the risk of developmental anomalies is much lower here than in later developmental phases (UNSCEAR 1986; Streffer 1987).

Radiation exposure during the main period of organ formation, however, may lead to severe macroscopic anatomic malformations, primarily of the vertebral column, the extremities and other parts of the skeleton. But other malformations, like hydrocephalus and cleft palate, occur as well. These malformations have by and large been found in animal experiments, yet also in some human cases, namely after irradiation of pregnant women suffering from cancer, at doses of several Sievert (UNSCEAR 1986; SSK 1985).

In Hiroshima and Nagasaki, malformations, in particular reduced circumferences of the cranium (microcephalus), were found in persons exposed to radiation doses above about 0.5 Sv in utero during the main period of organ formation. Based on data from animal experiments, the "spontaneous" rate of malformations during the main period of organ formation is assumed to double at radiation doses of about 0.2 Sv. There is obviously a threshold dose for these effects, at a level of about 0.05 Sv. Hints about this dose level have been gained from experiments with rats and mice. Experience, mainly from diagnostic exposures to radiation, indicates that the respective threshold dose for human beings is higher (SSK 1985).

Formerly, it was assumed that the fetal period was relatively insensitive to radiation. Yet recent analyses of data from Hiroshima and Nagasaki revealed that there is high radiosensitivity with respect to the induction of developmental anomalies of the central nervous system, primarily in the early fetal period. The rates of these anomalies are statistically significantly higher for persons irradiated in utero than for persons not exposed to radiation. In particular, during the eighth to the fifteenth week p.c, but still up to the 25th week p.c., heavy mental retardations may be induced (UNSCEAR 1986). By heavy mental retardation we understand that the persons concerned are not able to formulate simple sentences, to do simple calculations, and often cannot take care of themselves. Epidemiological data from Hiroshima and Nagasaki suggest that such developmental anomalies may already occur at doses above about 0.05 Sv. But the number of observed cases is extremely low for this low dose range and was found to be still lower after recent dose estimations, since some cases had to be classified into higher dose ranges (Otake et al. 1988). Based on the dosimetry of 1965 (abbreviation: T65DR), three or four cases of retardation have been observed for the two lowest dose classes respectively. According to the dose estimations of 1986 (abbreviation: DS86), there are only two cases for each of these classes (fig. 2.20).

Up to today, it could not be proved beyond doubt whether there is a threshold dose for an induction of this effect. Nevertheless, a threshold dose is assumed on the basis of

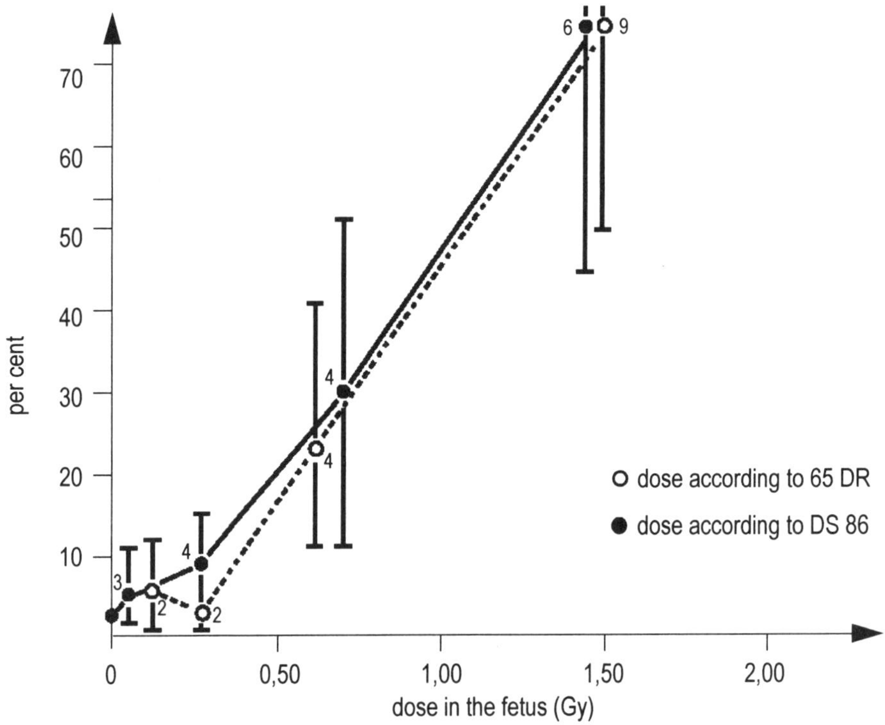

Fig. 2.20 Heavy mental retardation in persons from Hiroshima and Nagasaki who were irradiated in utero (exposure 8–15 weeks p.c., 130 exposed pregnant mothers) (Otake et al. 1988)

data from animal experiments, and this is supported by reflections that here we are faced with a mechanism which is basically different from that of carcinogenesis or mutagenesis. Available epidemiological data do not allow a definitive decision on this question. Yet they suggest a threshold dose as well, namely of the order of 0.01 to 0.1 Sv (Otake et al. 1988). This extremely important question needs further study.

Present experimental data and clinical experiences suggest that there is high radiosensitivity during the whole period of prenatal development. Particular attention should be paid to effects of exposure during the early fetal period. With advancing development during the fetal period, radiosensitivity decreases with respect to developmental anomalies, (though not with respect to induction of leukemia and cancer, which are discussed later).

2.2.5 Stochastic Effects

2.2.5.1 Hereditary Defects

Up to now, no increased rates of mutations of human gametes due to ionizing rays have been observed, which would serve to quantitatively assess a radiogenetic risk. Even in studies related to descendants of survivors of the bombing of Hiroshima and Nagasaki, no such statistically significant effects have been found. Only a trend to increased rates of mutations of the gametes has been observed in 20'000 children whose fathers or mothers or both parents were exposed to radiation during the bombing of Hiroshima and Nagasaki (Schull et al. 1981).

Hence, quantitative assessment of a radiogenetic risk is based on animal experiments, mainly with mice. In these experiments, the gonads are locally irradiated, and easily identifiable characteristics, like color of the fur, size of the ears, color of the eyes, development of cataracts etc., are analyzed. Changes in these characteristics are due to dominant but also to recessive mutations. Very often, linear dose response relations are ascertained for these effects, so that the mutation rate per Gray or Sievert can be calculated from the slope of such dose response relations.

The majority of results of experiments with mice was gained from radiation doses ranging above 1 Gy. Usually, the animals were irradiated at high dose rates. Two assumptions are necessary to use these experimental results in an assessment of the radiation risk for human beings (Ehling 1987):

- There is a linear dose response relation without threshold between dose and number of induced mutations.
- The frequency of radiation-induced mutations is about the same for gametes of mice and of human beings under comparable circumstances of irradiation.

Our present knowledge about the action mechanism of ionizing rays and about the emergence of mutations in gametes which lead to hereditary defects leaves virtually no doubt that there is a dose response relation without a threshold dose. Hence we must presume that mutations can also be induced at small doses and at low dose rates. It has been shown experimentally that mutations in mice can be induced by a chronic gamma-irradiation at a dose rate of 0.01 Gy per day over 300 days in the same way as by fractionated irradiation of 0.1 Gy per day over 60 days (Ehling 1987).

Today, based on empirical data and theoretical reflections, the basic assumption of radiation-induced mutation rates independent of the species is considered to be highly probable, although not absolutely sure. Apart from that, additional aspects have to be taken into account in the use of data from animal experiments for an assessment of a radiogenetic risk. Studies about the frequency of radiation-induced dominant and recessive mutations in different stages of development of the gametes showed that, for irradiation at equal doses, the mutation frequency is about three times higher in spermatocytes, spermatids, and spermatozoa, that is, in the maturing or mature stages, than in spermatogonia, that is, in the parent cells. Nevertheless, experimental data gained from irradiation of spermatogonia are used for a quantification of the radiogenetic risk. This is explained by the fact that spermatogonia exist and are exposed during the whole generation time of a human life, whereas spermatocytes, spermatids and spermatozoa have a comparably very short lifetime (UNSCEAR 1982). Development of mature spermatozoa from spermatogonia takes 72 to 74 days for human beings. During this period, all stages up to the spermatozoon are gone through. Total exposure of spermatogonia is, as a rule, so much higher that it is the frequency of their mutation which is of decisive importance for an assessment of the radiogenetic risk for human beings.

For reasons of feasibility, only males are irradiated in most radiogenetic experiments. If we want to approach real situations, in which normally both partners are irradiated, a correction is needed. Experimental data show that with respect to mice, the mutation rate after irradiation at low dose rates is higher for exposed spermatogonia than for exposed oocytes (UNSCEAR 1977; UNSCEAR 1988). UNSCEAR (United Nations Scientific Committee on Effects of Atomic Radiation) has taken this result into account in its risk assessment by setting the radiosensitivity of oocytes to only 40% of the respective value for spermatogonia. The BEIR-committee (Biological Effects of Ionizing Radiation), on the other hand, assumes equal radiosensitivity for male and female gametes.

For an assessment of the radiogenetic risk, mainly two methods are employed, namely direct and indirect assessment (UNSCEAR 1982; UNSCEAR 1986).

For a *direct assessment* of the radiogenetic risk, the rates of spontaneous and radiationinduced dominant mutations of an organ system are studied by means of animal experiments. The radiogenetic risk, that is, the number of mutations to be expected at a given radiation dose, may be calculated from these mutation rates according to the following equation (Ehling 1987):

$$\text{risk} = \begin{pmatrix} \text{mutations} \\ \text{per gamete} \\ \text{per dose} \end{pmatrix} \times \begin{pmatrix} \text{multiplication factor} \\ \text{for the total of} \\ \text{dominant mutations} \end{pmatrix} \times \text{dose} \tag{2.14}$$

As a basis for this kind of assessment, primarily the induction rates (mutations per gamete per unit of dose) for cataracts in mice with a dominant mode of inheritance were determined. Cataracts with a dominant mode of inheritance are, of course, only a small portion of the total of dominantly transmissible diseases. A survey of all known dominantly transmissible diseases describes altogether 934 well-defined anomalies with dominant transmission mode (McKusick 1983). 23 of them lead to opacities of the eye lens. From this relation 934/23, we derive a multiplication factor of 41, which is used to determine the total number of dominant mutations. For the mutation rate, hence for the first factor in the given equation, a mean value of 0.45 to 0.55 x 10^{-4} mutations per gamete per Gy has been found after single external irradiation with cesium-137. Data about the mutation rate for anomalies of the skeleton due to irradiation have been obtained in an analogous way (Ehling 1987). They yield the same risk values within the normal biological range.

With the help of such data, expected numbers of dominant mutations after exposure to radiation can be estimated for defined populations. In that way, a radiogenetic risk has been determined for descendants of survivors of the bombing of Hiroshima and Nagasaki (table 2.11). According to these risk values, a statistically significant increase in genetic damages due to the explosions of the atomic bombs was not to be expected, nor has it been observed.

Due to the increased number of studies in human genetics, we may expect the number of known dominantly transmissible diseases to increase. Since, on the other hand, probably all cataracts with a dominant mode of inheritance in human beings are known already, the multiplication factor of 41 is likely to be rather too low. Hence, the calculated radiogenetic risk is rather a little too low as well.

A new method permits us to simultaneously determine recessive mutations in irradiated animals, together with the rate of dominant mutations, hence mutations which actually will become manifest only in later generations. Although this does not allow us to calculate absolute rates of recessive mutations, since we do not know the multiplication factor here, we can establish the relation between dominant and recessive mutations. It is approximately 1:10 per gene for mice after exposure to radiation. If we assume this relation to hold for human beings as well, a respective risk assessment becomes possible (Ehling 1987).

In an *indirect assessment* of the radiogenetic risk, instead of absolute mutation rates we determine the doubling dose for induction of hereditary defects, which is then transferred to human beings (UNSCEAR 1977; BEIR 1980; UNSCEAR 1982). The doubling dose is that radiation dose which, on average, induces the same number of mutations of some kind as occur "spontaneously" in one generation of the studied organisms, hence

	due to cataracts	due to anomalies of the skeleton
mutations / gamete / Gy	$0{,}45 - 0{,}55 \times 10^{-4}$	$10{,}1 \times 10^{-4}$
multiplication factor for the total number of dominant mutations	41	5
genetically significant dose in person Sv	$1{,}1 \times 10^{4}$	$1{,}1 \times 10^{4}$
expected numbers for dominant mutations	20 - 25	56

Table 2.11 Calculated radiogenetic risk values for 19'000 descendants of survivors of the bombing of Hiroshima and Nagasaki, gained according to the method of direct assessment, namely by observing cataracts and skeleton anomalies in irradiated mice (Ehling 1987)

that dose which doubles this number. The doubling dose for sparsely ionizing radiation at a low dose rate amounts to 1 Gy for mice, according to the analysis of experimental data by UNSCEAR (UNSCEAR 1977; UNSCEAR 1982). All further information about "spontaneous" genetic defects needed for an assessment of the radiogenetic risk is gained from studies in human genetics. The risk (number of mutations to be expected at a given dose) can then be determined according to the following equation:

$$\text{risk} = \frac{\text{spontaneous hereditary burden}}{\text{doubling dose}} \times \text{hereditary share} \times \text{dose} \qquad (2.15)$$

The "spontaneous" hereditary burden is the number of observable "spontaneous" mutations which adjusts itself when the total number of "spontaneous" mutation events is in equilibrium with the number of eliminated mutations. The hereditary share is that share of a genetically caused disease which is considered to be due to hereditary factors (the sum of genetically caused diseases is the spontaneous hereditary burden). The value of the hereditary share, ranging from 0 to 1, is estimated on the basis of empirical data. Since mutations with dominant mode of inheritance always become manifest in the first generation, with no interaction of non-hereditary factors needed, the hereditary share for this case is 1. Table 2.12 shows the radiation-induced mutations to be expected under these circumstances at a radiation dose of 1 Gy per gamete, in one million live births, according to calculations of UNSCEAR (UNSCEAR 1977; UNSCEAR 1982).

classification of hereditary diseases	spontaneous hereditary burden	effect of 1 Gy per generation first generation	equilibrium
autosomal-dominant and X-bound mode of inheritance	10.000	1.500	10.000
recessive mode of inheritance	2.500	rare	very low increase
chromosome aberrations structural modifications	400	240	400
numerical modification (e.g. trisomy)	3.000	rare	very low increase
irregular mode of inheritance	90.000	450	4.500
total frequency	105.900	2.190	14.900

* figures were given only if an equilibrium is reached within five to ten generations.

Table 2.12 Estimated radiation risk for different hereditary diseases after irradiation at the doubling dose of 1 Gy per gamete (sparsely ionizing radiation at low dose rates), in one million live births (Ehling 1987)

The advantage of an indirect assessment of the radiogenetic risk lies in the fact that for different types of hereditary diseases, expected numbers can be calculated, once the representative doubling dose for all genetic defects has been determined. Direct assessment, on the other hand, can yield more accurate values, due to better experimental data for the evaluation of selected dominant mutations.

In the above-mentioned studies concerning children of survivors of the bombing of Hiroshima and Nagasaki, the authors derived a mean doubling dose of 1.56 Gy from the observed trend (Schull et al. 1981). Therefore, we may assume that application of the doubling dose determined in experiments with mice (1 Gy) does at least not lead to an underestimation of the radiation risk for human beings.

Yet a number of open questions remain, the answering of which is necessary for a more realistic assessment of the radiogenetic risk for humans. In particular, the dose effect relation for low doses and low dose rates has to be studied further, to increase the reliability of extrapolations. In addition, it must be determined whether the examined genes are representative for the whole genome or only for a certain class of mutated genes. Finally, the basic assumption that mutation rates are the same for mice and humans, has to be corroborated.

2.2.5.2 Induction of Malignant Diseases

In the case of induction of malignant diseases, it is far less possible to extrapolate results from animal experiments to humans than it is in the case of genetic damages. Yet, while there are virtually no empirical data for human beings with regard to an assessment of the genetic radiation risk, an estimation of the risk of leukemia and cancer can rely on some epidemiological studies. Among them are, primarily, studies concerning (UNSCEAR 1977; BEIR 1980; UNSCEAR 1988):

- survivors of the bombing of Hiroshima and Nagasaki,
- persons occupationally exposed to high radiation,
- patients exposed to high medical irradiation.

Since a malignant disease caused by radiation (leukemia or solid tumors) does not differ from one occurring "spontaneously", we cannot identify an individual case of such a disease as being due to radiation simply by considering its manifestation or its clinical development. Epidemiological studies can only provide statistical data for an assessment of the radiation risk to humans. For this purpose, radiation-exposed collectives of persons are compared with collectives of persons not exposed to radiation yet comparable to the latter with regard to further risk factors for leukemia and cancer (e.g. age, sex etc.). In some cases, and in particular with respect to survivors of the atomic bombing, it was even possible to determine dose dependences by comparing frequencies of cancer among persons of the same collective, which were, however, exposed to different doses.

The common approach of these statistical analyses is to compare the frequencies with which the studied disease occurs in different collectives and to thereby ascertain whether and to what extent rates of leukemia and/or solid tumors are correlated to irradiation. These data will yield a certain probability of getting cancer after a certain exposure to radiation. They will also allow us to determine the probability with which a certain actually existing case of cancer could have been induced by irradiation of some known quantity (NIH 1985). The reliability of such studies is impaired, however, by the fact that these diseases appear with a considerable latent period (five to ten years for leukemias and several decades for solid tumors) and are thus less amenable to recollection.

Since leukemias occur relatively soon after irradiation and with a particularly high relative risk, a comparably large set of data is available for these diseases. It is found that certain types of leukemias (acute and chronic myeloic leukemias as well as acute lymphatic forms) appear increasedly after irradiation. On the other hand, no increases due to radiation have been observed for another form (chronic-lymphatic leukemias) (NIH 1985; Preston et al. 1987). Studies of survivors in Hiroshima and Nagasaki and studies about collectives of patients irradiated because of Bechterew's disease suggest that for radiation doses of above 0.2 Sv a statistically significant increase in the rate of leukemia has to be expected for the total population (all ages). Lower radiation doses did not yield statistically significant increased rates of leukemia in epidemiological studies of large parts of the population (UNSCEAR 1977; BEIR 1980; Boice & Fraumeni 1984; UNSCEAR 1988).

From what we know today, however, this does not mean that radiation doses lower than 0.2 Sv do not cause leukemias. But if they do, the number of cases then becomes so small that the variability of the risk of "spontaneous" leukemia superimposes the additional risk due to radiation in a way that the latter is no longer identifiable within the variation range of the rate of "spontaneous" leukemia (BEIR 1980; Boice & Fraumeni 1984; Streffer 1985). The same holds with respect to the rates of solid tumors.

The relative risk of dying from leukemia after irradiation at a dose of 4 Gy has been observed to be increased by about the factor 16 in survivors in Hiroshima and Nagasaki (fig. 2.21). The radiation dose is mainly due to gamma-radiation with a quality factor of 1. Hence the dose equivalent in Sv has to be set equal to the energy dose in Gy. The relative risk of all types of cancer except leukemias, on the other hand, was increased by only the factor 2 for the same radiation dose. When the relative risk is calculated, the "spontaneous" rates are set to one (Preston et al. 1987).

There are, however, some solid tumors for which the observed increase in the relative risk is considerably higher than the mean value of 2; including, particularly, breast cancer in women (BEIR 1980; Boice & Fraumeni 1984). Female breast tissue has shown to be particularly radiosensitive (ICRP 1977). This strongly depends on the age at which the exposure occurs (Boice & Fraumeni 1984). The risk is crucially higher for women who were younger than 30 or even younger than 20 years of age at the time of exposure than it is for older women. This age-dependence has been observed both in women in Hiroshima and Nagasaki and in women whose breast tissue was exposed to relatively high radiation by frequent X-rays. Besides the bone marrow (induction of leukemia) and the breast tissue, the lungs and the epithelia of the bronchia are also particularly radiosensitive tissues with respect to induction of tumors (NIH 1985; UNSCEAR 1988).

Recent studies in Japan show that only in the last few years, tumors of the gastro-intestinal tract and also of the liver increasingly occur among survivors of the atomic bombing. These data suggest that there may be considerable latent periods (of several decades) for radiation-induced solid tumors. Thus it is to be assumed that malignant tumors may be induced in basically any human tissue by ionizing rays, but that different tissues and organs show very different radiosensitivity (BEIR 1980; NIH 1985).

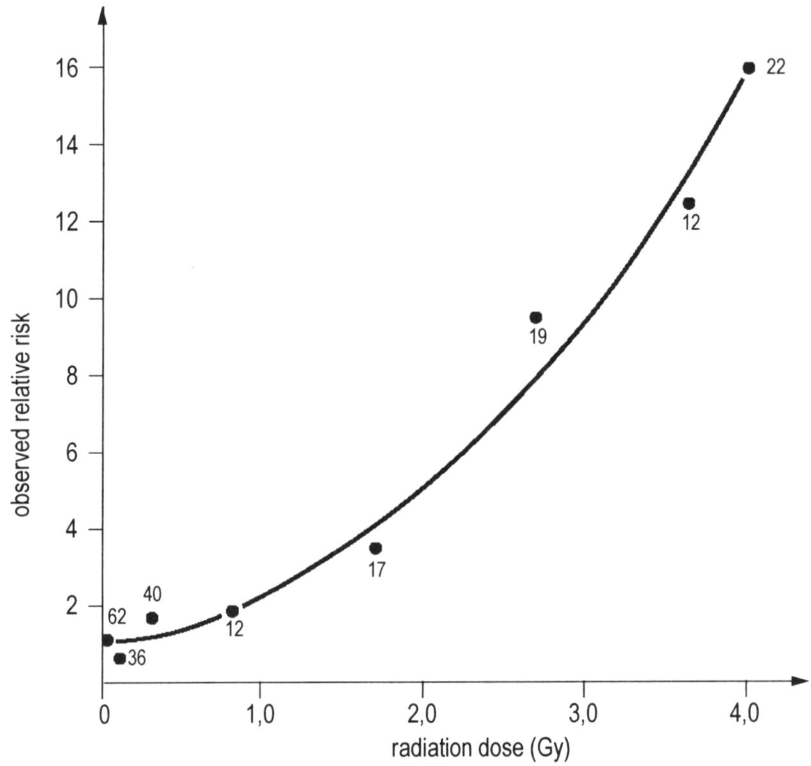

Fig. 2.21 Relative risk of leukemia as a function of the radiation dose (organ dose) on the basis of data from Hiroshima/Nagasaki (figures next to the observation points ● denote the number of cases) (Preston et al. 1987)

A considerable dependence of the radiation risk on the age at the time of exposure has not only been found for breast cancer but for other malignant diseases as well. In general, a particularly high risk is observed for radiation exposure of children and adolescents. Such data has increasingly been gathered in recent years among survivors of the bombing of Hiroshima and Nagasaki. This age-dependence is of importance in establishing risk factors. It plays a particular role in the evaluation of limits for occupationally exposed persons and for the population at large, since no exposures during childhood have to be taken into consideration for occupational exposure (UNSCEAR 1988).

The extended studies of prenatal exposure to radiation show very different results. Stewart and Kneale, in particular, have observed increased death rates due to leukemia and solid cancer during the first years of life after prenatal exposures to relatively low radiation doses (10–40 mSv) (Stewart & Kneale 1968). Some other authors found similar results. Studies concerning the survivors in Hiroshima and Nagasaki, on the other hand, did not yield higher rates of malignant diseases for prenatal exposure to radiation than for postnatal exposure (UNSCEAR 1986). To be on the safe side, we assume a higher risk of induction of leukemia and cancer for exposure during the late fetal period than for postnatal exposure.

With respect to most types of cancer, the so-called relative-risk model best describes empirical data and is hence used for risk assessment. Only the empirical data for leukemias and bone tumors are better represented by the absolute-risk model (NIH 1985; Kellerer 1990). According to the absolute-risk model, the rate of diseases increases after a latent period against the "spontaneous" rate, and then decreases again to the "spontaneous" value after a certain time (fig. 2.22). It must be assumed that by then all radiation-induced cases of illness have become manifest. According to the relative-risk model, the rate of malignant diseases in the irradiated collective also increases above the "spontaneous" rate after a latent period (fig. 2.22). Yet in this model, this increase is then maintained throughout the lives of the persons concerned, in a way that the relation between the rates for irradiated and not irradiated persons remains constant (BEIR 1980; Kellerer 1990; UNSCEAR 1986). This means that radiation-induced tumors mainly appear at an age when "spontaneous" tumors become manifest as well. Risk assessment on the basis of the relative-risk model is pessimistic and certainly marks an upper limit for a risk factor. For a more exact risk assessment on this basis, it is necessary to continue epidemiological studies as long as possible. The projected figures for cases of illness to be expected in future, which are included in the assessment, may then be verified.

2.2.6 Effective Dose

Because of different radiosensitivity of individual tissues and organ systems as to induction of malignant diseases, the International Commission on Radiological Protection (ICRP 1977) established different weighting factors for individual organs and tissues (table 2.13). As was briefly discussed in section 2.1.3, the so-called effective dose equivalent has been defined for a quantification of the total radiation risk at small doses. It is called 'effective dose' here, in accordance with an increasingly common simplified usage. The effective dose evaluates the total stochastic radiation risk, hence induction both of hereditary defects and of malignant diseases. The genetic risk (hereditary defects) is assumed to cover 25% of the total risk. Since the sum of all weighting factors is 1.0, a weighting factor of 0.25 (table 2.13) results for the gonads. The remaining 75% are distributed among the other organs and tissues, allowing for the respective risk of induction of malignant diseases. The effective dose is calculated by multiplying dose equivalents in the individual organs and tissues with the respective weighting factors and summing up the resulting products.

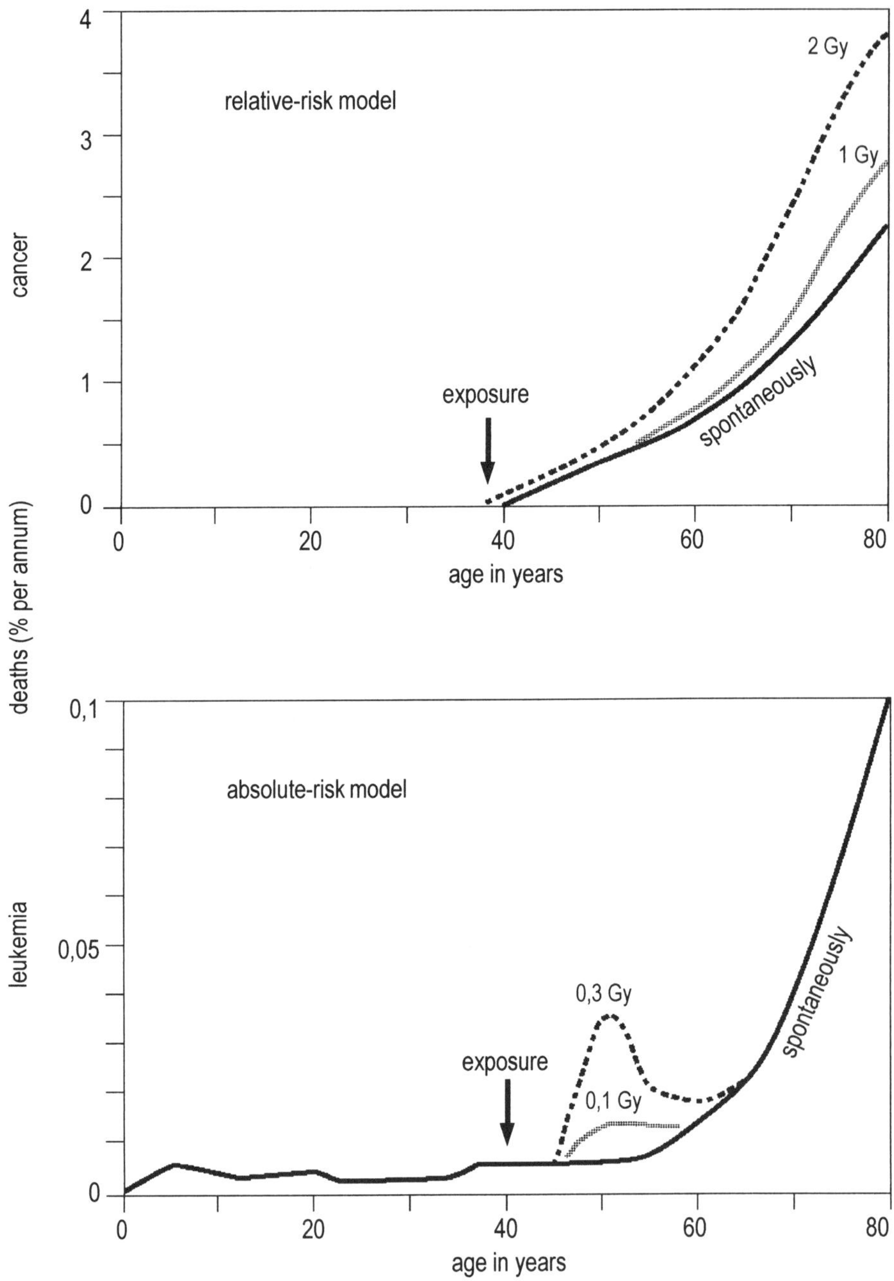

Fig. 2.22 Risk models
The curves show the age-dependent probability per annum of dying from cancer or leukemia, both "spontaneously" and after irradiation.

organ / tissue	W_T	$H_{eff} = W_T \times (H_T)_{1\,mSv\ homogeneously}$
gonads	0,25	0,25 mSv
breast tissue	0,15	0,15
bone marrow	0,12	0,12
lungs	0,12	0,12
thyroid	0,03	0,03
bone surface	0,03	0,03
remaining organs	5 x 0,06	0,30
sum (H_{eff})		1,00 mSv

Table 2.13 Effective dose (H_{eff}) and shares of exposures of individual organs (H_T) due to a homogeneous whole-body irradiation (dose: 1 mSv) (W_T is the weighting factor for the risk of stochastic damages to individual organs)

The main utility of the effective dose lies in its ability to provide a uniform basis for an assessment of the radiation risk, even with respect to an inhomogeneous irradiation, e.g., with respect to selective exposures of individual organs to radiation from incorporated radioactive substances. A critical objection is that weighting factors take into account just the mortality risk and not the morbidity risk with regard to leukemia and solid tumors. If this was allowed for, the sum of the weighting factors would, of course, still be 1, so that this would only shift the weights among the different organs. In particular, the weighting factor for the thyroid, for which there is a particularly wide gap between mortality risk and morbidity risk, would increase. Hence the weighting factors for other organs would have to decrease correspondingly. A further objection is that the weighting factors for some tissues and organs have to be corrected on the basis of recent epidemiological data (Preston et al. 1987; Shimizu et al. 1987).

Since the sum of the weighting factors equals 1.0, the whole-body dose equivalent is identical to the effective dose for a homogeneous whole-body irradiation. Yet for an inhomogeneous irradiation, which affects some organs or tissues with special intensity, these high local irradiations particularly influence the effective dose. This holds also, and particularly, for exposure to natural radiation, since irradiation of the respiratory tract by radon and its radioactive daughter products is especially high. Today, the effective dose is commonly used for limits in laws, regulations and recommendations.

2.2.7 The Small-Dose Problem

Determining an effective dose and evaluating the radiation risk with the help of the specific concept of the effective dose equivalent makes sense only if stochastic radiation effects are assumed to follow a linear dose response relation without threshold dose. Yet there is no definitive scientific proof for this assumption, it is rather based on plausibility considerations. We will end this section by discussing this issue.

In various epidemiological studies a statistically significant increase in malignant diseases was observed only at doses above the range of some tenths of a Sievert to one Sievert, hence at doses far above those which are found in the environment and for which limits through environmental standards are needed. Radiation risks at lower doses, which are of importance with respect to exposure of the population in the vicinity of nuclear facilities (some 10–100 μSv/a) and even with respect to occupational exposure (some mSv/a to some 10 mSv/a), have therefore to be derived by extrapolation from empirically determined risks at higher radiation doses. These extrapolations extend over large dose ranges and are therefore subject to considerable uncertainties. Moreover, the statistically significant epidemiological data concerning increased occurrences of leukemia and cancer do not only refer to exposures to high radiation doses but also to exposures at high dose rates. As exposure at the work-place and, to a still greater extent, in the environment takes place at low dose rates, this results in further uncertainties in the use of epidemiological data for risk assessment.

Often, reduction factors of 2 to 3 for the risk factor are employed, to take into account the fact known from general experience that the risk is relatively lower at low dose rates and in a low dose range (UNSCEAR 1977; ICRP 1977; UNSCEAR 1988). As to the induction of malignant diseases, epidemiological data may often be described both by a linear-quadratic dose response relation and by a strictly linear one. For some cases, also a strictly quadratic dose response relation is considered (figure 2.23). From what is known today, we may assume that a linear dose response relation best describes available data about induction of breast cancer and tumors of the thyroid, whereas a linear-quadratic dose response relation yields better results for other malignant diseases (NIH 1985). A risk assessment for low doses, however, is generally based on a linear dose response curve, for the sake of safety and simplicity, whereby the risk is rather overrated than underrated.

A dose response relation without threshold dose is plausible if the damaging development is initiated by a unicellular process, that is, if the malignant transformation of one cell suffices to induce such a disease. Yet the probability of induction is low. Experiments in vitro showed that the probability of transformation following an X-ray dose of 10 mGy is about 10^{-3} per cell (UNSCEAR 1986; Streffer 1985). As to the induction of leukemias,

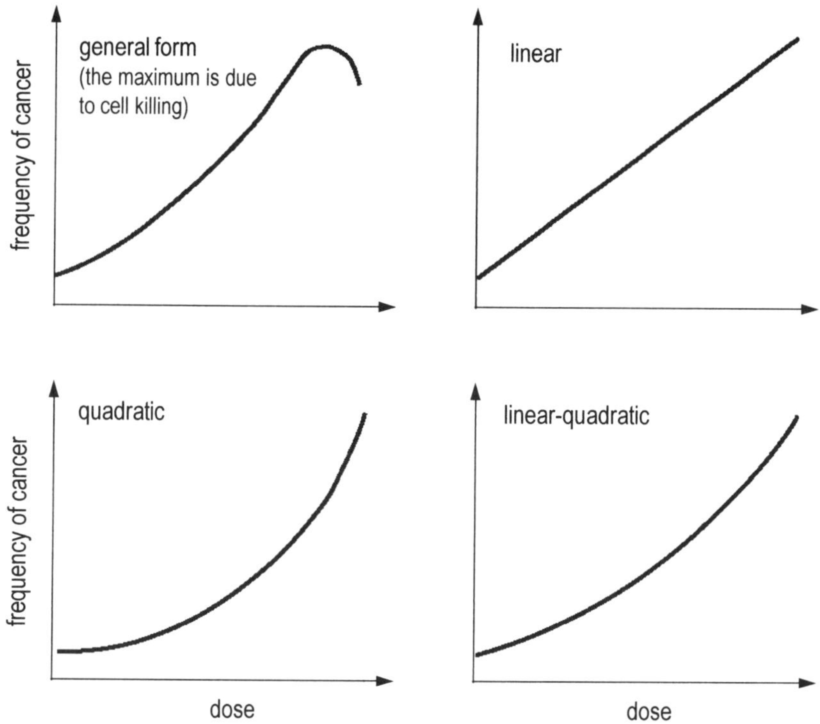

Fig. 2.23 Different types of dose response curves for the frequency of cases of cancer

it has been estimated that a dose of 2 Gy leads to a probability of an eventuation of only 10^{-8} per competent cell of the bone marrow[12], hence to about 10^{-10} per competent cell, if we extrapolate the respective value for 10 mGy. If each cell of the bone marrow had this competence, the probability would amount to only 10^{-16} per cell. The probability that a transformed cell induces a disease must therefore be extremely low.

The idea that one single transformed cell may lead to leukemia or solid tumors is based on the assumption (supported by experimental results) of a monoclonal growth of a tumor. So, in the case of tumors, often identical cytogenetic characteristics or the expression of the same specific gene products are found in each cell. Although this renders the assumption that there is no threshold dose for the induction of this radiation effect considerably plausible, this has not yet been proved so far. Sometimes the hypothesis is put forward that low radiation doses may stimulate repair processes in the genome. Epidemiological studies are not useful for a clarification here, because of fundamental statistical problems and because of the multitude of other factors influencing the frequency of cases of cancer. The same holds for animal experiments. Instead, intensified studies about the induction mechanism of tumors due to radiation are necessary for a solution of this problem (UNSCEAR 1986; Streffer 1985).

As far as the mechanism of oncogenesis is understood today, exposure to radiation leads at first to a malignant transformation of cells (initiation of oncogenesis). In analogy to cell transformation by chemical substances, we assume that changes in the DNA, that is, in the genetic information of the cell, are essentially involved in this transformation process. Yet, which of the radiation-induced damages of the DNA characterized in an earlier section may lead to cell transformation is still unknown. In this context, processes of cell proliferation and DNA repair are of importance (fig. 2.24), however, these cannot be assessed in detail so far. In-vitro experiments have shown that malignant transformations may even occur following radiation doses below 0.01 Sv (UNSCEAR 1986). It is only by an enormous cell proliferation, though, that a clinically manifest tumor with some billions of cells may then emerge from a few cells or one single malignant cell (Streffer & Müller 1984). A tumor weighing one gram comprises about one billion cells. The steps necessary in such a development are denoted as promotion of oncogenesis.

Little is known so far about the processes following the initiation of oncogenesis, which may take many years or even decades (UNSCEAR 1986). If the assumption is correct that there is no threshold dose, there must be a certain, if small, probability that any malignant cell may lead to a manifest tumor. This is the basis for mechanistic reflections about the monoclonal growth of the tumor.

If we proceed on these assumptions, including the assumption of no threshold dose, a risk assessment is possible even for small radiation doses (fig. 2.25). Mathematical extrapolation then leads to risk factors of about 200 to 1'000 deaths due to leukemia and solid tumors after whole-body irradiation of one million people at a dose of 0.01 Sv, that is, the risk factor is 2×10^{-2} to 10^{-1} per Sievert[13] (UNSCEAR 1977; BEIR 1980; UNSCEAR 1986). This means that a lifetime radiation dose of 0.01 Sv, which would result from complete exploitation of the limit of 0.3 mSv/a, would increase the observed "spontaneous" death rate of the German population due to leukemia and cancer by one or several parts per thousand (Streffer 1985; Jacobi 1988). But the exposure of the population in the vicinity of nuclear facilities in Germany amounts to only about 10 μSv/a for normal operation, design basis accidents included. This is four to five orders below the dose range for which epidemiological data are available. Such an extrapolation, and with it the prognosis of a number of additional deaths from cancer due to nuclear facilities, can therefore hardly be supported from a scientific point of view and should be omitted.

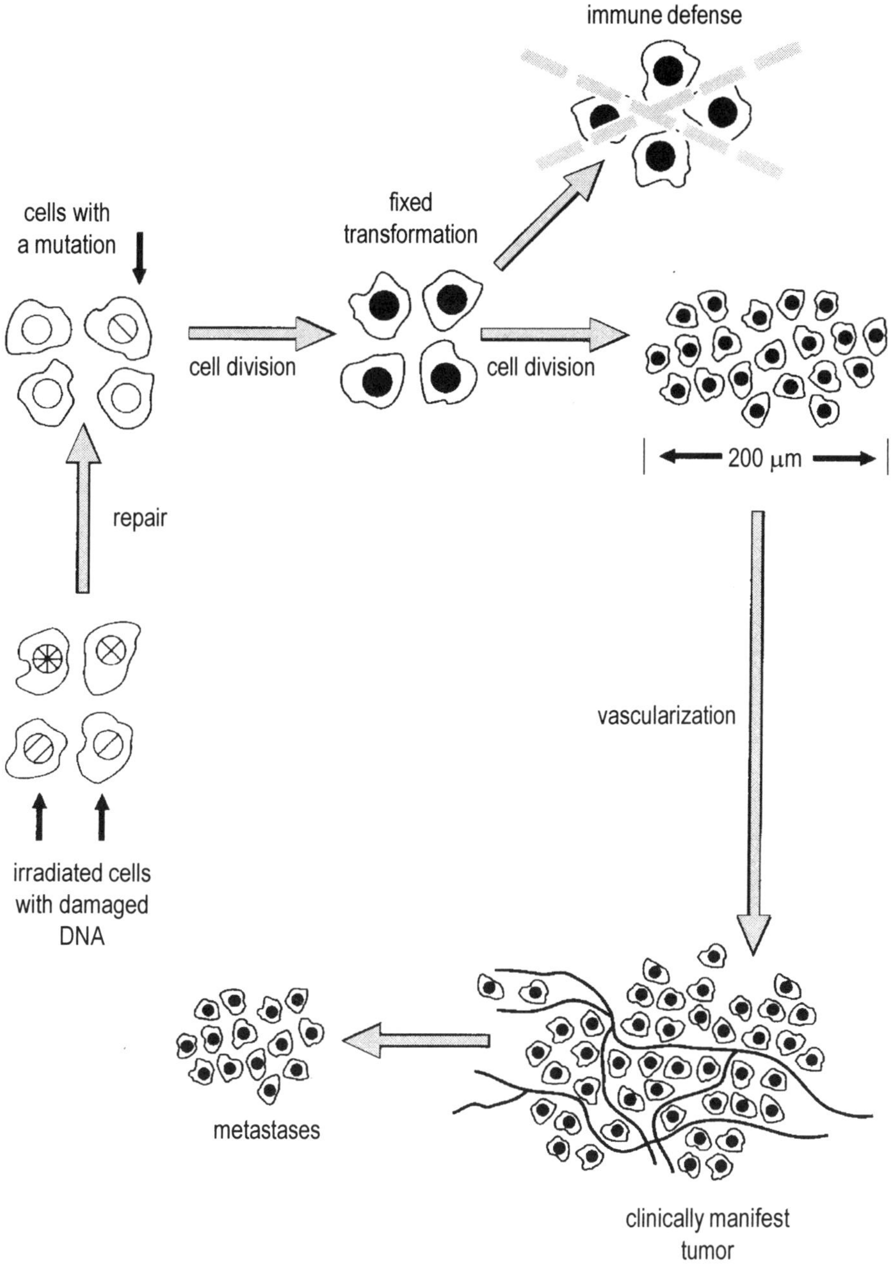

Fig. 2.24 Mechanism of oncogenesis due to ionizing rays

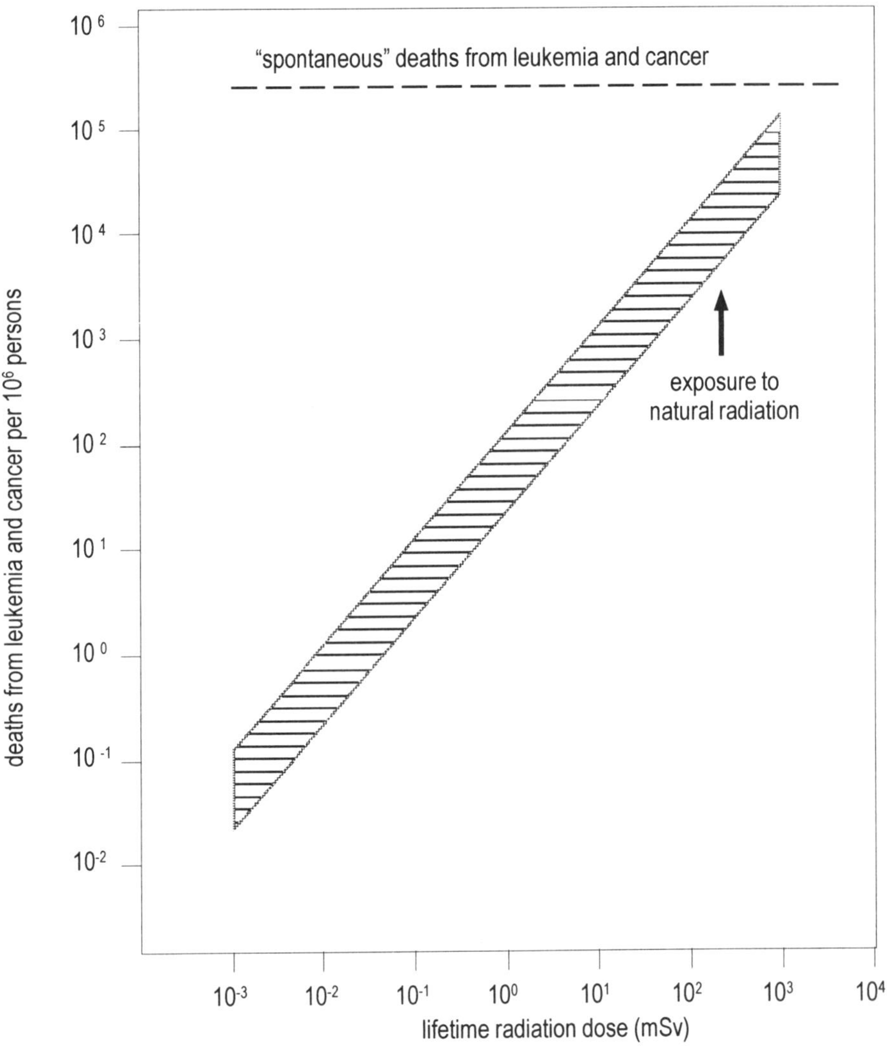

Fig. 2.25 Rate of induced leukemias and cancers as a function of the radiation dose, for a very low dose range

Repeated reports about an increased number of cases of childhood leukemia in the vicinity of nuclear power plants in Germany have not been proved by epidemiological methods so far. The observed rates of leukemia are of an order found also in regions without nuclear facilities. Several studies, however, have found increased rates of childhood leukemia in the vicinity of reprocessing installations in Great Britain. In the vicinity of such facilities, so-called clusters of cases of leukemia appeared; that is, the number of cases of leukemia was higher than it would have been expected for this population (Roman et al. 1987; Darby & Doll 1987), although absolute figures were very low.

In a comprehensive study, mortality from leukemia and other cancers in England and Wales was compared to mortality from cancer in the vicinity of nuclear installations (Cook-Mozaffari et al. 1989a). 70 districts which, with some part of their areas, belong to the immediate surroundings of nuclear facilities (range of 16 km), were compared to 330 other districts far from such installations. During the years 1969–1978, the relative risk for persons of the ages from 0 to 24 years in the vicinity of nuclear installations was increased by 15% with respect to mortality from all kinds of leukemias, and by 21% as regards lymphatic leukemias. Moreover, it was found for the same collective of persons during the same period that mortality from Hodgkin's disease was increased by 24%. For persons of the ages of 25 to 64 years, on the other hand, mortality from lymphatic leukemias was reduced by 14%.

Since the annual radiation doses received by children living near nuclear facilities are essentially lower than those for which, according to all radiobiological knowledge, a measurable increase in the rate of leukemia would have been expected, the question arises whether other parameters might play a role in this context as well. The same working-group therefore studied the surroundings of further nuclear installations which have been operating only since 1978 or later, as well as regions in which nuclear facilities are planned (Cook-Mozaffari et al. 1989b). The result was that for persons of the ages of 0 to 24, the relative risk of leukemias is increased also in regions where nuclear installations are planned or are under construction. The values for the relative risk are comparable to those which are observed in the vicinity of existing nuclear facilities. There is no statistical significance, though, since the number of cases is so low.

The relative risk of falling ill with Hodgkin's disease was also significantly increased in regions where nuclear installations were planned or under construction. In that case, the relative risk was even higher for planned facilities than for already existing ones. Findings concerning the value for all malignant lymphomata were similar. In the respective studies, again a higher value was found for lymphatic leukemias than for leukemias in general. Moreover, mortality from lymphatic leukemias was reduced here again for persons of the ages from 25 to 64 years.

These studies suggest that the increased risk of leukemia and lymphomata in the vicinity of nuclear facilities is not connected to a regional contamination by radioactive substances but that these findings are due to other factors. Yet so far, available studies gave no hints about the nature of such factors. There are, however, studies about other Scottish regions with considerable fluctuations of the population, like in the originally rural nuclear sites, which also show an increased mortality from leukemias (Kinlen 1988). The author assumes that, due to such movements of the population, in particular due to immigration of new population groups, infectious agents, e.g. viruses, might play a role in an additional induction of leukemias. This assumption is also adopted and considered to be possible by Cook-Mozaffari et al. A clarification of this issue will be the task of further studies.

Just recently, an epidemiological study carried out in Sellafield showed a correlation between the occurrence of childhood leukemia and fathers being exposed to increased radiation before procreation (Gardner et al. 1990). Whether there is any causality here, remains an open question for the time being. Extended studies about children of survivors of the atomic bombing did not suggest such a correlation.

The broad range of risk factors, as it is shown in fig 2.25, reflects the uncertainty of these estimations. The dissimilarity of the exposed populations from which epidemiological data were gained as to their "spontaneous" cancer risks, and reference to different populations of comparison clearly contributed to this variation. A major reason for this range is the revision of results from studies about survivors of the atomic bombing in Japan, which yielded higher risk factors (UNSCEAR 1988; Preston et al. 1987; Shimizu et al. 1987). This increase is due to three reasons:

- New calculations concerning dosimetry after the explosions of the atomic bombs have led to new dose estimates.
- In the last few years, new data have been collected about rates of leukemias and tumors. Particularly, new cases of illness or death have been observed in persons who were exposed when they were young.
- These results have, for the first time, been consistently evaluated according to the concept of the relative risk. Thereby, the radiation risk for still living exposed persons is projected into the future. Cases of death from cancer to be still expected in future are determined with the help of age-dependence of the cancer risk in not exposed persons, assuming a constant increase in the relative risk due to radiation.

While the revision of dosimetry only marginally influenced the risk factor, the other two aspects have more distinct consequences. They result in a risk factor which is higher than the formerly assumed value of 2×10^{-2} per Sievert by about the factor 5. This value is valid for high radiation doses and high dose rates. Taking a reduction factor of 2 into consideration for sparsely ionizing radiation at low doses and low dose rates, we obtain a value of about 5×10^{-2} per Sievert.

Once again, stimulating effects of ionizing radiation in the low dose range (mSv to 100 mSv) on biological processes are discussed in literature. A survey of these biopositive effects, which are denoted by the term "hormesis", is given in Luckey 1980. Explanations for this theory start from the assumption that life on earth has developed under conditions of ionizing radiation and is still continuously exposed to this radiation today. It has often been shown experimentally that ionizing radiation stimulates cell proliferation in plants and also in mammalian tissues, including human tissues, whereby they possibly stimulate growth and also healing of wounds. Additionally, resistance of mammals to radiation (surviving acute radiation effects) may be increased by previous irradiation (0.5–1.0 Sv) under defined conditions during a short period. The immune system may possibly be stimulated under certain conditions. Yet there is no experimental evidence so far that a stimulation of the immune system results in a reduction of the carcinogenic effect of ionizing radiation. Recent experimental studies discuss a phenomenon which is called "adaptive response" and which is possibly due to stimulation of repair and recovery processes.

Even if sometimes the possibility of positive mutations is mentioned in principle, this is certainly of no importance with respect to a hormesis effect. The vast majority of observed mutations (even mutations caused by natural radiation) have negative consequences. Nor are there experimental hints that "positive" mutations might emerge in gametes or soma cells. In terms of a conservative risk assessment, as is usual for radiation protection, biopositive effects of ionizing radiation with respect to stochastic damages can therefore not be taken into account so far.

2.3 DISCUSSION OF METHODS AND ERRORS

This section will once again comprehensively and critically discuss the present state of knowledge. In doing so, it will focus on possible errors and on fundamental methodological limits.

2.3.1 Observations about Dosimetric Errors

A critical examination of dosimetry has to distinguish between directly measurable quantities and quantities derived by means of models from measured values.

Directly measurable physical quantities which describe a state, as for instance temperature, have a clear, i.e. true, value, which, however, can never be measured perfectly. There is, on the one hand, an inevitable statistical scattering of measured values around the true value. If the measurement is repeated often enough, the mean value of all measurements approaches the true value as close as is desired. On the other hand, systematic errors may occur in measurements, due to imperfect or erroneous measurement conditions. Such errors can only be eliminated by analyzing measurement conditions.

Yet when radioactivity is measured, we have, moreover, to deal with the fact that radioactive decay is itself a statistical process, so that the measured number of radioactive disintegrations in a given time is actually not the same for different measurements. If measurement time is prolonged, a single measurement approaches the statistical mean value for the number of disintegrations per unit time, hence the "true" value, as closely as desired. A systematic error in determining radioactivity may, for example, occur with respect to radioactivity of one single radionuclide which has to be chemically separated from a mixture of several radionuclides before measurement. Systematic errors may creep into such a chemical procedure, which then, of course, cannot be eliminated by frequent repetition of separation and measurement. This was, e.g., the case with the determination of strontium-90 in a mixture of radionuclides as it occurred as fallout in Germany after the Chernobyl accident. Only a few very experienced laboratories were able to carry out reliable measurements here.

As to biological quantities, which are measured in living organisms, there is usually no single true value but a range of values, dependent on individual differences. For example, the share of an ingested amount of radionuclides accumulating in a single organ may be influenced by the individual size of the organ; and the pace at which it is eliminated may depend on individual metabolic particulars. One analogue familiar to everybody is the different development of body weight which different persons show even with very similar nutrition.

Aside from directly measurable quantities there are also those which have to be calculated from other measured quantities by means of models. Models are simplified descriptions of a condition which is too complex to be understood or even to allow a feasible description. One example is the distribution of radionuclides in an organism. Such a distribution is described by subdividing the organism into a small number of compartments, which are treated as if they were homogeneous. If – e.g., in the case of distribution of radioactive iodine – the gastro-intestinal tract is regarded as a compartment, its individual elements, e.g. the stomach, the small intestine, the colon, actually behave so similar, compared for instance with the thyroid, that a combination in one compartment is justified. Of course, true values would still be better approximated if these individual elements were regarded as separated compartments with different storage capacity. The example indicates that we may continue in that direction as far as we desire by choosing ever smaller compartments, but that we will soon reach limits of feasibility – even if the necessary data were available – without gaining a significant increase in accuracy. So, in modeling, we will always have to respect the following principle: as close to reality as possible, but at the same time as simplified as necessary. In many cases, however, approximation of reality is not so much limited by the feasibility of the model but by availability of data.

Thus, methodologically different errors may occur at any stage of dosimetry. In the following, we will, on the one hand, convey an idea of the accuracy attainable in radiation dosimetry, on the other hand we will point out gaps in our knowledge, which are of importance in this context and which might realistically be expected to be closed by research efforts.

The first stage of dosimetry deals with measurements of the intensity of external radioactive sources, hence measurements of radioactivity. With qualified personnel and appropriate equipment, such measurements are so exact today that they meet any demand, with the exception of measurements of radionuclides which have to be chemically separated at first. This requires experience and, above all, time. Yet this is the case for only relatively few radionuclides, namely those which do not emit γ–radiation so that they cannot be detected in a mixture by means of γ–spectrometry. Among the radionuclides which are produced in a nuclear facility and which contribute significantly to exposure of the population to radiation or might contribute in case of malfunction are, primarily, strontium-90, iodine-129, and plutonium. Here, improvement of analytical techniques is desirable.

Local and personal doses can be measured with similar accuracy and reliability, as far as they are from γ–emitters. As to pure β–emitters, doses can, in most cases, only be calculated from measured source intensities. α–emitters have to be considered only as internal radiation sources, due to their short range, so that local or personal doses in the usual sense are of no relevance here. Incidentally, measurement of neutron doses, which are of no importance in the protection of the population against radiation, involves considerable uncertainties today and is therefore subject to research.

The next step of dosimetry is a model calculation, namely calculation of organ or body doses from the measured values of source intensities or local or personal doses. The model uses physical data, e.g., scattering cross-sections for different types of rays in the organism, and biological data, in particular anatomic ones. Physical data are well-known. As to biological data, differentiation according to age and sex leads to a wide variation, which we try to meet by definition of a reference individual. With the help of a human phantom, corresponding with the defined reference individual, we can experimentally verify the calculated values for organ and body dose, and make sure that the obtained result is still reasonable in spite of inevitable model simplifications. The approximation of reality through calculation models is improved particularly by drawing on methods of computer tomography. All in all, we can say that not much of the accuracy with which source intensities or local or person doses are measured is lost in calculation of organ and body doses for penetrating γ–radiation. Therefore it should be possible under most circumstances relevant for radiation protection of the population to determine values of organ exposures with an accuracy of a factor 2.

Internal dosimetry is generally less reliable than external dosimetry, not because of fundamental biological and physical differences between external and internal exposure, but because of four basic difficulties in calculations. One is the very inhomogeneous distribution of certain radionuclides, e.g. iodine, strontium and plutonium isotopes, in the human organism, for biokinetic reasons. The second difficulty is due to the fact that some organs are at the same time radiation sources and radiation targets. This makes itself felt more, the lower the penetrating power of the radiation emitted by the respective radionuclide is. Third, age-dependence of biokinetic data is only partly known for many radionuclides. And fourth, the mathematical anthropomorph phantoms underlying the calculations leave room for improvement (e.g., by series of computer-tomographic imaging).

Strontium-90, as a pure β–emitter, is a good example of the combination of very selective accumulation, namely in the bones, with radiation of low penetrating power, which virtually leads only to exposure of that organ in which the radionuclide is accumulated. This results in dose gradients in this organ which are difficult to calculate; and neighboring organs are affected in the contact zone to an extent which is difficult to calculate as well. The most important example of the other extreme is cesium-137, which spreads uniformly through all organs and which then, as a γ–emitter of medium energy, also leads to a virtually uniform irradiation of the whole body. For those radioactive iodine isotopes which are γ–emitters and which are accumulated in the thyroid, the calculation was described in an exemplary way, together with the related possibilities of errors, in section 2.1.5.

For those few radionuclides with long half-lifes which are produced in large amounts during nuclear fission and which thus mainly determine the dose caused by emissions of nuclear facilities, distribution data are so well-known that there is no need to make great scientific efforts to generally improve them. Here again – apart from the difficulty to gain human biological data – we are faced with the basic problem of biology that data vary within natural ranges, which we try to allow for by definition of reference values. Yet besides true individual variabilities, which may, for instance, be caused by physical or genetic disposition, there are also systematic differences, primarily dependent on age and sex. As to age-dependence of distribution, there is a certain data base available for some radionuclides, but all in all there is still need for further research. The same holds for the prenatal transition to embryo or fetus.

Let us assume, with respect to internal dosimetry, that uncertainties concerning the major part of known radionuclides are very high, but that these radionuclides altogether contribute only very little to exposure of the population to artificial radiation. With regard to those few radionuclides which are crucial for this kind of exposure, uncertainty is clearly lower, so that uncertainty in dosimetry is not much higher for internal than for external irradiation. This holds primarily for individuals whose metabolic characteristics are within the normal scattering range around the reference organism. With respect to the infantile organism, there are valuable approaches taking its characteristics into consideration; but little is known so far about other individually specific features.

In a last step, we have to deal with the transition to dose equivalent or to effective dose. Quality factors allowing for a transition from energy dose to dose equivalent are obviously only very roughly represented by the three approximative values 1, 10 and 20 employed in German radiation protection regulations. This rather rough approximation is, however, not due to lack of basic knowledge or lack of data, but has been established in the interest of feasibility.

One factor influencing the effective dose is the relative risk of stochastic damages faced by single organs or organ groups. Of course, this can be only a rather roughly estimated figure. So, the effective dose is not the basis of an assessment of health risks. Determination of risk coefficients by means of epidemiological studies will always be based on organ doses. The effective dose, on the other hand, is a precious aid in exposure management, which permits an intercomparison of different exposures as well as their summation, if need be.

2.3.2 Discussion of the Scattering Range of Risk Factors

Exposure and impact are decisive with respect to risk. It is the stochastic radiation impacts which are of pivotal importance for an evaluation of radiation risks, and among them, in particular, the induction of malignant diseases. So, what degree of certainty or uncertainty do assessments of these risk values have?

Epidemiological studies are the most important basis for ascertaining the risk of radiation cancer, and among them, primarily, studies concerning the survivors of the atomic bombings of Hiroshima and Nagasaki. In addition, there are studies focusing on collectives of persons who were exposed to rather high radiation due to their occupational activities or due to medical applications, and collectives exposed to high natural radiation. For a final assessment of risk factors, only Japanese data have been taken into account, since the respective group is the largest available collective for such studies, since all age groups are represented, and since data refer to a relatively homogeneous whole-body irradiation. There is a total of about 120'000 persons considered in the study, and for 75,991 of them, dose estimations on the basis of the new dosimetry system of 1986 (abbreviation: DS86) are available. Here, dosimetry system means the theoretical calculation of dose distribution within the range of an exploding atomic bomb. An older dosimetry system dates from the year 1965 (abbreviation: T65DR). Preconditions for a generally valid assessment of the radiation risk are less favorable for other collectives of exposed persons, e.g. for those 14'000 persons who, in the first postwar decade, were irradiated with X-rays at high doses because of Bechterew's disease, since dose distribution is very inhomogeneous here and since only certain age groups are represented.

The frequency distribution of the different "spontaneously" occurring types of cancer shows distinct national differences. The frequency of gastrointestinal tumors, in particular tumors of the stomach, is significantly higher among the Japanese population than among people in Western Europe, and the reverse holds for tumors of the female breast. Therefore the question arises whether risk values referring to Japan may be transferred to populations with another spectrum of "spontaneous" malignant diseases. This question has been thoroughly discussed, and intercomparisons of tumor rates in Great Britain and in Puerto Rico led to the conclusion that such a transfer is perhaps problematic for individual types of tumors, but that national differences are probably irrelevant in ascertaining the total risk for all malignant diseases, since their different tendencies counterbalance each other. Values referring to the total risk of radiation cancer, obtained from epidemiological studies in Japan, can therefore be transferred to other populations with a sufficient degree of certainty.

Risks are obtained by comparing frequencies of deaths from cancer in collectives exposed to different levels of radiation. By 1985, about 40% of 75,991 survivors of the atomic bombing had died due to old age. The causes for their deaths were determined and included in the epidemiological study. The risk assessment, however, does not only take into consideration persons who have already died, but starts from the assumption that the "relative risk" of radiation-induced cancer (with the exception of leukemias) remains constant during a whole life, hence continues to be of relevance for persons who are still alive. This projection into the future has been included into considerations about the risk values, which made them increase considerably. Only future deaths will show how justified such a way of proceeding is.

With regard to persons who were irradiated because of Bechterew's disease, and for whom there is an already considerably longer time of observation, some indications suggest that the relative risk might not be constant but might decrease in the course of time. In that case the relative risk for persons still alive might be lower than it had been for those persons who have already died.

It was mentioned already that statistically significant increases of cases of cancer have been observed among the survivors of the atomic bombing only for radiation doses above 0.2 Gy (there was a total of 300 to 500 additional deaths from leukemias and cancers due to the atomic bombs). Table 2.14 lists the values of the "relative risk" of additional lethal cases of cancer in Japan in relation to dose ranges. Furthermore, it lists the 90% confidence intervals and the respective number of deaths. The relative risk is significantly increased both with respect to leukemias and with respect to other types of cancer from the fourth dose range on (0.20–0.49Gy). The step from this dose range to estimates within the range of the dose limit for the German population (0.3 mSv per annum or about 0.02 Sv per lifetime) thus required extrapolation not only over a large dose range but also, over an enormous range of temporal distribution of exposure. For these parameters, in this range, no empirical knowledge is available about the character of the dose effect relation. The large confidence intervals reflect the uncertainty of the values. This holds in particular for leukemias because of the very small number of cases, but we can assert for the whole study that, due to the smallness of the effect, the number of persons exposed is still too low to allow the derivation of truly solid statistical values.

Finally, we have to consider the fact that irradiation due to the explosions of the nuclear bombs took place at extremely high dose rates. As compared with that situation, radiation exposure of the population living in the neighborhood of nuclear installations during normal operation, and even in accident situations like that of Chernobyl, does not only take place at low doses but also at very low dose rates. So far, we cannot be sure whether and to what extent this might reduce radiation-induced cancer in human beings, as this is the case with some other radiation effects. For the collective of patients with Bechterew's disease, lower risk factors were found after fractionated irradiation than for the survivors of the atomic bombing after acute irradiation, so that such a reduction is to be presumed. As to the induction of leukemia, it was actually found that values for relative risk, related to 1 Gy in a low dose range (< 0.50 Gy) are lower than in a higher dose range (> 0.50 Gy). This does not, however, hold for other types of cancer (table 2.15). The values in table 2.15 are taken from table 2.14 and related to 1 Gy.

Apparently, recovery processes, which lead to a reduction of the risk of radiation-induced cancer at low dose rates, play a significant role. This role is different for different types of cancer, though, and never provides a complete explanation of the phenomenon. Data from animal experiments confirm this high variability. UNSCEAR suggested a range of 2 to 10 for the reduction factor. If the reduction factor is assumed to be in the range of about 2, this is obviously an assumption on the safe side. With such a reduction factor, we get a risk factor of 5×10^{-2} per Sievert for exposure to sparsely ionizing rays, at a low dose and at a low dose rate.

dose range (Gy)	leukemia			cancer except leukemia		
0,01 - 0,05	0,99	0,68 - 1,40	33	1,06	1,0 - 1,12	1394
0,06 - 0,09	0,61	0,25 - 1,22	5	1,08	0,98 - 1,19	341
0,10 - 0,19	1,08	0,61 - 1,82	11	1,06	0,97 - 1,16	410
0,20 - 0,49	1,79	1,18 - 2,68	23	1,12	1,03 - 1,21	529
0,50 - 0,99	4,15	2,76 - 6,19	24	1,36	1,23 - 1,51	273
1,0 - 1,9	8,01	5,34 - 11,94	24	1,66	1,45 - 1,90	158
> 2,0	18,57	12,09 - 28,20	21	2,05	1,66 - 2,50	67

Table 2.14 Estimated risk values (persons not exposed: 1.0) in relation to absorbed organ doses (each second column lists the 90% confidence intervals and each third column lists the deaths)

If the risk factor were essentially higher, exposures to different natural radiation would have to become manifest in different types of cancer. For a mean exposure to natural radiation at an effective dose equivalent of 2 mSv per annum, a total effective dose equivalent of about 140 mSv results over a period of 70 years. Considerations based on the newly estimated risk factors would then lead to a share of 4% of all deaths from cancer and a share of about 20% of deaths from leukemia resulting from exposure to natural radiation. An exposure to natural radiation which is increased by 50–100% should then lead to a significant increase in the number of deaths from cancer and leukemia, for which there is no evidence so far, though. These considerations show that the listed risk factors will certainly not lead to an underestimation of the risk of cancer due to ionizing radiation. Even overestimation is possible for the case of exposure to sparsely ionizing rays and at low dose rates.

	dose < 0,50 Gy	dose > 0,50 Gy
leukemia	2,94	5,02
cancer except leukemia	1,32	1,29

Table 2.15 Estimated relative risk values (persons not exposed: 1.0), related to 1 Gy, in low and high dose range

Studies based on animal experiments and studies of cellular biology can help to clarify the mechanism of carcinogenesis due to radiation and can thus contribute to an answer to the question whether we may rightly assume that there is no threshold dose. Such studies may also help to find out whether the assumption of a reduction factor at low dose rates is justified. Yet they do not allow for an extrapolation to assess risk factors for human beings, since the spontaneous rates of mortality from cancer differ very much from animal species to animal species and even from animal phylum to animal phylum. Apart from many other factors, phenomena of promotion of tumors, e.g. by hormones, the regulation of which is very different for different mammalians, play a considerable role.

In addition, it should be mentioned that there are no epidemiological data for the radiogenetic risk, since no statistically significant increase in genetic damages was observed in descendants of the survivors of Hiroshima and Nagasaki. Radiogenetic risk factors used in radiation protection are based on very extensive experiments with mice. Although their transfer to human beings is theoretically rather well-founded, there remains a basic uncertainty regarding such transfers. The risk factor derived in that way is of the order of 10^{-3} to 10^{-2} per Sievert. The fact that the survivors of the atomic bombing do not show statistically significant increases in hereditary damages is in keeping with this value.

ANNOTATION

1 incorporated = taken into the body, for example ingested from food.

2 An ion pair is a positively charged atom or molecule and a free electron.

3 1 R = 2.08 x109 ion pairs per cubic centimeter of air, which is the same as 2.58×10^{-4} C/kg (the unit Roentgen is defined especially for absorption in air).

4 An approach towards appropriate models of biological effects of chemicals could perhaps be provided by an understanding of structure activity relations, which, however, is still in its infancy.

5 Stochastic risks are above all the cancer risk and the risk of genetic mutations. It is not the severity of damage which depends on the extent of exposure here, but the probability of its incidence. This is discussed in more detail in 2.3.5.

6 ICRP is the International Commission on Radiological Protection, founded in 1928.

7 This exposition has been adopted, with the authors' permission, from a work by Henrichs, Müller-Brunecker and Paretzke (Henrichs et al. 1983).

8 The expression 'somatic' is derived from the Greek word soma = body. We distinguish, for instance, somatic cells and germ cells (gametes), and somatic and genetic effects.

9 A stochastic event is a random event.

10 By "spontaneous" diseases we understand here those diseases that affect persons who were not additionally exposed to radiation. They may be due to endogenous causes, and also to other exogenous causes, like, e.g., exposure to natural radiation.

11 These cataracts have to be distinguished from those which only show in the following generations, as a consequence of a radiation-induced mutation of the gametes.

12 Cell of the bone marrow which is transformable and may induce leukemia. The fraction of competent cells in the total number of cells of the bone marrow has been assumed to be 10-6.

13 This means: if a collective is irradiated at 1 Sv per person, 2–10% of the persons will die from cancer (in addition to the 20% which will normally die from cancer)

LITERATURE

Alper, T. (1979): Cellular Radiobiology, Cambridge, Cambridge University Press.

Altmann, K.I., Gerber, G.B., Okada, 5. (1970): Radiation-Biochemistry, New York, Academic Press.

BEIR (1980): The Effects of Exposure to Low Levels of Ionizing Radiation, Washington, D.C., National Academic Press.

Boice, J.D., Fraumeni, J.F. (1984): Radiation Carcinogenesis: Epidemiology and Biological Significance, New York, Raven Press.

Bond, V.P., Fliedner, T.M., Archambeau, J.O. (1965): Mammalian Radiation Lethality: a Disturbance in Cellular Kinetics, New York, Academic Press.

Chadwick, K.H., Leenhouts, H.P. (1981): The Molecular Theory of Radiation Biology, Berlin, Springer.

ChemG 1982 (Chemikaliengesetz): Gesetz zum Schutz gegen gefährliche Stoffe.

Cook-Mozaffari, P.J., Darby, S.C., Doll, R., Forman, D., Hermon, C., Pike, M.C., Vincent, T. (1989a): Geographical Variation in Mortality from Leukaemia and Other Cancers in England and Wales in Relation to Proximity Nuclear Installations, 1969–1978, Brit. J. Cancer 59, 476–485.

Cook-Mozaffari, P.J., Darby, S.C., Doll, R. (1989b): Cancer Near Potential Sites of Nuclear Installations, Lancet II, No 8672, 1145–1147.

Darby, S.C., Doll, R. (1987): Falbut, Radiation Doses Near Dounreay, and Childhood Leukaemia, Brit. Med. J. 294, No 6572, 603–607.

Dikomey, E., Franzke, J. (1986): Three Classes of DNA Strand Breaks Induced by X-Irradiation and Internal ß-Rays, Int. J. Radiat. Biol. 50, 893–908.

Dunning, D.E., Schwarz, G. (1981): Variability of Human Thyroid Characteristics and Estimates of Dose from Ingested I-131, Health Phys. 40, 661–675.

Eckerman, K.F., Kerr, G.D., Raridon, R. (1980): Organ Doses from Isotropic and Cloud Sources of Gamma Rays, Health Phys. 39. 1054.

Ehling, U.H. (1987): Quantifizierung des strahlengenetischen Risikos, Strahlen, Strahlenther. Onkol. 163, 283–291.

Elkind, K.F., Sutton, H. (1960): Radiation Response of Mammalian Cells Grown in Culture. I. Repair of X-Ray Damage in Surviving Chinese Hamster Cells, Radiat. Res. 13, 556–593.

Feinendegen, L.E. (1977): Das Strahlenrisiko bei Kernreaktoren und radioaktivem Müll, Öff. Gesundh.-Wesen 39, 584–598.

Fischer, A. (1981): Schilddrüsengewichte im saarländisch-pfälzischen Raum – Ein Beitrag zur Endemie des Kropfes, Dissertation, Universität des Saarlandes, Homburg.

Gardner, M.J., Snee, M.P., Hall, A.J., Powell, C.A., Downes, S., Terrell, J.D. (1990): Results of a Case-Control Study of Leukaemia and Lymphoma among Young People Near Sellafield Nuclear Plant in West Cumbria, Brit. Med. J 300, 423–434.

Generoso, W.M., Shelby, M.D., Serres de, F.J. (1980): DNA Repair and Mutagenesis in Eukaryotes, New York, Plenum Press.

Glöbel, B. (1978): Die Ökologie von stabilem und radioaktivem Jod und Bedeutung für die Beurteilung des Strahlenrisikos bei der medizinischen Anwendung, Habilitationsschrift, Universität des Saarlandes, Homburg.

Hall, E.J. (1978): Radiobiology for the Radiologist, Hagerstown/Maryland, Harper and Row.

Hanawalt, Ph.C., Cooper, P.K., Ganesau, A.K., Smith, Ch.A. (1979): DNA Repair in Bacteria and Mammalian Cells, Annu. Rev. Biochem. 48, 783.

Hellstem, P., Keller, H.E., Weinheimer, D. (1978): Thyroid Jodine Concentration and Total Thyroid Iodine in Normal Subjects and in Endemic Goitre Subjects, J. Endocrinol. 9, 351–356.

Henrichs, K., Kaul, A. (1982): Age Dependent Values of Specific Absorbed Fractions and Specific Effective Energy for the Dosimetry of Internal Emitters, Rad. Prot. Dosim., Vol. 3, No. 1/2, 71–73.

Henrichs K., Müller-Brunecker, G., Paretzke, H.G. (1983): Zur Strahlenexposition der Schilddrüse bei Inkorporation von Jod-Isotopen, GSF-Bericht, GSFS-960.

Hoffmann, F.O. (1973): Parameters to be Considered when Calculating the Age Dependent I-131 Dose to the Thyroid, Bericht des Instituts für Reaktorsicherheit, Köln (jetzt GRS), IRS, W-5.

ICRP (1975): Report of the Task Group on Reference Man, ICRP Publication, ICRP 23, Oxford, Pergamon Press.

ICRP (1977): Recommendations of the International Commission on Radiological Protection, ICRP Publication, ICRP 26, Oxford, Pergamon Press.

ICRP (1979): Limits for Intakes of Radionuclides by Workers, ICRP Publication,

ICRP 30, Oxford, Pergamon Press.

Ishihara, T., Sasaki, M.S. (1983): Radiation-Induced Chromosome Damage in Man, New York, Alan R. Liss, Inc..

Jacobi, W (1975): The Concept of the Effective Dose – A Proposal for the Combination of Organ Doses, Radiat. Environ. Bioph. 12, 101–109.

Jacobi, W., Paretzke, H.G., Ehling, U.H. (1981): Strahlenexposition und Strahlenrisiko der Bevölkerung, Kapitel II, GSF-Bericht, GSF-S-710.

Jacobi, W. (1987): Strahlenschutz in Forschung und Praxis, Stuttgart, G. Thieme, vol. XXIX.

Jacobi, W. (1988): Strahlenexposition und Strahlenrisiko der Bevölkerung durch den Tschernobyl-Unfall, Phys. BI. 44, 240–246.

Karhausen, L., Pages, J.P., Vacca, G., Piepz, A., Visscher, M. (1973): Metabolisme de l'iode chez l'enfant et l'adolescent dans une région de la communauté, Bericht der Europäischen Gemeinschaften, EUR 4964 f. 18/86.

Kaul, A., Roedler, H.D. (1980): Radioiodine: Biokinetics, Mean Dose and Dose Distribution, Radiat. Environ. Bioph. 18, 185–195.

Kellerer, A.M. (1990): The New Estimates of Radiation Risks, Vol. 55, No 4, 198–203.

Kiefer, J. (1981): Biologische Strahlenwirkung, Berlin, Springer.

Killough, G.G., Eckerman, K.F. (1983): Internal Dosimetry, in: Radiological Assessment, J.E. Till & H.R. Meyer (eds.), U.S Nuclear Regulatory Commission, Washington, NUREG/CR-3332, 7–88.

Kinlen, L. (1988): Evidence for an Infective Cause of Childhood Leukaemia: Comparison of a Scottish New Town with Nuclear Reprocessing Sites in Britain, Lancet II, No 8624, 1323–1326.

Kocher, D.C. (1983): External Dosimetry, in: Radiological Assessment, J.E. Till & H.R. Meyer (eds.), U.S. Nuclear Regulatory Commission, Washington, NUREG/CR-3332, 8.1–8.52.

Kramer, R., Drexler, G. (1982): On the Calculation of the Effective Dose Equivalent, Rad. Prot. Dosim. Vol. 3, No 1/2, 13–24.

Luckey, T.D. (1980): Hormesis with Ionizing Radiation, Boca Raton, Florida CRC Press, Inc..

Matthies, M., Eisfeld, K., Paretzke, H.G., Pröhl, G., Wirth, E. (1982): Simulation des Transfers von Radionukliden in landwirtschaftlichen Nahrungsketten, GSF-Bericht, GSF-S-882.

McKusick, V.A. (1983): Mendelian Inheritance in Man, 6th ed., Baltimore, John Hopkins University Press.

NIH (1985): Report of the National Institutes of Health ad hoc Working Group to Develop Radioepidemiological Tables, NIH Publication No. 85 2748, U.S. Department of Health and Human Services, Washington, D.C..

Nothdurft, W. (1985): Knochenmark, in: Handbuch der Radiologie, Berlin, Springer, vol. XX, 235–264.

Otake, M., Yoshimaru, H., Schuh, W.J. (1988): Severe Mental Retardation among the Prenatally Exposed Survivors of the Atomic Bombing of Hiroshima and Nagasaki: A Comparison of the T6SDR and D586 Dosimetry Systems, Radiation Effects Research Foundation, Hiroshima, RERF TR.

Pampfer, S., Streffer, C. (1988): Prenatal Death and Malformations after Irradiation of Mouse Zygotes with Neutrons or X-Rays, Teratology 37, 1–9.

Patterson, M.C., Bech-Hanssen, N.T., Smith, P.J., Mulvihill, J.J. (1984): Radiogenic Neoplasia, Cellular Radiosensivity and Faulty DNA Repair, in: Radiation Carcinogenesis, J.D. Boice & J.F. Fraumeni (eds.), New York, Raven Press, 319–336.

Preston, D.L., Kato, H., Kopecki, K.J., Fujita, Sh. (1987): Cancer Mortality among A-Bomb Survivors in Hiroshima and Nagasaki, Life Span Study Report 10, Part 1, 1950–1982, Radiation Effects Research Foundation, Hiroshima, RERF TR 1–86.

Pröhl, G., Friedland, W., Paretzke, H.G. (1986): Intercomparison of the Terrestrial Food Cham Models FOOD-MARC and ECOSYS, GSF-Bericht, GSF 18/86.

Roedler, H.D. (1977): Strahlenexposition des Patienten durch Radiopharmaka – Grenzen der Genauigkeit von Dosisberechnungen, Dissertation, FU Berlin.

Roman, E., Beral, V., Carpenter, L., Watson, A., Barton, C., Ryder, H., Aston, D.L. (1987): Childhood Leukaemia in the West Berkshire and Basingstroke and North Hampshire District Health Authorities in Relation to Nuclear Establishments in the Vicinity, Lancet, 17–22.

Schuh, W.J., Otake, M., Neel, J.V. (1981): Genetic Effects of the Atomic Bombs: A Reappraisal, Science 213, 1220–1227.

Shellabarger, C.J., Chmelevsky, D., Kellerer, A.M. (1980): Induction of Mammary Neoplasms in the Sprague-Dawley Rat by 430 keV Neutrons and XRays, J. Natl. Cancer 1.64, 821–833.

Shimizu, Y., Kato, H., Schull, W.J., Preston, D.L., Fujita, Sh., Pierce, D.A. (1987): Comparison of Risk Coefficients for Site-Specific Cancer Mortality Based on the D586 and T6SDR Shielded Kerma and Organ Doses, Life Span Study Report 11, Radiation Effects Research Foundation, Hiroshima, RERF TR 12–87.

Snyder, W.S., Fisher, L., Ford, M.R. (1969): Estimates of Absorbed Fractions for Monoenergetic Photon Sources Uniformly Distributed in Various Organs of a Heterogeneous Phantom, MIRD Pamphlet 5, J. Nucl. Med. 10, Suppl..

Sonntag von, C. (1987): The Chemical Basis of Radiation Biology, London, Taylor & Francis.

Stather, J.W., Greenhalgh, J.R. (1983): The Metabolism of Iodine in Children and Adults, NRPB – R 140, Didcot.

Stewart, A., Kneale, G.W. (1968): Changes in the Cancer Risk Associated with Obstetric Radiography, Lancet, No 7534, 104–107.

SSK 1985 (Strahlenschutzkommission): Wirkungen nach pränataler Bestrahlung, Bundesminister des Innern (ed.), vol. 2, Stuttgart, Gustav Fischer Verlag.

StrlSchV 1989 (Strahlenschutzverordnung): Verordnung über den Schutz vor Schäden durch ionisierende Strahlen.

Streffer, C. (1969): Strahlen-Biochemie, Heidelberger Taschenbücher 59/60, Berlin, Springer.

Streffer, C., Müller, W.-U. (1984): Radiation Risk from Combined Exposure to Ionizing Radiations and Chemicals, Adv. Radiat. Biol. 11, 173–210.

Streffer, C., van Beuningen, D. (1985): Zelluläre Strahlenbiologie und Strahlen-Pathologie, Handbuch der Radiologie, Berlin, Springer, vol. XX, 1–39.

Streffer, C. (1985): Mechanismen der strahlenbedingten Kanzerogenese: Aspekte zellbiologischer und tierexperimenteller Untersuchungen, in: Strahlenschutz in Forschung und Praxis, Stuttgart, G. Thieme, vol. XXVIII, 34–47.

Streffer, C. (1987): Risiko nach Strahlenexpositionen während der pränatalen Entwicklung des Menschen, in: Strahlenschutz in Forschung und Praxis, Stuttgart, G. Thieme, vol. XXVIII, 34–47.

Trott, K.R. (1985): Strahlenwirkungen auf die Abdominalorgane, in: Handbuch der Radiologie, Berlin, Springer, vol. XX, 69–100.

UNSCEAR 1977 (United Nations Scientific Committee on the Effects of Atomic Radiation): Sources and Effects of Ionizing Radiation, United Nations, New York.

UNSCEAR 1982 (United Nations Scientific Committee on the Effects of Atomic Radiation): Ionizing Radiation: Sources and Biological Effects, United Nations, New York.

UNSCEAR 1986 (United Nations Scientific Committee on the Effects of Atomic Radiation): Genetic and Somatic Effects of Ionizing Radiation, United Nations, New York.

UNSCEAR 1988 (United Nations Scientific Committee on the Effects of Atomic Radiation): Sources, Effects and Risks of Ionizing Radiation, United Nations, New York.

3

DEVELOPMENT OF ENVIRONMENTAL STANDARDS FOR RADIATION AND OTHER NOXAE

Reasons for the anthropocentrism of environmental protection have been given in chapter 1. At the same time, we have pointed out that this basic idea must not result in a protection of nature only to the extent which is useful and usable for human beings living today. Nature in its diversity has to be preserved also as a habitat, and for future generations. In this sense, environmental standards have to be regarded as instruments to protect humans and their environment against harmful changes and influences. Moreover, environmental standards may also serve to reduce already existing pollution, thus restoring the environment to a less polluted state.

Important foundations for the establishment of environmental standards are found in the natural sciences and medicine concerning the impacts of noxae on the environment – and on the organisms living in it – as well as on human health. The improvement of this knowledge, but also the increasing sensitivity of people in industrial countries to possible dangers from noxae which could harm environment and health, have crucially shaped the development of environmental standards.

In the following sections, we will describe the historical development of environmental standards for ionizing radiation and radioactive substances as well as for some other representative environmental noxae. The development of environmental standards followed the increase in scientific knowledge about the diffusion of noxious substances in the environment as well as the possible ways that animate and inanimate nature, including the earth's atmosphere and the stratosphere, are exposed to and affected by them. Frequently the risks from these noxae have first been recognized at the work-place. The experiences gained there have then been used in the development of environmental standards.

3.1 HISTORICAL DEVELOPMENT OF RADIATION PROTECTION STANDARDS

Noxious impacts of ionizing radiation were observed shortly after the discovery of X-rays and of natural radioactivity, at the end of the 19th century. Becquerel himself, the discoverer of natural radioactivity, noticed that a radium preparation, which he carried in the pocket of his suit, led to reddening and damaging of the skin. Severe skin damages were also found in radiologists; leukemia and skin cancer also appeared, which had obviously been induced by ionizing radiation. At the end of the twenties, the American scientist Muller observed in his studies of the fruit fly *drosophila melanogaster* that X-rays produce genetic mutations (Streffer 1988).

All these effects demonstrate that some protection is needed when handling ionizing radiation or radioactive substances. At first, this meant protection of those persons who had to professionally deal with ionizing radiation or radioactive substances. To ensure effective radiation protection, it was necessary to be able to carry out an exact physical dosimetry. Biological procedures were employed for this purpose; so, the skin erythema dose was defined. This meant the „amount of rays“ that, with hard radiation, at a focused distance of 23cm from the skin, after one week produced a light reddening of the skin, after three weeks a brownish discoloration, and after six weeks a distinct brown color of the skin, on an area of 6 _ 8 cm in size. In 1924, the „Physikalisch-Technische Reichsanstalt“ (former German national institution of physics and technology), in cooperation with the German Society of Radiology, for the first time introduced a physically defined unit of dosage, based on the ionization of the air. This unit of dosage „Röntgen“ eventually became internationally accepted, it was declared the internationally binding unit of dosage for X-rays at the second International Congress of Radiology in 1928. This unit of dosage was also employed for the first establishments of tolerance doses and dose reference values. In 1953, at the seventh International Congress of Radiology in Copenhagen, the energy dose with the unit of dosage „rad“, and the dose equivalent with the unit of dosage „rem“ were internationally introduced. The present radiation protection regulations, which are indicated in Gray or Sievert today, are still based on these dose quantities (Muth 1980).

The tolerance doses and the recommended dose reference values were initially based on the early units of dosage. So, in 1925, a tolerance dose was recommended which equaled 1/20'000 skin erythema dose per hour; this resulted in a daily tolerance dose of 0.25 Röntgen for a normal workday. In 1928, at the second International Congress of Radiology, the International X-Ray and Radium Protection Commission, forerunner of the present International Commission on Radiological Protection (ICRP), was founded, which crucially influenced radiation protection and corresponding regulations. In 1934, at the fourth International Congress of Radiology in Zurich, a tolerance dose of 0.2 Röntgen per day was recommended for occupationally exposed persons. At the same time, a lower tolerance dose of 0.1 Röntgen per day was introduced as a reference value in Sweden and in the U.S. Based on the works of the International Commission founded in 1928, in 1950, at the sixth International Congress of Radiology in London, the ICRP was founded, which, as its first measure, recommended a maximal admissible dose of 0.3 Röntgen per week for occupationally exposed persons. With improved knowledge about the biological effects of ionizing radiation and about technological possibilities of radiation protection including dosimetry, in 1956, on the occasion of the eighth International Congress of Radiology in Mexico City, a maximal admissible annual dose of 5 Röntgen (i.e. 0.1 Röntgen per week) for occupationally exposed persons was introduced (Muth 1980). This dose reference value, now called dose limit, has remained valid up to this day insofar as it is approximately the same as the present dose limit of 50 mSv per annum. In addition, an occupational lifetime dose of 400 mSv has recently been introduced. With the increasing application of ionizing radiation in medicine and technology, and because of exposure to environmental radiation due to the atmospheric nuclear tests, the necessity of introducing dose limits for the general public as well was increasingly acknowledged. The discussion about such values primarily focused on the genetic effects of ionizing radiation.

Subsequently, discussions in the U.S. and in Great Britain led to a so-called genetic population dose. In the fifties, it was estimated that exposure to natural radiation over a period of thirty years (the time of one generation) amounts to about 5 Röntgen for the general population. Furthermore, it was assumed that during this period there is an additional exposure for medical purposes to a dose of about 3 Röntgen (whole-body dose or genetically significant dose). On the basis of these estimations and taking the genetic risk for future generations into account, a commission of the U.S. National Academy of Sciences (the Committee on Biological Effects of Atomic Radiation, BEAR-committee) recommended that exposure to man-made radiation should not exceed 10 Röntgen for a period of 30 years. This dose should be indicated as whole-body dose or gonad dose, because of the genetic effects (BEIR 1980).

In 1957, a committee of the U.S. National Council on Radiation Protection and Measurements worked on the problem of how to limit radiation exposure of the general population. This committee recommended that the tolerance dose for exposure to man-made radiation, excluding medical exposure, should not be higher than mean exposure to natural radiation. In addition, the principle was formulated that exposure to man-made radiation should be kept as low as technically achievable. The BEAR-committee recommended to split up the tolerance dose for the general population (now 10 rem instead of Röntgen, for a period of 30 years) into 5 rem for medical exposure and 5 rem for non-medical exposure to radiation. – The BEAR-committee was later renamed the BEIR-committee (Committee on Biological Effects of Ionizing Radiation). – In chapter 2, the correlations between the dose quantities of ion dose, energy dose, and dose equivalent have been discussed in detail.

The further development was essentially shaped by the ICRP. One basic recommendation, on which our present radiation protection regulations are largely based, was published in 1977. In this recommendation, the ICRP at first established three principles of radiation protection (ICRP 1977):

- Any application of radiation which does not entail some benefit should be omitted (justification of an irradiation).
- The exposure of individual persons to radiation resulting from applications should not exceed certain limits (dose limits). This does not comprise exposure of a person to medical irradiation and exposure to natural radiation.
- In addition, any exposure to radiation should be kept as low as reasonably achievable in consideration of the state of science and technology, and taking the specific circumstances into account (optimization of radiation protection).

In the English version of the ICRP-recommendation of 1977, this third principle is called ALARA-principle (**a**s **l**ow **a**s **r**easonably **a**chievable). Accordingly, the dose limits are not supposed to be interpreted as tolerance doses which may be exhausted. The ICRP rather thinks that the mean dose for occupationally exposed persons should only amount to about one tenth of the dose limit per annum.

The establishment of dose limits orients itself by the two types of damage discussed in chapter 2, which are characterized with the help of the dose effect relation, namely stochastic and non-stochastic radiation effects. Dose limits are established for individual organs and tissues as well as for the effective dose equivalent defined and described in chapter 2. The dose limits for individual tissues and organs (irradiations of parts of the body) are such that the threshold doses for non-stochastic effects in these organs and tissues are not exceeded by an exposure to radiation at the dose limit level.

The dose limit for the effective dose equivalent has been set to 50 mSv per annum (5 rem per annum) for occupationally exposed persons. This value hence equals the former dose limits for the whole-body dose or the genetically significant dose for occupationally exposed persons. Recently, an additional occupational lifetime dose of 400 mSv has been introduced. The dose limit orients itself by the stochastic radiation effects (genetic and carcinogenic radiation effects). The connected risk, which is derived by means of extrapolation from higher dose ranges, was compared to risks in other occupations (see section 2.2.5). In these comparisons, the risk for persons occupationally exposed to radiation was found to be within the mean range of occupational risks. It was thus classed as tolerable, all the more so as risk assessments for other occupations often do not consider long-term effects, e.g. carcinogenic risks for workers in chemical plants.

The ICRP-recommendation of 1977 also establishes dose limits for individual persons of the general population. The effective dose equivalent has been set to 5 or 1 mSv per annum here. These values have been adopted by the European Communities (today EU) into their EC basic norms, where the dose limit for persons of the general population was set to 5 mSv per annum. This dose range is far below the threshold doses for non-stochastic effects. Thus, only stochastic effects are relevant for risk assessments. Even for these effects, limits are in a range in which risks can only be assessed by extrapolating calculation. The EU member states are bound to respect these dose limits in their value regulations, yet they may well establish lower dose limits. Germany has made use of this possibility when establishing dose limits for the general population.

The German radiation protection regulation sets the dose limit for the effective dose equivalent in the surroundings of nuclear facilities to 0.3 mSv per annum at the locations of most unfavorable effect. Radioecological models are employed for calculation, which take into consideration external irradiation and internal irradiation after ingestion of radioactive substances. They start from the extreme assumption that the person in question dwells at the place with the highest radiation dose, and that all food is produced at the location of most unfavorable effect, with the highest possible radioactive contamination. These calculations are done for emissions via air and water. For both paths, a respective dose limit of 0.3 mSv per annum is conceived. Both exposure to external radiation, e.g. soil radiation, and exposure to internal radiation, e.g. due to ingestion, are considered. Since the places of most unfavorable effect are not the same for the paths air and water, and an individual can only ingest the assumed package of food once, an addition of both values can be ruled out. The place of most unfavorable effect with respect to the air path is usually at the fence of the installation. To determine the ingestion dose, it is assumed that all food is produced there. The place of most unfavorable effect (highest radioactivity) for the water path is a different location. To determine the dose for the water path, it is assumed that all food is produced there or at a place where the respective radioactivity is transferred to. So the ingestion dose is calculated separately for the air path and the water path and is included into the overall calculation for the respective path. These extremely conservative calculation models yield radiation exposures which do not occur in reality. The calculated radiation exposures are design values, which may only be reached under most unfavorable conditions, for individual persons, in individual years.

The value of 0.3 mSv per annum is based on a recommendation of the SSK (German commission on radiation protection), which advises the German government on radiation protection issues. Its reflections about establishing this dose limit were focused mainly on exposure to natural radiation, similar to considerations of the BEAR-committee in the U.S. Yet the SSK did not orient itself to the total value of mean radiation exposure but to the variation range of the natural radiation which the majority of the German population is exposed to. According to most recent studies, this variation range is about the order of 0.5 mSv for the effective dose equivalent. Exposures largely exceed this value in cases of extreme concentrations of radon in houses. Hence the dose limit of 0.3 mSv per annum is below the variation range of exposure to natural radiation. Thus, an individual change of habits, including a change of place of abode within Germany, may lead to changes of exposure to natural radiation which exceed the range given for the dose limit in the vicinity of nuclear installations.

For 90% of the German population, terrestrial radiation leads to exposures varying within the range of 0.2–0.7 mSv per annum (whole-body dose). For the remaining 10%, radiation doses are considerably higher. A flight from Germany to New York at about 11 km altitude exposes a passenger to a radiation dose of about 0.03–0.04 mSv. Finally, we will point to the fact that for the majority of the population a considerable portion of the effective dose resulting from natural radiation sources is due to the inhalation of radon and its radioactive daughter products. Radon concentrations in the air, including the air in closed rooms, is highly dependant on the content of radioactive substances in the soil and on the nature of the foundations of the houses. The resulting effective dose varies for the German population within a far larger range than the dose limit of 0.3 mSv per annum.

On the basis of the risk values given in chapter 2, a single effective dose of 0.3 mSv would increase the carcinogenic risk by about one tenth of a thousandth of the „spontaneous" rate of death from cancer if we assume that there is no threshold dose and extrapolate linearly. A whole set of human habits entail risk values within that range or higher. The dose limit of 0.3 mSv per annum for the effective dose is in keeping with the principles of pragmatic consistency formulated in this study (section 1.3.3).

3.2 DECISION PROCESSES IN THE FORMATION OF SOME OTHER STANDARDS

This section will discuss four examples of environmental noxae in an exemplary way. Two cases are about health endangering noxae, namely about hazardous substances at the work-place in general, and about asbestos at the work-place and in the environment in particular. Limits at the work-place are, strictly speaking, not environmental standards, but they are nevertheless of great interest from a procedural point of view. The other two cases are about noxae with an ecological impact, namely chlorofluorocarbons (CFCs) and sulfur dioxide (SO_2).

3.2.1 Standards of Workers' Protection

As stated before, standards of workers' protection served as a basis for later establishments of environmental standards several times. A senate commission of the DFG, the MAK-commission, works out proposals as to the „maximal admissible concentration of a work material at the (...) work-place which, according to the present state of knowledge, (...) does usually not impair the health of the workers and does not molest them in an inappropriate way. (...) Relevant here are scientifically founded criteria of health protection, not technological and economic possibilities of practical realization" (MAK-Werte 1988, part I). These proposals orient themselves to threshold concentrations or threshold exposures known from toxicological research which are regarded as sufficiently reliable by the commission.

The list „MAK-Werte 1988" registers standards of workers' protection for 348 different substances. Another 36 substances are under investigation while this study is carried out; the information gathered, however, is not yet sufficient to justify the establishment of MAK-values in the latter cases. If there is substantial reason to suspect some work substance to be carcinogenic, the commission feels that they are not in a position to denote any concentration as unrisky. These substances are registered in a special list and classified according to the respective state of knowledge into:

A 1) substances which are carcinogenic for humans, as we know from experience; at the time of this study, this class comprises 18 substances

A 2) substances which, only in animal experiments, have clearly proved to be carcinogenic; at the time of this study, this class comprises 74 substances

B) substances for which there is substantial reason to suspect carcinogenic potential; at the time of this study, this class comprises 61 substances.

Classification into group B is only a preliminary one. This group is annually checked for whether a substance can be assigned to groups A 1) or A 2), or whether the suspicion of carcinogenicity has proved to be insubstantial. Apart from information about these 153 carcinogenic substances, for which the MAK-commission explicitly refuses to establish standards since there is no threshold value, special problems regarding the effects of further substances are discussed.

In ascertaining MAK-values, the commission proceeds as follows: Inspired by scientific publications, by voices from inside the commission, or by other expert circles, the commission deliberates on whether an MAK-value is to be set or altered. A respective intention is imparted to the expert public. In doing so, the commission asks for information to be given within the period of one year, which is then discussed with the informants, if need be. Evaluation of epidemiological studies as well as of animal and cell experiments leads to specifying that concentration of a substance which can be held unrisky with regard to typical working conditions. This concentration value marks the limit below which no impairment of a worker's health is to be feared. Explicitly mentioned is the necessity to base these considerations on sufficient experience about handling the substance. In case of conflict, further experts are consulted for a final deliberation, in agreement with different institutions. Among them are representatives of the committee of the German states on workers' protection and safety technology, of the Federal Association of German Industry BDI, of the head association of the industrial cooperatives, and of the national association of trade unions DGB. As far as possible, different individual sensitivities are also represented here.

If, for some substance, it is impossible to define a concentration which can still be regarded as unrisky, no MAK-value is suggested. The substance may be registered in the category of carcinogenic work materials. This may possibly result in a technical reference concentration (TRK) being defined by the Committee on Dangerous Substances (AGS), hence not by the MAK-committee itself. Once MAK-values are established, they are published together with the reasons that led to their establishment. These reasons include, above all, critical evaluations of the current state of knowledge.

To become binding standards, these values have to be published by the Federal Minister for Labor and Social Affairs as technical rules on dangerous substances.

Criteria of practicability and realizability do not play any role for the MAK-commission in their defining concentration levels which are regarded as unrisky. So, sometimes the reduced MAK-values cannot immediately be observed in some firms. All technological possibilities have then to be exhausted so as not to exceed the reduced values. If this proves to be impossible within the period of one year, a transition period has to be applied for.

In this context, the federal minister is advised by the AGS, which is constituted by representatives of the following five groups (§ 44 Abs. 1 GefStoffV of 26 August 1986):

- trade unions (7 representatives),
- employers' organizations (9 representatives),
- public authorities (9 representatives),
- scientists (12 representatives), and
- one representative of a consumers' organization.

This committee is, among other things and apart from its annual resolution on the inclusion of the revised MAK-list into the code of technical rules on dangerous substances, responsible for the establishment of TRK-values.

The AGS, or its forerunner, the Committee on Dangerous Work Materials, have established technical reference concentrations since the seventies. According to the legal definition of § 15 Abs. 6 GefStoffV, these are the concentrations of substances in the air at the work-place which are achievable with the current state of technology. Such TRK-values are defined only for those dangerous substances for which, at that point in time, no MAK-values based on toxicology and industrial medicine exist. Their establishment is essentially based on:

- the condition that concentrations of substances in the range of the TRK-value can be analytically determined,
- the consideration of the current state of process engineering and ventilation engineering, taking into account what will be achievable in the near future,
- the principle that TRK-values must not be contradicted by experiences of industrial medicine or toxicological findings (TRGS 102, part III).

Observance of these TRK-values is supposed to reduce the risk of health impairment, but it cannot guarantee its complete exclusion.

At the time of this study, the AGS regards 93 substances as definitely carcinogenic and classes them into the three danger groups

I (extremely endangering)
II (very endangering)
III (endangering).

„This classification according to the intensity of impact relies mainly on scientific knowledge. In exceptional cases, other aspects may also be attributed greater relevance, e.g. arguments from the field of labor policy, economy, or technology, as well as the spread of a dangerous substance“ (TRGS). Another 61 substances are suspected by the MAK-commission to be carcinogenic. These more than 150 different substances meet, in the TRGS 102 of March 1989, with a total of 31 TRK-values for 24 substances or groups of substances.

If the AGS decides to establish a TRK-value for a substance which the MAK-commission holds to be carcinogenic, a subcommittee of the AGS becomes active. It selects some firms which are assumed to observe modern safety regulations and asks them for information about actually measured concentrations in the air at the work-place (Beyersmann 1986). The results of these measurements are regarded as representing the actual state of technology in individual production sectors. Often they are subject to considerable fluctuations, fluctuations of up to twice the order of magnitude. Before an agreement is reached on a TRK-value which will be technically achievable in the near future in the individual production sectors, it is tested whether this value can be monitored by means of the current measurement technology. There are cases in which the target value is defined according to the level of lowest measurable concentration, or even below it. Finally, it is asked whether experience of industrial medicine or toxicological knowledge might oppose the suggested value (TGRS 102, 1989). In the case of beryllium, for example, an animal experiment became known according to which the currently valid TRK-value would have been too high, hence an examination of this result was urgently recommended. Incidentally, for beryllium, due to technical problems, a TRK-value higher than the generally valid one has been established for grinding shops (0.005 mg/m^3 instead of 0.002 mg/m^3), which will have to be checked by 1 July 1989.

If the subcommittee, perhaps in consultation with further experts, decides in a closed session on a TRK-value, this proposal is transmitted to the AGS. The AGS again decides in a closed session and perhaps in consultation with further experts, with the majority of the votes of its present members entitled to vote. Similar to other environmental and workers' protection standards, here the following also holds:

„The TRK-values require continuous adaptation to the state of technological development and analytical possibilities, as well as a check according to the state of knowledge of industrial medicine and toxicology.“ This obliges the employer „to strive for concentrations below, as far as possible, the technical reference concentrations, through continuous improvement of technological conditions and protection measures" (TRGS 102, 1989). This corresponds with the minimization principle in radiation protection.

For most TRK-values, the reasons elaborated by the AGS are published in the appendix to TRGS 102 (1989). They comprise statements about research results on the effect of substances and their spread, about the state of the respective analysis technology, and about results of concentration measurements.

Balancing considerations between what is achievable according to the state of technology (§ 15 Abs. 6 GefStoffV) and what is reasonable according to the constitutional principle of proportionality are made in the AGS, which decides with the majority of votes. It can be observed here, in the context of the respective balancing discussion, that the individual groups are ready to make compromises. So, representatives of the employees will also consider economic interests in their claim for optimal health protection, if otherwise they have to fear the loss of jobs. On the other hand, representatives of the employers – which are often the only experts – will also make their own suggestions about protection measures if there are no initiatives on the part of the employees' associations. The group of scientists will often have to make up their minds to advocate preventive measures before the harmfulness of dealing with certain substances is proved with the greatest possible scientific certainty. Thus, the respective social and political value concepts play an essential role. The administration occupies a mediator role in this net of interests; it is responsible for sufficient protection measures and at the same time must not admit any production constraints unless this can be justified by clear scientific facts (Beyersmann 1986).

3.2.2 Standards for Asbestos

It has been known for almost 90 years that asbestos has health-endangering properties. Like with ionizing radiation, we have to distinguish between stochastic and non-stochastic effects. Since the late sixties, the respective discussion has focused on an assessment of the carcinogenic effect of asbestos dust. The discussion about this risk of cancer has become topical primarily due to the ascertainment of high concentrations of asbestos in closed rooms, which led to closure and rehabilitation of public buildings. The high carcinogenic potential of this substance and the extensive spread of products containing asbestos led to intensive research. Today, asbestos is the epidemiologically best studied carcinogenic noxa apart from radiation and smoking. In nature, asbestos is found as a mineral in rock. If such rock erodes, asbestos is also emitted into the environment.

By asbestos, we mean certain crystalline silicates, which are found in workable concentrations in different deposits. According to their chemical composition or their crystalline structure, we distinguish between chrysotile and amphibole asbestos. The long and elastic fibers cannot be burned and do not rot; the raw material is cheap and easy to manufacture. These properties have led to an extensive use of asbestos; between 1950 and 1975, the use of asbestos multiplied by more than the factor ten.

Due to its favorable properties, asbestos has already been used for millennia, but its industrial use only began around 1880, with the exploitation of the chrysotile fields in Quebec.

In 1900, for the first time, Murray expressed a suspicion about the harmful effects of this substance on health. In an autopsy, he found a large number of asbestos fibers in the lungs of an asbestos worker who had died from fibrosis. A short time later, asbestos dust was declared to be health endangering (1902); the first study on deaths in the French asbestos industry during the period of 1890–1895 was published (1906).

After the notion of „asbestosis" had first appeared in a publication (in 1927), the knowledge about this disease was summarized, together with the appeal for a reduction of the exposure to dust, in a report to the British Parliament in 1930. In Germany, asbestosis was acknowledged as an occupational disease in 1936. In the mid-thirties, studies about a possible link between asbestosis and lung cancer were published; this suspicion was confirmed by the first animal experiment with exposure to asbestos dust in 1941. This resulted in an extension of the catalog of occupational diseases by asbestosis in connection with lung cancer (1943).

At the end of the thirties, the interlinked appearance of asbestosis and a form of cancer of the pleura, the mesothelioma, was observed. After the first case studies in the early fifties, the supposed connection between exposure to asbestos dust and the induction of mesothelioma was confirmed by epidemiological studies in the vicinity of an asbestos mine in 1960. In 1976, this tumor was also included in the list of occupational diseases (Umweltbundesamt 1980).

In 1970, for the first time, an appendix with the category „carcinogenic work materials" (among them asbestos) was included into the MAK-list; shortly afterwards, the establishment of an MAK-value for asbestos was demanded. Respective considerations were focused on the avoidance of asbestosis, a dose-dependent disease with a threshold value. In 1972, the commission announced a discussion on this issue, but then declared in the MAK-list of 1973 that no MAK-value could be defined since all types of asbestos (in particular crocidolite) could induce malignant tumors. Thus, no concentration could be regarded as unrisky. However, a preliminary technical reference value was suggested. With this first establishment of a TRK-value, the MAK-commission set foot on new ground. For it was only months later that the Committee on Dangerous Work Materials (today: Committee on Dangerous Substances AGS) defined the concept of the technical reference value.

The TRK-values for asbestos of 1973 orient themselves to the fibrogenic effect (induction of benign tumors of the connective tissue) of the mainly industrially used chrysotile. Simultaneously, the carcinogenic potential of all types of asbestos is pointed out. The values were revised in 1976, due to a prolonging of the primarily assumed exposure time from 35 to 50 years as well as the progress in knowledge about the importance of the fibers for the pathogenic effects of asbestos; they were extended to amosite and adopted by the AGS or the federal minister of labor into the TRgA 102 (technical rule on dangerous work materials). Another three years later, the values were extended to all types of asbestos and revised anew. One of the essential reasons for this was the assessment that, according to available results of measurements of actual asbestos concentrations at exposed work-places, the set reference values are achievable in foreseeable time and can be observed. Not least for economic reasons, however, older facilities were permitted to produce at double the concentration values during a limited transition period (TRGS 102, 1988). On the basis of various EU directives (then EC directives), the values were lowered again in 1986; an explicit strategy of a minimization which is feasible in practice was then pursued. At the time of this study, these limits are being reexamined; the aim is to avoid the use of asbestos altogether, if possible.

Concluding, it has to be summarized that within thirteen years the work-place limit for chrysotile fine dust of 0.15 mg/m^3 was lowered for the asbestos fine dust of crocidolite to 0.025 mg/m^3 and for the other asbestos fine dusts to 0.05 mg/m^3.

Noxious substances known from the work-place are partly also found in the environment. Since it is assumed that there is no threshold dose, these substances may mean a certain, if low risk to the population, even if their concentrations are measured to be several orders of magnitude lower here. Since the mid-seventies, respective suspicions with respect to exposure to asbestos have been taken into consideration.

The federal law on immission protection of 1974 (BImSchG), among other things, imposes on certain operators of plants the duty to make provisions, according to the state of technology, to limit emission. The TA Luft, which contains detailed regulations to this law, makes concrete the legal concept of „state of technology" numerically, with the state of technology implying, according to the legal definition of § 3 Abs. 6 BImSchG, an up-to-date procedure whose practical appropriateness to limit this emission of harmful substances seems to be guaranteed.

The TA Luft of 1974 contains, for the first time, a limit for emissions of asbestos. It assigns asbestos to the class of those substances which are subject to the strictest requirements as to the state of technology to be observed. The TA Luft of 1983 then assigns asbestos to the newly constituted group of substances with the greatest carcinogenic potential, for which a value is established at the border of the technologically feasible. During a transition period lasting until 1986, exceeding this value is tolerated as far as observance is not yet technologically achievable. At the same time, a minimization rule which heeds the principle of proportionality is formulated.

The emission limits for asbestos corresponding with the respective state of technology were set to 20 mg/m^3 in 1974, according to technological progress they have amounted to 0.1 mg/m^3 since 1986.

3.2.3 Risk Assessments

In some cases of establishing environmental standards, there are approaches to conduct quantitative risk assessments concerning carcinogenic substances for which no threshold of effect is known, and to orient limits for exposure of the population to these risk assessments. This section will report on two such approaches.

In 1981, the German federal government ordered the BGA to lay the foundations for assessing the effects of asbestos immissions on health under consideration of the most recent information. In that very year a respective report was presented „making use of (...) all available sources of knowledge and methods of evaluation" (Aurand & Kierski 1981).

Since there were no studies on the effects of respectively small exposures to asbestos, the report, starting from the assumption that there is no threshold dose, calculated by linear extrapolation a dose effect relation for immissions of asbestos observed in the environment, on the basis of the dose effect relation for tumors due to asbestos at the work-place (mainly in smokers). The method of linear extrapolation is chosen because of the precaution principle, since it yields the most conservative (most pessimistic) assessment. According to unanimous expert opinion, the so calculated risk is probably higher than the real risk. In comparison with other tolerated causes of death, and with reference to the assessment that a risk of the order of 1 : 1'000'000 per annum is acceptable (a concept common in American literature as „virtually safe"), it is finally recommended that a value clearly below 1'000 fibers/m^3 should be aimed at. This way of proceeding is controversial (Fischer & Meyer 1984).

Referring to this, the BGA pointed out that 200 fibers/m^3 are also found in areas with clean air, and that a general reduction of immissions to such values would probably require considerable additional expenses of labor and energy from the national economy, which would be higher the more rapidly the goal of complete elimination was aimed at. That would result in new risks due to accidents and pollution, so that, on the whole, an increase of the total risk had to be feared. In 1983, the LAI was asked by the conference of the environment ministers to develop „assessment standards for carcinogenic air pollution – for example in the form of reference values" (LAI-Bericht 1987), in order to obtain the correlation between lung cancer and general air pollution.

Starting from the assessment that the demand for a zero risk is virtually unrealizable, it is seen as the objective of environmental health protection to minimize the emission of carcinogenic substances. In addition, advanced assessment standards for the acceptability of a risk shall enable further focused measures to limit pollution by carcinogenic substances.

To clarify the question of what should be seen as an acceptable residual risk, a comparison is made with other man-made risks and with the fatality rate from natural catastrophes. A lifetime risk of 1 : 1'000 referring to all cases of death from all carcinogenic substances seems to be acceptable in consideration of these already given death probabilities. This risk is of the same order as the one originating from ionizing radiation emitted by nuclear plants if the dose limit of 0.3 mSv per annum at locations of most unfavorable effect is exhausted. Like with that noxa, however, a minimization rule respecting the principle of proportionality is also conceived for carcinogenic pollutions.

On the basis of this arbitrarily chosen total risk of 1 : 1'000, it can be calculated for individual carcinogenic substances, with the help of the „unit risk" defined by WHO experts, which concentrations of this substance in breathing air have to be observed so as not to exceed this risk. For this purpose, a relevance series of carcinogenic air-polluting substances for which the establishment of assessment standards seems urgent, has been worked out. A risk quantification relying on several studies has been made for the seven most important substances and groups of substances according to this series (among them also asbestos), the results of which have been correlated with each other (taking the measured average annual exposures in conurbations into account).

Finally, the assessment values for the studied substances are calculated, with the sum of cancer risks „assigned“ to the concrete values of the individual pollutants being required to by no means exceed the value of 1 : 1'000. The chosen risk limit is, however, supposed to be exhausted by only 70%, so as to maintain a safety reserve for taking remaining imponderabilia into consideration.

These assessment standards obtained on the basis of risk considerations are sometimes heavily criticized with the argument that for hardly any of these substances the scientific basis is solid enough for a risk quantification of sufficient accuracy. In reaction to that, the WHO decided on the occasion of its setting air quality guidelines (1987) to give estimated risk values instead of guideline values for a number of carcinogenic substances. For nine substances and substance groups, they determine, with the help of the „unit risk“, the cancer risk due to lifelong exposure to a concentration of 1µg/m^3.

3.2.4 Chlorofluorocarbons (CFCs)

Only initial steps have been made so far to establish standards for the protection of the ozone layer and for an avoidance of the greenhouse effect. The existing regulations concerning CFCs pursue a strategy of emission reduction. In particular, they emphasize substitution for CFCs in all fields of their present use, thereby respecting the precaution principle. The example of the CFCs will serve here to describe how a global environmental problem is dealt with and what influence scientific findings may have on political decisions.

In 1974, for the first time, the American scientists Rowland and Molina pointed out the danger posed to the stratospheric ozone layer by the increasing use of CFCs. At two international conferences in 1978, mathematical models were presented which enabled calculation of reductions of stratospheric ozone concentrations, taking the currently available measurement results into account. About four years later, these depletion models had to be revised so as to yield lower ozone losses. Distinct ozone losses in the stratosphere above the Antarctic were reported for the first time in 1984; one year later, scientists of the British Antarctic Survey informed the public about this „ozone hole“. An increase in skin cancer due to unfiltered UV-radiation was prognosticated as a most important consequence of a reduction of the ozone concentration also above populated regions of the earth. The influence of the stratospheric ozone concentration on the greenhouse effect is probably rather negligible. The CFCs, however, are themselves effective greenhouse gases.

In 1975 the German federal government had a study carried out which analyzed the situation_s scientific and economic aspects. Simultaneously, it intensified promotion of research in the field of atmospheric processes. Similar studies were also carried out in other countries, taking into consideration danger to the environment, economic effects, and effects on the labor market.

At the beginning of 1977, German industry agreed on the objective to reduce the share of CFCs in aerosol propellants by 30% within a few years. Also in the international context it was preferred to proceed via voluntary self-restriction of industry towards a reduction of an emission of CFCs. The EU (then EC) suggested a search for new possible substitutes for these substances. Also, it was demanded that there should be incentives to reduce leakages, above all in cooling units, and not to increase production capacities.

This phase began one year after the first hints about the harmfulness of CFCs and lasted about four years. Then the measures taken were evaluated and further strategies were developed. Particular emphasis was placed on the improvement of scientific knowledge. The German federal government was asked by the Committee on the Interior of the German national parliament to produce several reports and studies, among other things on

- possibilities to renounce the use of CFCs and to replace them, and
- financial consequences and other effects of such a renunciation for the respective industrial branches.

On the basis of these studies, the German government holds a total ban of CFCs to be expedient but at the same time stresses the necessity of a differentiated way of proceeding, since a substitution for CFCs would require unjustifiable expenses and could even entail new safety risks in several fields of their application.

At the EU level, an extensive package of measures was discussed in 1981, by means of which a significant reduction of CFC-emissions in the different sectors of their application should be achieved. This program was also adopted by the EU council (then EC council) by the end of 1982.

On the basis of improved scientific model prognoses, it was first declared at the EU level that neither a tightening of the previous policy nor additional measures were needed according to these data. In a statement to the press of July 1984, the Industrial Community on Aerosols published under the catchphrase „ozone hysteria", among other things, the following admonition: the German authorities concerned with environment protection should avoid rash measures when issuing administrative regulations, and should consider the current state of knowledge. Thereby, both unnecessary unnerving of the environment-conscious consumer and considerable economic consequential costs due to rash measures could be avoided (Deutscher Bundestag 1989, (inquiry commission on „provisions on the protection of the earth's atmosphere")).

The UBA (Federal Environmental Protection Agency) reply to this statement was that the restrictive use of CFCs was a reasonable and imperative precaution.

The EU commission (then EC commission) continued its previous strategy of voluntary self-restriction of industry, and, in accordance with the relevant European industrial associations, in 1985 presented three behavior codices for a reduction of CFC-emissions. In spring of the same year (22 March 1985), the „agreement on protection of the ozone layer" was signed in Vienna. It obliged the signatories to take measures against activities which are likely to damage the ozone layer.

The Technical Ordinance „Clean Air" (TA Luft) of 1986 issued emission values for the two particularly widespread and long-lived CFC11 and CFC12, based on the respective state of technology. The act on emission limitation of highly volatile halocarbons (1986) limited the use and the loss of certain solvents in plants which are not subject to the licensing duty of the BImSchG and hence not to the TA Luft either.

The policy statement of the German government of 18 March 1987 pointed at the increasing global endangering of the earth's atmosphere and announced that the government would advocate a ban of CFCs in aerosols. Both chambers of the German national parliament also urged such a ban. The government indicated that it attached greater relevance to the enforcement of international measures than to national prohibitions. Moreover, a national interdiction to produce CFC-containing sprays had to accord with EU law, which would result in a delay. Shortly, a more rapid and more effective way was followed. Shortly thereafter, the chemical industry sent the responsible national minister a declaration of self-restriction to reduce the use of fully halogenized CFCs in sprays.

In reaction to the unexpectedly marked ozone hole, an international agreement was reached within a few years. On September 16, 1987, the Montreal protocol on substances leading to a depletion of the ozone layer was signed. This agreement, following and concretizing the Vienna agreement, obliged the signatories to freeze production and consumption of eight substances and to reduce some of them gradually by 20% or 30% annually.

In the same year, the German Bundestag constituted the inquiry commission on „provisions for the protection of the earth's atmosphere", which was supposed to comprehensively take stock of the current situation and make suggestions for precautionary and defending measures to protect humans and the environment. One year later, the German government constituted an advisory board on climate in order to intensify climate research, and initiated several new research programs. In autumn 1989, the draft of a regulation was submitted according to which the use of CFCs in the different individual fields of application should be prohibited in temporally graded steps.

Concluding, it has to be said that over the years, balancing considerations on advantages and disadvantages of a ban of CFCs have had different results, and have changed in accordance with scientific prognoses. When the scientific models on ozone depletion were revised so as to yield lower effects, at first the focus was on the costs of introducing substituting materials; yet when the marked ozone hole appeared, these considerations receded into the background again.

3.2.5 Environmental Standards for SO_2

Sulfur dioxide is a very pungent-smelling gas, even if heavily diluted, the biological effect of which has been known and made use of for a long time. In Homer's „Ulysses" it is mentioned as a disinfecting fumigant, and until very recently it was used to control vermin and to disinfect sick-rooms.

Since sulfur is an element of animal and vegetable proteins, it is contained in all fossil fuels. In particular, during their combustion for the purpose of producing energy, SO_2 is emitted in large amounts into the environment; in 1986, 2.2 million metric tons of sulfur dioxide were emitted in West Germany (VCI, Umweltbericht 1988/89).

In humans, sulfur dioxide causes irritations of the mucous membranes, especially the respiratory tract. But even for particularly sensitive persons (children, asthmatics), effects are no longer found below certain concentrations. We have therefore to assume the existence of a threshold value. The MAK-value amounts to 5 mg/m^3, the maximal immission value recommended by the VDI-commission „preservation of clean air" to protect health of the general population amounts to 0.3 mg/m^3 (VDI-Richtlinie 2310, 1988).

During a nation-wide smog situation in January 1985, mean day concentrations were measured which were almost twice the level of the trigger value for smog alarm. The expert council for environmental issues mentioned that well-documented increased morbidity and mortality were at least partly due to the high immission values (Der Rat von Sachverständigen für Umweltfragen 1987, Rdnr. 1751).

The sensitivity of plants towards air pollution was already known in the last century under the name of „hut smoke damage". While the VDI (1988) defines immission limits for differently sensitive plants between 0.05 and 0.12 mg/m^3, a damage limit of 0.02 mg/m^3 is discussed for stocks of fir trees in climatically favored locations (Wentzel 1987). Thus, SO_2 is an example of a noxa to which plants clearly react more sensitively than humans.

The report „Transboundary Pollution – the Impacts of Sulfur in the Air and in Precipitation", which Sweden presented at the first environmental conference of the United Nations, for the first time drew global attention to the problem of acidification of waters.

A purposive fight against air pollution by SO_2 was started only after a smog situation in December 1962, when a significant increase in mortality was observed in the German state North Rhine-Westfalia (Knoepfel & Weidner 1985). In reaction, the highly industrialized and densely populated state built up its own measuring and monitoring net, and created a legal basis for the enforcement of traffic restrictions when smog limits are exceeded. This strategy was also adopted in the enactment of the federal law on immission protection, leading to a nation-wide legal basis for traffic restrictions. The consequent regulations which concretize the BImSchG aim, among other things, at a reduction of the high SO_2 immission values in concentrations below the level which is harmful for humans. The TA Luft of 1974 established an immission limit to be observed in the licensing of plants (0.14 mg/m^3, still valid at the time of this study); the third BImSchV enforced the use of low-sulfur fuels by a limitation and repeated reduction of the admissible sulfur content. A counter-effect, however, resulted from the policy of coal priority, which has been pursued since the seventies and according to which the sulfurous coal from the German Ruhr District has to be preferred to less expensive and qualitatively better imports (Knoepfel & Weidner 1985). So, the „century contract" signed in 1980 guarantees that the use of hard coal will increase up to 1990 and will remain constant up to 1995.

Figure 3.1 below shows the development of emissions on German territory during the years 1966–1986, and contrasts it with the development of immissions in the

Ruhr District during the same period. It is a striking fact that immission values in the Ruhr District decreased by almost 30% during the first ten years, although nation-wide emission values did not decrease at all during this period.

This development may be explained by the different possible measures to reach a pollution reduction. Essentially, there are two ways:
- diluting the concentration by widespread dissemination
- reducing emissions by use of modern retention technology.

In the seventies, in its attempt to improve the air quality in polluted areas, North Rhine-Westfalia preferred, with its policy of high smokestacks, the strategy of dilution of emission to that of its reduction. This policy was expected to yield a regional relief from immission by about 25%. Preference for this strategy is partly due to technological and financial considerations, where the generally bad economic and labor market situation of those years led to a particularly emphasized objection of „economic injustifiability“ against supplementary claims according to § 17 BImSchG. Fears that such a strategy would only shift the immission problems to more distant regions were said to lack substantiation (Knoepfel & Weidner 1985).

Despite an emission situation which had changed only a little, the SO_2-pollution in the Ruhr District decreased from 170 microgram per cubic meter in 1965 to 80 microgram per cubic meter in 1980, due to the „policy of high smokestacks“ (Wentzel 1987). Yet it is exactly this far-reaching dissemination of pollutants which is responsible for the considerable forest damages in areas far from industrial regions, as well as for the acidification of waters, for example in Scandinavia.

Environmental damages and a generally growing environmental concern led to a turn in German clean air policy. While up to then it had implicitly been assumed that limits for human health would also sufficiently protect ecosystems (plants, animals), it was now understood that the environment itself had to be protected against harmful anthropogenic influences. This – with respect to humans – precautionary environment protection contains „the beginning of a dichotomous thinking about goods worthy of protection and about limits. (...) Now, the authority specifying standards systematically distinguishes between limits and reference values which protect human health, and limits and reference values which are supposed to ensure the viability of rather delicate plants and animals“ (Der Rat der Sachverständigen für Umweltfragen 1987, Rdnr. 1683). A continuing anthropocentric orientation of the law on preservation of clean air is proved by the fact that animals, plants etc. are protected only if their endangering will mean impaired general welfare or if their vicinity will be affected in an unacceptable

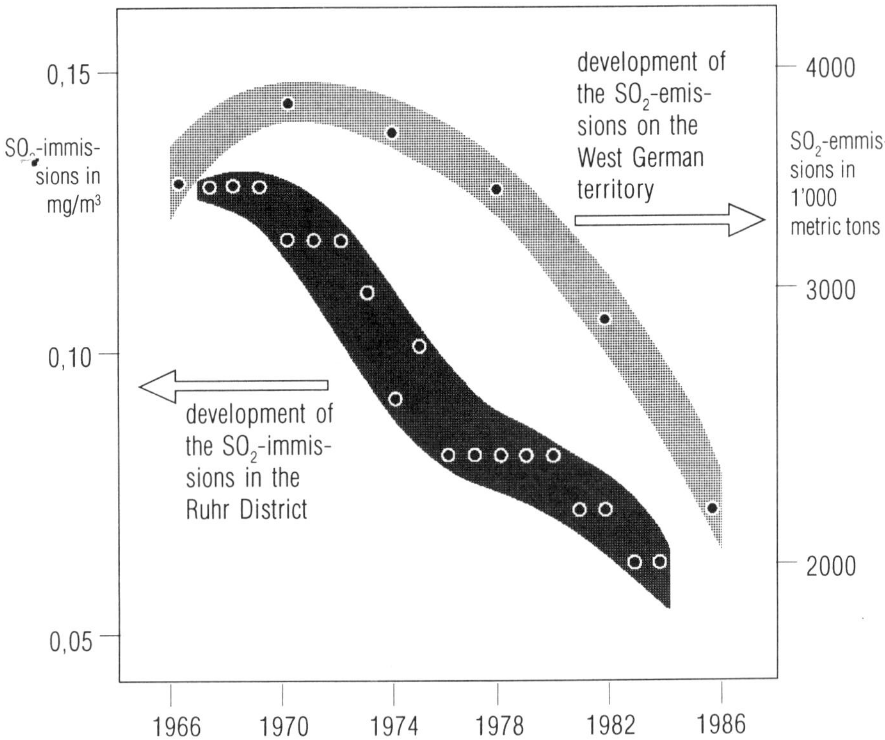

Fig.: 3.1 Comparison of the developments of SO_2-emissions and SO_2-immissions during the years 1966–1986.

a) Emissions of sulfur dioxide during the years 1966–1986 in the German Federal Republic; a further decrease to 1'000'000 metric tons is prognosticated to be reached by 1995.
(Source: Vierter Immissionsschutzbericht der Bundesregierung. Bundesdrucksache 11/2714 of 28 July 1988).

b) Surface-related annual mean values of sulfur dioxide immissions during the years 1967–1984 in the Ruhr District.
(Source: Umweltgutachten 1987 des Rates von Sachverständigen für Umweltfragen).

way (BImSchG and TA Luft). But now the focus is more on emission reduction, flue gas desulfurization is intensively advanced. Moreover, in the context of a harmonization of state regulation on smog, limits were set to more stringent values, and governmental measures were made dependent on the fact that smog exceeded the limits at only two measurement points (act on smog of the German state Bavaria 1985).

Both the new kind of forest damages which has received public attention under the catchword „Waldsterben", and the increasing acidification measured in Scandinavian waters, triggered a wave of new research. At the international level, the realization of a correlation between SO_2-emissions and „acid rain" led to the Geneva convention on far-reaching transboundary air pollution.

A few years later, several nations agreed on a reduction of their SO_2-emissions on the occasion of the Ottawa conference on „acid rain". Many other nations joined this „Ottawa-Club" in the following years. In 1985, the ECE executive organ of the Geneva Clean Air Convention signed a supplementary agreement in Helsinki, which obliged the signatories to reduce their annual sulfur emissions or the respective transboundary sulfur exports by 30%, by 1993 at the latest.

The German government abandoned its reserves against containment of transboundary air pollution in 1982, at a conference in Stockholm, under the pressure of the public debate on „Waldsterben". Since „Ottawa", Germany has been among those countries which announced and realized the largest emission reductions. Even at the level of the EU, Germany has taken the lead in clean air activities (Weidner 1987).

The first German pilot facility for flue gas desulfurization started to operate in 1975. Respective commercial facilities for hard coal power plants started to operate in 1977 and 1979. Since later installation of such facilities can only be demanded within the scope of „economic justifiability" (BImSchG), respective measures are highly subsidized. In the individual case, equipment with supplementary facilities is achieved by making licenses for new plants conditional on improvements of old ones (Knoepfel & Weidner 1985). Apart from that, the individual German federal states strive for further agreements with individual operators of plants. In 1982, the first agreement with a corporation was reached to gradually have all its major lignite power plant blocks equipped with desulfurization facilities by 1987.

The regulation on large-scale firing plants (1983) established emission values which have to be observed during the operation of these plants. The respective values, among them values for sulfur dioxide, aim first of all at prevention of harmful environmental impacts on forests. For the use of different fuels, they vary, according to the state of the respective technology, between 5 mg and 2'000 mg per cubic meter of flue gas. The emission reduction was achieved primarily by applying these limits also (in temporally graded steps) to already licensed (older) plants. The desulfurization facilities needed to reach these values led to a reduction

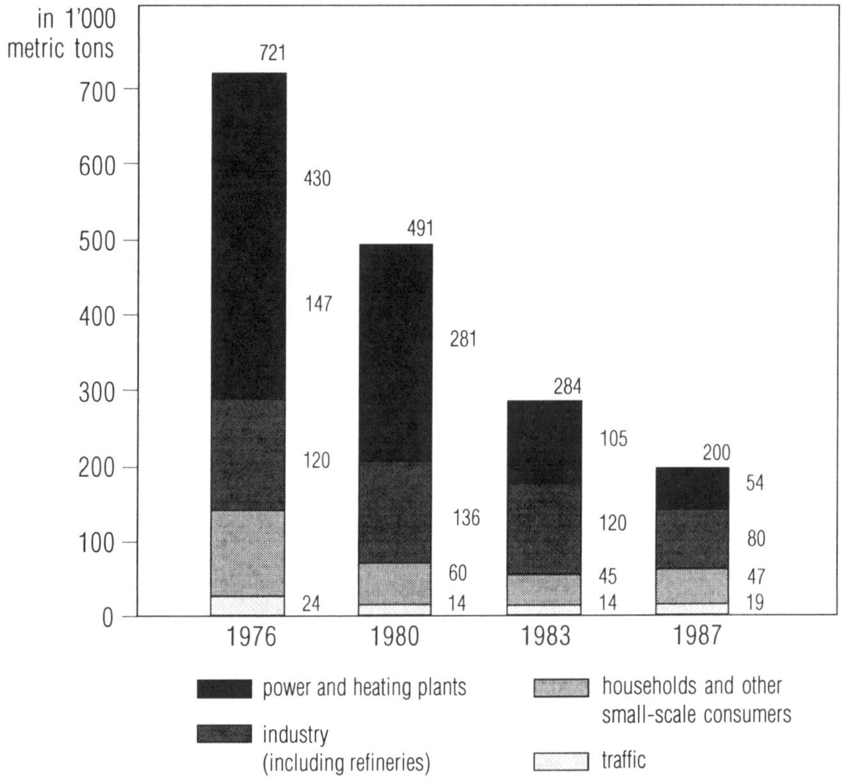

Fig. 3.2 The decrease of sulfur dioxide emissions during the years 1976–1987 in Bavaria. Source: Bayerische Staatsregierung 1989

of sulfur emission from power plants from about 1.6 million metric tons per annum to 0.485 million metric tons per annum (Rat der Sachverständigen für Umweltfragen 1987). This entailed investment costs amounting to about 10 billion German marks. A proposal of the state government of Hesse (1983) to enact a law on levying a charge on sulfur dioxide emissions, which was to support the thirteenth BImSchV, was abandoned.

The TA Luft of 1974 was modified in 1983, not least to the end of protecting the forests. Thereby, the existing immission values were adapted to the improved state of knowledge and were differentiated, for the first time, according to different goods worthy of protection. The values concerning pollution by SO_2 remained essentially unchanged. Yet immission protection was considerably refined by reducing the reference area for immissions from 16 km^2 to 1 km^2, and the possibility to hide local pollution behind mean values was considerably restricted.

The revised version of the TA Luft of 1986 focused on emission limitation and on the possibility of enforcing requirements on older installations; regulations concerning immission were more or less adopted from the 1983 version. Together with the thirteenth BImSchV, this TA Luft now establishes emission regulations for virtually the whole range of industrial facilities, which orient themselves to the state of technology and go far beyond the requirements of protection of human health. The amended TA Luft facilitates the issue of supplementary regulations concerning older plants in case of noxious environmental impacts on ecosystems which are worth being protected. In addition, the possibility formulated in the TA Luft of 1983 that new installations may compensate for their obligations to reduce emission by rehabilitation measures in neighboring installations was extended to older installations as well (TA Luft 1986).

The amended version of the BImSchG of 1985 invalidates the, up to then, weighty objection that supplementary regulations were „economically unjustifiable“ in individual cases. Such binding administrative acts are now merely subject to the interdiction of being unproportional (§ 17 Abs. 2 BImSchG), which is usually ascertained according to economic criteria. Another legal basis which was lacking so far is now provided for, namely the basis for the already mentioned compensation regulations (§ 48 Nr. 4 BImSchG).

Further measures to reduce pollution by SO_2 were taken by amending the third BImSchV and revising the first BImSchV. Both acts regulate the admissible contents of sulfur in fuels.

3.3 COMPARATIVE CONSIDERATION

Pollutants, including radioactive substances, are emitted from technological installations with waste air or waste water. Via the media air, water and soil, they spread in the environment. By transfer, often via rather long chains, e.g. of food, pollutants reach the organisms or places where they then exert their harmful effects. These damages or changes may roughly be classified into the following categories:

- changes due to physico-chemical environmental damages; e.g. damages to the ozone layer by CFCs, accumulation of CO_2; acidification of waters.
- ecological damages to plants and animals (collective damages), e.g. formation of reactive toxic agents by sulfur oxides; killing of insects by toxic substances, damages to plants by SO_2 and others.
- damages which induce diseases in humans (individual damages).

The discussed developments of environmental standards refer to pollutants which are examples of these categories. It shows that usually a noxa is classed in only one of these categories, so that the establishment of environmental standards is oriented to the action mechanisms of these critical damages. Low SO_2-immissions, for instance, impair vegetation; hence environmental standards for SO_2 refer to damages to plants, although damages to human health only occur at higher concentrations.

Ionizing radiation may basically induce damages in every living organism. Yet it requires considerably higher radiation doses to induce measurable irreversible non-stochastic damages in plants and animals, with the exception of mammals, than in humans.

Dose limits (whole-body dose or effective dose) are such that only stochastic damages (genetic mutations and cancer) have to be considered for a risk assessment. So always only very few individuals will be affected at low doses. Environmental standards in the field of radiation protection are therefore oriented to potential damaging effects on human health. The dose limits for individuals of the general population, but basically the limits at the work-place, too, are such – as has already been mentioned – that non-stochastic damages are avoided if radiation doses meet these limits or remain below them. On the other hand, stochastic damages will still be possible at these exposures, since it is assumed that there are no threshold doses governing these dose effect relations. These principles (avoidance of non-stochastic effects) have often also served as foundations for environmental standards concerning chemical substances.

As has been already discussed in chapter 2, the exposure level is decisive for a potential radiation risk. As regards nuclear plants, exposure is not so much due to direct radiation emitted from the plant but for the most part due to radioactive substances which are disseminated in the environment. As a result of processes of radioactive decay, these substances may irradiate humans externally, or they may be ingested from food or inhaled from breathing air and thus irradiate humans internally after radioactive decay. The dispersion of radioactive substances in the environment, transportation via air, e.g. bound to aerosols, via food chains, as well as internal transportation in humans, all follow basically the same mechanisms and processes as chemical noxae do. These processes are determined by physical, physico-chemical and chemical properties of the substances, and are independent of the radioactivity of a substance. The radioactive substances emitted from nuclear plants are often of a simpler chemical nature than complex organic emissions from chemical plants, but their atmospheric transportation frequently happens after they have been attached to aerosols. In such cases, dissemination will depend on the properties of the aerosols. Distribution processes in food chains are of a far more specific nature. Nevertheless, basic mechanisms are similar here as well, even if there may be differences in detail in individual steps for specific substances and for specific radioactive materials. Of course, there are differences also in transfer rates between different compartments of the environment and of humans. So, ecological transportation models for distribution calculations are basically identical for chemical noxae and for radioactive substances.

But there is far more certainty in determining exposure of critical molecules and of compartments in human cells and tissues to ionizing radiation than to chemical substances. When radioactive substances decay in a cell, all compartments or volume elements are irradiated with the same probability. Thus it is irrelevant to localize the radionuclide in the intracellular space, except for tritium.[1] With chemical noxae, however, the harmful substance has to reach the critical molecule to be able to affect it by means of a chemical reaction. Thus, apart from knowledge about the distribution of the substance among human tissues and organs, knowledge about the intracellular distribution of the chemical substance, including kinetic processes, is needed as well.

The damaging effects of ingested noxae in human tissues are often activated or inactivated by enzymatic processes. So, the carcinogenic nitrosamines from amines and nitrite can only be built in the human organism. The originally inert polycyclic hydrocarbons are activated in the tissue by enzymatic reactions so as then to react with the DNA and other biomolecules. These kinetic processes considerably complicate an assessment of the exposure of critical molecules to chemical noxae. Finally, it should be mentioned that microdosimetric considerations have yielded the finding that energy transfer through ionizing radiation always occurs in „energy packages“ (clusters of ionizations), the data of which are relatively well known. On the other hand, reactions of chemical noxae with biomolecules, e.g. with DNA, are to a larger extent singular processes.

Today, we assume that changes of the genome due both to ionizing radiation and to chemical substances may lead to genetic mutations (changes in germ cells) or to cancer (changes in soma cells). An establishment of environmental standards is particularly problematic for these genetically harmful agents, since with respect to their dose effect relations it is assumed, based on the present concept of their action mechanism, that there is no threshold dose (see chapter 2). Today, a number of changes of the DNA following exposure to radiation or to chemical substances are described in detail. Yet, so far we do not know which specific processes or changes of the DNA are responsible for the emergence of a stable mutation or the malignant transformation of a cell. Once a mutated germ cell or a cancer cell has been formed, the development of a sick child with a hereditary defect, or of cancer in the exposed person is, with a high probability, independent of the question whether the primary cellular event was due to irradiation or to a chemical substance.

These analogies and parallels between the impacts of ionizing radiation and those of a multitude of environmental noxae, the limits for which are oriented to the effects on human health, make clear that dose limits for ionizing radiation are embedded in the comprehensive context of environment protection and environmental standards. There are numerous cases where radiation protection was assigned the position of a pacesetter which other fields of environment protection followed. This advancing of radiation protection is no doubt due to the fact that assessments of doses or exposures are easier to establish here, and biolocical-medical radiation effects are better known than is the case with noxious chemical substances. Therefore, dose effect relations can be established in a more comprehensive way for ionizing radiation than for other toxic agents in the environment. Finally, the exposure to natural radiation provides a reference point for dose limits for which there is none or only a very limited one for chemical noxae.

As has already been said, the so to speak simple case of setting standards consists of establishing an exposure limit on the basis of an experimentally and/or epidemiologically proven dose effect relation from which a „No observable effect level" (NOEL) can be derived. In practice, this will lead to a threshold value of effect. This threshold value is then set as a limit – usually with safety reserves for the general population. This has been the case for all MAK-values for the workplace, and dose limits concerning non-stochastic effects, for irradiations of parts of the body, are established in such a way as well.

Even if there is no example of an exposure limit for a chemical substance being set to a higher level than its known NOEL, the limit must nevertheless not simply be associated with the threshold value.

If no threshold values are known for some carcinogenic and mutagenic chemicals, there are usually no limits even now – at least as far as environmental concentrations are concerned – but only minimization rules. Minimization rules without supplementing limits provide a situation of imprecise regulation. This situation is due to two reasons. First, sufficiently reliable dose effect relations are lacking for most carcinogenic and mutagenic chemicals so far. Second, the establishment of an exposure limit would also require the establishment of an acceptable risk here, which, when German radiation protection standards were established, was facilitated by the possibility to relate such values to natural exposure and to set limits within the range of this natural exposure. But for carcinogenic and mutagenic chemicals, further development will go, like in radiation protection, in the direction of explicit acceptance of a risk. In the U.S., a first step in this direction was made in 1986. The EPA issued guidelines for the „mutagenicity risk assessment", according to which scientific foundations are to be created which allow of establishing limits on the basis of acceptable risks. Even if it is not of prior importance to change the present practice in Germany, the core of which is, as has been said, a minimization rule (supplemented by technical reference concentrations for work-places, which reflect the state of technology), also in Germany, we will have to seriously get prepared to face the question about limits for carcinogenic and mutagenic chemicals.

This creates a new situation for toxicology. On the one hand, it is necessary to supply foundations for rationally balancing considerations on levels of acceptable risks from chemicals and hereupon-based decisions about such levels which will meet with consensus. On the other hand – and this will probably be the greater problem – great efforts are needed to acquire, at least for the most important chemicals, as extensive knowledge about the mutagenic and carcinogenic risk for the population as we have at our avail for ionizing radiation. The extent of this task can be vaguely imagined if one recognizes that now there are about 100'000 chemicals in circulation, and for only about ten of them is this required risk assessment already completely achieved. At present, 600 substances, hence less than 1% of the total number, have been selected for more specific examination, among them 60 chemicals of particular priority. In view of this situation, we will probably have to constrain ourselves for the time being to establishing limits in an exemplary way, as it were, for a few toxicologically relatively well-explored carcinogenic chemicals. For the rest, we will perhaps – at least for the time being – have to content ourselves with less certain toxicological foundations and hence with limits which have rather the character of reference values.

It would probably be a reasonable way to ascertain a general value for an acceptable range of risk if, with respect to various exposures, we referred to the range of exposure to natural radiation. In radiation protection, the basic limit for the general population of 0.3 mSv per annum is justified on the grounds that an additional exposure to a respective level of radiation and hence also the connected additional risk for various groups of the population will be unavoidable unless large parts of Germany with high natural exposure are declared as uninhabitable. Therefore it is pragmatically consistent to also permit exposure to such a level of additional radiation at the fence of a nuclear plant, which serves the general welfare.

This argumentation can be generalized: it would also be consistent to accept another risk of the same order, which is linked to corresponding benefits. In other words, the risk of cancer or mutation connected to an exposure to radiation of 0.3 mSv per annum could basically also serve as acceptable risk for chemicals. The respective value for the additional risk of death from cancer is of the order of 10^{-5} per annum (of 10^{-3} for the whole life) if all cases of death are referred to, and of the order of 10^{-4} per annum if only „spontaneous" deaths are regarded. That is the same order of magnitude as the order of internationally discussed values for acceptable risks for the population. Here, we would have an exemplary application of a risk-risk comparison as it will be basically discussed in the following chapter 5, and as it was already once practically applied by the LAI to a real problem, namely to limits for carcinogenic air pollutants. A particular advantage would be that such considerations do not refer to some life risk or technological risk whatsoever, but to a risk which, through establishment of an environmental standard, has also been accepted by the German Federal Constitutional Court.

Of course, possible coincidence of different exposures has always to be taken into account when limits are established on the basis of risk assessments. If coincidences are sufficiently probable, the respective risks have to be summed up and safety rebates have to be made. A certain restriction of such coincidences results, however, from the location of technological facilities, in particular when the environmental standard refers to the place of most unfavorable effect, as is the case with nuclear plants.

ANNOTATION

1 Only in tritium decay, a β-radiation emerges whose mean range is smaller than the cell diameter. Hence localization of tritium within the cell is of relevance

LITERATURE

Aurand, K., Kierski, W.-S. (ed.) (1981): Gesundheitliche Risiken von Asbest. Eine Stellungnahme des Bundesgesundheitsamtes, Berlin, D. Reimer.

Bayerische Staatsministerien für Ernährung, Landwirtschaft und Forsten sowie für Landesentwicklung und Umweltfragen (ed.) (1989): Wald in Gefahr – Eine Information der Bayerischen Staatsregierung, Bayreuth, E. Mühl.

BEIR (1980): The Effects of Exposure to Low Levels of Ionizing Radiation, Washington D.C., National Academy Press.

Beyersmann, D. (1986): Verfahrens- und Argumentationsmuster bei der Festlegung von Grenzwerten im Ausschuß für gefährliche Arbeitsstoffe (AgA), In: Grenzwerte – Interdisziplinäre Untersuchungen zu einer Rechtsfigur des Umwelt-, Arbeits- und Lebensmittelschutzes, G. Winter (ed.), Düsseldorf, Werner-Verlag, 86–94.

Deutsche Forschungsgemeinschaft (ed.) (1988): Maximale Arbeitsplatzkonzentrationen und Biologische Arbeitsstofftoleranzwerte 1988, Mitteilung XXIV der Senatskommission zur Prüfung gesundheitsschädlicher Arbeitsstoffe, Weinheim, VCH.

Deutscher Bundestag, Referat Öffentlichkeitsarbeit (ed.)(1989): Schutz der Erdatmosphäre: Eine internationale Herausforderung; Zwischenbericht der Enquête-Kommission des 11. Deutschen Bundestages „Vorsorge zum Schutz der Erdatmosphäre“, Bonn.

Fischer, M., Meyer, E. (ed.) (1984): Zur Beurteilung der Krebsgefahr durch Asbest, München, MMV.

ICRP (1977): Annals of the ICRP, Publication 26, Recommendations of the International Commission on Radiological Protection, Oxford, Pergamon.

Knoepfel, P., Weidner, H. (1985): Luftreinhaltepolitik (stationäre Quellen) im internationalen Vergleich, Berlin, Ed. Sigma Bohn.

Länderausschuß für Immissionsschutz (LAI) (1987): Arbeitsgruppe „Krebsrisiko durch Luftverunreinigungen“, Beurteilungsmaßstäbe für krebserzeugende Luftverunreinigungen, Berichtsentwurf.

Muth, H. (1980): Zur Geschichte der Dosiseinheit „Röntgen“ (R), Schriftenreihe Deutsches Röntgenmuseum 2.

Der Rat von Sachverständigen für Umweltfragen (1987): Umweltgutachten 1987, Vorabdruck.

Streffer, C. (1988): Strahlentoxikologie, in: Gift – Geschichte der Toxikologie, M. Amberger-Lahrmann & D. Schmähl (eds.), Berlin, Springer, 127–166.

Türck, R. (1988): Luftgüteleitlinien der Weltgesundheitsorganisation, in: Grenzwerte und Risikobetrachtungen in der Umwelthygiene, K. Jander & E. Lahmann, Stuttgart, G. Fischer, 105–116.

Umweltbundesamt (ed.) (1980): Luftqualitätskriterien; Umweltbelastung durch Asbest und andere faserige Feinstäube, Berlin, E. Schmidt. Verband der Chemischen Industrie e. V. (ed.) (1989): Umweltbericht 1988/89, Stand Juni 1989, Frankfurt.

Verband Deutscher Ingenieure (VDI) (1988): Kommission Reinhaltung der Luft, Band 1, Berlin, Beuth.

Weidner, H. (1987): Luftreinhaltepolitik in Europa: Ein 17-Länder Vergleich, Berlin, Edition Sigma.

Weinmann, W., Thomas, H.-P. (1989): Gefahrstoffverordnung mit Chemikaliengesetz, Textausgabe mit Erläuterungen, Köln, C. Heymanns.

Wentzel, K.F. (1987): Waldschäden – was ist wirklich neu? In: Patient Wald, Gesellschaft für Strahlen- und Umweltforschung mbH (ed.), München, Biehl, 19–28.

World Health Organization (1977): International Agency for Research on Cancer, IARC-Monographs on the Evaluation of Carcinogenic Risk of Chemicals to Man: Asbestos, Vol. 14, Lyon.

4

INDIVIDUAL, SOCIAL AND GOVERNMENTAL RISK ASSESSMENT

We feel and are exposed to risks at all times; we are either aware of these risks ourselves or we hear about them, we assess them more or less realistically and react (or do not react) to them. How this happens does not only depend on our information level, our sensitivity towards hazards, our general value attitudes and our specific attitudes, but also on social processes of influencing, the dynamics of public opinion and collective convictions.

If certain risks go beyond „objectively" determinable tolerance limits (lack of „acceptability") or „intersubjectively" perceived thresholds of endangerment (lack of „acceptance"), there is usually the call for governmental action, be it by means of legal regulations or of politically-administrative measures.

This chapter studies the individual, social, juristic and politically-administrative processes that take place when risks (radiation risks in particular) are evaluated and limited. In doing so – similar to the previous chapter – it confronts with reality certain scientific and normative maxims of assessing risks (chapters 1 and 2), and at the same time it represents the empirical starting point for proposals on a more rational evaluation and limitation of risks (chapters 5 and 6).

4.1 FORMATION OF INDIVIDUAL ASSESSMENTS ABOUT AND ATTITUDES TOWARDS RISKS

When individual perceptions, assessments, rejections or acceptances of risks are discussed in public, one could easily get the impression that these are uncommon, even dramatic experiences and decisions for the individual. However, that is by no means the case. Nothing is more customary than dealing with risks, i. e. evaluating, putting up with or avoiding personal costs, hazards, disadvantages and damages that will possibly arise if one wants to achieve certain advantages, benefits, values and satisfaction. Every little child learns that.

Typical of this is the experience of having successfully performed something, the endeavor, which is found universally in young children, to experience one's own ability in accomplishing tasks that could either succeed or fail. Naturally, the more difficult the task is, the more the danger of failure is subjectively perceived, the greater is the emotional experience of pride when it is accomplished. Therefore, risk-option models (Heckhausen 1980) have played a pivotal role in the intensively pursued research on individual development, personal formation and situative manifestation of the achievement motivation ever since the early fifties. Summarizing the most important results of these investigations and of other fields of psychological research relevant for risk, choice and decision behavior, five general conclusions can be drawn:

(a) Perceived risks are by no means only negatively experienced as triggers of avoidance behavior, they are also, and often at the same time, characteristics of the attractiveness of an objective. For it often holds that the higher the risk of failing when carrying out an action, the more positively its result is experienced.

(b) Since risks often have to be assessed in complex, nontransparent and dynamically changing problem situations, people acquire a number of typical heuristics in the course of their cognitive development in order to reduce the complexity of situations and their personal uncertainty. We will discuss these „cognitive mechanisms“ in detail later.

(c) In daily choice and decision behavior, risks are not calculated as objective probabilities, but as subjective expected values, and results or consequences of decisions are anticipated not only in their actual useful effect, but also in the emotionally associated experiences. Not facts, but personal interpretations are decisive for acceptance or rejection of risks.

(d) Already in early childhood, relatively stable differences among individuals manifest themselves in the way persons either rather tend to be out for risks or to avoid them. For some persons, these dominant tendencies (hope for benefit against fear of loss) have a relatively general character, hence are independent of the respective field of action, but for many people, there is a large individual range. This range may reach as far as to specific fears (e.g. fear of flying), the causes of which lie in the biography of the individual and thus in the personal learning history.

(e) The influences of groups often lead to significant intensifications or diminutions of individual prejudices towards specific risks so that we have to start from the assumption that individual risk evaluations are often superimposed by processes of social judgments.

The latter regularity is of particular practical importance when certain risks become matters of public attention and when, at the same time, the individual foundation of experience and knowledge for assessing the risks is comparatively limited. That is presently the case with many scientific-technological innovations. A good example of this is radiation protection, which has been in the center of intensive socio-political discussions for the past few years.

Due to the significance of the problem in terms of energy economy, there are comparatively many cognitive-scientific and socio-psychological studies on the development of individual assessments and attitudes concerning radiation risks. Therefore, this topic serves as a particularly useful example for the psychological analysis of human behavior with respect to scientific-technological risks in general.

Since public opinion on peaceful use of nuclear energy has systematically been recorded for a long time, it can be ascertained by means of the available polling results that the public attitude towards radiation risks has changed significantly in the course of the past ten years. All in all, the fear against nuclear power plants has become greater and the readiness to accept the risks involved has decreased so that the social acceptance of nuclear power has diminished.

This is the continuation of a trend that started from a positive appraisal of nuclear technology in the fifties and led, with little fluctuations, to increasingly negative assessments on the part of a growing section of the population; yet, on average, no dramatic tendencies of rejection are identifiable so far. However, the disaster of Chernobyl has rapidly accelerated the development of negative attitudes: the number of people expressing a disapproving opinion increased immediately after this event, in West Germany, for example, temporarily from 35% to 70%. But also in the U.S., more than half the population is now opposed to nuclear power and the building of new nuclear power plants (OTA 1984; Van der Pligt 1985). Although radical opinions have also intensified during this process and the connected potential of protest has increased, the average attitude profile is differentiated within itself. A relatively large percentage of those who generally oppose nuclear energy only demand that no new nuclear power plants are built, but not, at the same time, that nuclear power as an energy source is given up immediately.

Even if we consider the entire spectrum of opinions with its differentiations, there is altogether a relatively strong contrast between the attitudes of a large number of people against nuclear power on the one side and the advocating of this energy source by experts and official authorities as particularly environmentally friendly and safe, on the other. How can this contradiction between public opinion and expert assessment be explained? This question has emphatically been examined for the past few years in a number of countries including West Germany (Renn 1984; Slovic et al. 1982; Slovic 1987), but so far, no satisfying theory explaining this phenomenon has been found (Kasperson et al. 1988). Until now, there are only a number of hypotheses that are empirically more or less proved.

4.1.1 Hypotheses on Differences between Rational and Intuitive Assessments

A statement that is very general but is often heard to explain the discrepancy between assessments of the lay public and of experts refers to the difference between irrational and rational thinking. The rejection of nuclear power is seen as the result of evaluation processes that are not in accordance with rational criteria and are mainly caused by uncertainty, anxiety and dramatizing news coverage on individual accidents.

To justify this view, it is often noted that lay persons develop firm opinions on and against nuclear power without, however, even having the special knowledge needed to judge. Therefore, it was long thought to be plausible that the acceptance of nuclear power would improve linearly with increasing knowledge about it (Inglehart 1984). Several more recent studies prove, however, that there is no such simple correlation. At least among lay persons, the knowledge levels of proponents and opponents of nuclear energy are not significantly different from one another (Levi & Holder, n. y.). Only on the expert level, the anticipated differences to the lay public appear.

Another common assumption, which is tightly connected with the knowledge hypothesis, is that contradictory information leads to personally experienced uncertainty when the individual is not familiar with the matter. This uncertainty causes anxiety, and „anxiety is the worst adviser" – as a German saying goes, for judging difficult situations. For it generally holds that the more anxiety a phenomenon causes, the less people seem capable of observing accurately, thinking sensibly and using appropriate methods to describe, understand, predict and control the phenomenon.

This thesis seems to describe an important aspect of how lay persons form their assessments of nuclear energy. It has proved in the meantime that proponents and opponents of nuclear energy are not different in their disposition to anxiety; they do differ, however, in the intensity of specific experiences of fear of nuclear energy and of the associated risk of accidents (Levi & Holder, n. y.).

This state of anxiety is increased by a wide-spread loss of credibility and trust which experts and government officials have experienced in the meantime. All recent studies establish this sometimes dramatic decrease in confidence on the part of the lay public and its close connection to their intensified opposition to nuclear power. The American social psychologist Albert Bandura attributes this statistical regularity to a central mechanism: „People who distrust the judgmental efficacy of risk analysists or their impartiality are likely to be swayed more by their own intuition than by the analysists' probability calculations" (Bandura 1986, 138).

That „own intuition", of course, does not mean reliable and valid assessment processes. It rather refers to the forming of opinions in situations of personally experienced uncertainty and threat. Insofar as subjective problem definitions play as important a role as do the influences of relevant reference persons and reference groups.

Since assessments about the benefits and risks of nuclear power compared to other energy sources deal with extremely complicated matters, according to all available scientific results, lay persons form their assessments with the intention of radically reducing the complexity of the problem and thereby minimizing their own uncertainty. This is done by building up convictions and expectations that are based on subjective probability assessments about the risks of nuclear power.

Now the theoretically and practically important question is, in what way and by what criteria can we test and measure how „correct", how appropriate or at least how reasonable such assessments are?

4.1.2 Criteria of Rational Assessments and Decisions

In decision theory, different criteria are used for evaluating assessments, all being subject to the concept of rational thinking and deciding. In other words: subjective probability assessments and risk evaluations are classified as more or less valid according to their deviations from the results of rational assessments. In general, despite a number of differences in details, assessments about nuclear energy can then be regarded as *rationally founded*,

- if risk estimations are carried out within the scope of clearly defined cost-benefit analyses;
- if risks are determined as the multiplicative function of the probability of occurrence of harmful incidents and of the expected damage;
- if the criteria of damage seem precise and reasonable;
- if the assessment is reached methodically and the arguments used are logical

so that the subjective risk evaluation corresponds to a great extent with the objectively observed or scientifically predicted relative frequencies of the probability of occurrence of a damage and the costs involved, and thus proves itself valid.

A criteria catalog that is so sophisticated is by no means regarded as artificial by experts, because the human way of thinking and judging is seen by many cognition scientists as analogous to that of a statistician (or of a computer that has been programmed accordingly). At the same time, it is also assumed that without a sufficient individual knowledge base, with incomplete or incorrect information and with affective prejudices, the intuitive assessments of humans deviate heavily from this ideal model of decision theory. This often seems to be the case when lay persons evaluate nuclear energy. Therefore, in the following we will not study subjective probabilities and risk assessments *the way they should be*, but only those which really occur and which need scientific explanation.

4.1.3 Formation of Intuitive Assessments

Compared to scientific procedures or idealized subject models, daily risk perceptions, i. e. the assessment and evaluation of risks by lay persons, are by no means conscious, methodically strictly controlled and logically performed processes but often impulsive judgments, conclusions and evaluations against the background of available knowledge and personal convictions, so that criteria of consistency and coherence among different components of an assessment are often violated. Against the theoretical background of cognitive psychology, such assessments are the result of a subjective processing of the available information by means of stored knowledge and the operational routines governed by this knowledge. Thus, for example, excellent chess-players are not different from mediocre ones in their general intellectual capacities or in their talent for systematically analyzing complex situations, but mainly in the amount and the quality of memorized chess constellations, which allow them to act rapidly and successfully (Chase & Simon 1973).

The idea, however, that such everyday assessments were always, or at least frequently, detached and separate processes of knowledge-based information processing would be wrong. This is rather the exception. In many cases, intuitive assessments are formed by building up opinions and consequent subjective assessment tendencies over fairly long periods of time. Within this more or less stable reference system, a statement on current problems or questions can then be given relatively quickly. For most people, this has the enormous advantage that they avoid the unpleasant experience of subjectively losing control, the development of cognitive dissonances, and the experience of conflicting ambiguities, which are almost always connected with the state of ignorance or with not being able to reach a judgment. The less specific knowledge a person has, the smaller an individual's ability is to assess a complex matter, and the more affect laden the problem is, the more the person will fall back upon solid convictions and on very general cognitive strategies. The assessments are then mostly made very quickly and are extremely general and usually as clear-cut as they are one-sided.

4.1.4 Cognitive Heuristics in the Formation of Intuitive Assessments

So it would be wrong if one attempted to regard the process of risk assessing by lay persons even as merely analogous to a rational decision procedure. Instead, depending on the respective way the question is posed, momentary mental models are formed, which already include, in a simplified form, the decisive elements of the assessment that is to be developed. Such mental models are fragmentary, they are based on the momentarily accessible knowledge, they replace missing information with general convictions, are strongly influenced by the respective social context, and form the basis for relatively vague, intuitive assessments. In these models, certain heuristics are used, which can be regarded as rules of thumb or plausibility rules for a non-logical way of thinking. These heuristics are very general, always applicable and effective, but at the same time they are extremely prone to producing assessment mistakes, as the following examples illustrate[1]:

(a) Usually, the information that is available and accessible to the person making an assessment is thought to be necessary and sufficient. So, when concrete, actual mental models are formed, easily retrievable memories, vivid ideas, ever recurring personal experiences and easily comprehensible opinions of others play a particularly accentuating role, because the connected information is always present. Due to the amplification effect of the media and of relevant social reference groups (Kasperson et al. 1988), actual or expected catastrophes, dramatic accidents and frequently recurring problem situations gain a particular importance for the formation of individual assessments on nuclear energy. In a way, they represent the easily accessible mental material for assessment formation. Whatever is not present at the very moment is not taken into account in a self-evident manner and is often not even missed („availability").

(b) The subjective confidence in the correctness of one's own assessment is often significantly higher than its reliability and validity, measured against objective criteria. Incidentally, this tendency to overrate one's own ability to judge holds both for lay persons and for experts. That has to do with the fact that positive evidence in favor of one's own opinion is taken into account more often and to a greater extent than negative evidence („surplus of confidence").

(c) Two interconnected events are generally considered to be more likely than each of both events on its own. For example: (A) Long-haired people take part in demonstrations. (B) Bank employees take part in demonstrations. (C) Long-haired bank employees take part in demonstrations. Of which of the three groups do more people take part in demonstrations? (C) is chosen more often than (A) and (B), because, in an intuitive judgment, the basic rates of the three events are not taken into account appropriately („basic-rates trap").

(d) Assessment formation is often heavily influenced by the dominance of current reference systems. Naive thinking is always integrated in current contexts, which can partly be manipulated by phrasing the question or the problem that is to be decided on in different ways. For example, whether losses or gains are emphasized in an identical situation often leads to completely diverging decisions. Imagined losses are generally rated as more substantial than imagined gains („contextual embeddedness"). This tendency is illustrated by an example by Tversky and Kahnemann (1985), which has become very well known:

> In California, a certain type of Asian influenza is feared to spread. Without suitable precautions, 600 people are likely to die. There are two alternative medical schemes available:
> - Scheme A: 200 people will be saved by this scheme;
> - Scheme B: 1/3 probability that all 600 people will be saved; 2/3 probability that nobody will be saved.
>
> With this phrasing, 72 % of the test participants chose scheme A and only 28 % chose scheme B.
>
> In a second version, the phrasing was changed as follows:
> - Scheme C: 400 people will die;
> - Scheme D: 1/3 probability that nobody will die; 2/3 probability that 600 people will die.
>
> In this case, scheme C was only chosen by 22 % and 78 % chose scheme D.

Why do the same situations result in such different subjective assessments, depending on the phrasing of the problem? The different context is decisive here. Imagined losses are, as mentioned earlier, evaluated as more substantial than imagined gains. Figure 4.1 illustrates the functional relations.

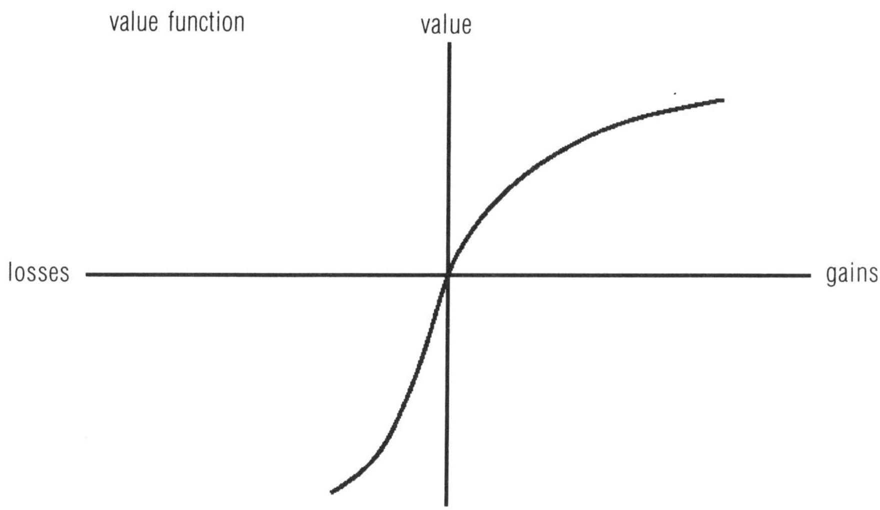

Fig. 4.1 Hypothetical function of risk evaluation, depending on subjectively perceived (or expected) losses and gains (according to Tversky & Kahnemann, 1985, 111).

In accordance with the regularity depicted in figure 4.1, most of the test participants in the study of Tversky and Kahnemann found it hard to accept the explicitly mentioned death of 400 people in scheme C, even though they quietly put up with it in scheme A. This explains the apparently paradoxical result that with imagined gains, efforts are taken to minimize the risk, but with imagined losses, a higher risk seems acceptable if there is the chance of avoiding losses.

(e) Assessment processes are often very impulsive, because there is a strong tendency to reach quick certitude when uncertainty is cognitively experienced. This „fiat tendency" furthers the latent readiness for self-confirming wishful thinking and for the creation of cognitive illusions („fiat tendency").

The above-mentioned examples of rules of thumb employed to quickly cope with difficult assessment problems have been studied intensively in recent years[2] and have been referred to theoretically and practically when the evaluation of nuclear power by lay persons was scientifically investigated.[3]

4.1.5 Relevant Factors in Intuitive Risk Evaluation

Although stereotype information processing in the course of solving complex assessment problems can be described well by means of the mentioned heuristics and by some others that are not referred to here, the resulting assessments and evaluations do not depend on these heuristics alone. Additionally, they are determined by the respective contents dominating in information processing. So, we get a second class of differences between rational and intuitive assessment formation. Lay persons usually cannot consider risks in an abstract, formalized or even quantified form; instead, they orient themselves by qualitative characteristics. In special studies, a number of very general points could be identified which rather strongly influence the assessment of the risk of nuclear power and its subjective acceptance. Such qualitative attributes of risk or loss are, for example:

- the number of deaths and – not so often – the number of injuries to people's health that are expected during normal operation of nuclear power plants, without any catastrophe occurring. This factor plays a much greater role for experts than it does for lay persons;
- the potential of catastrophes, which is, particularly for nuclear power, deemed very severe. It is indicated by the expected number of casualties in cases of very serious accidents. If lay persons are asked, for example, how many additional deaths they fear to happen in a particularly disastrous year, compared to the average values, the presumptions vary for most risk sources from one to three times the average value; for nuclear power, this value is multiplied by 100;
- the dreadfulness of the anticipated consequences in case of a catastrophe; i. e., the fear to lose the chance of taking action to limit the losses; the far-reaching destruction of the nearby environment and hence of the living conditions of many people, as well as damages to nature that are regarded as hardly repairable or even irreversible in the long term;
- the perceived loss of the chance to control and to direct in order to actively influence the causes, the probabilities of occurrence, the courses and the consequences of cases of damage, mainly those that concern the person in question himself/herself. Since these personal means of control are considered to be very large in situations of driving a car, skiing, or mountaineering, the risk of these activities is subjectively underestimated, if individual assessments are compared with objectively ascertainable facts;
- the feeling of being involuntarily exposed to a danger source, without personal decisions on the acceptance of the risk having any influence. However, recent studies of the correlations between voluntarily accepting a risk or having it involuntarily imposed on oneself on the one side and the summary risk assessment on the other side have shown that this factor is only of little importance (Jungermann & Slovic 1991), although it does have great plausibility and is therefore again and again highlighted in the respective literature

as being particularly significant. It has to be considered here that in empirical studies, voluntariness or involuntariness of taking a risk is often confounded with subjectively perceived controllability or uncontrollability of sources of danger and damaging consequences. There is no doubt that the individually experienced control possibilities are of great importance for risk assessment, which is sometimes falsely attributed to the factor „voluntariness".

The results reported by Jungermann (1982), according to which the subjectively perceived risk could be predicted almost precisely (with a multiple correlation coefficient of .95) with the help of the predictors „dreadfulness", „severity of a damage", „potential of catastrophes" and „estimated average number of casualties", show which significance such qualitative categories have for the assessment formation of lay persons. The risk of nuclear power is rated highest here, even though the lowest fatality rate per year is ascribed to it.

The estimated number of deaths caused by nuclear power plants, however, strongly varies depending on whether lay persons are asked to give their estimation for an „average year" or for a „disastrous year". This is shown by the following two diagrams (4.2 and 4.3), where the statistical number of casualties caused by traffic accidents was given as a guiding figure.

Figures 4.2 and 4.3 illustrate a general tendency to overestimate risks, where the assessment of the relative significance of various damage sources is comparatively good for average years (the correlation between the estimated and the real numbers is .87) while this is not the case for disastrous years (correlation: .39). O. Renn, who carried out the study, writes: „For a normal year, our 'common sense' seems to produce a relatively correct estimation of the losses. However, if there are no intuitive standards, as in the case of rating catastrophic developments, relatively constant rates of losses are assumed without making distinctions according to the type of risk, or else extremely high numbers are chosen" (Renn 1984, 151).

The relatively realistic estimation of various loss probabilities for normal years in the study of Renn (see also Daamen et al. 1986) should, however, not be overinterpreted. There are two main reasons for this warning: first, a comparison of different empirical studies demonstrates that the results are by no means consistent. This also holds for the estimation of nuclear power as a risk source (Daamen et al. 1986). The second reason why we should be careful when theoretically interpreting the results shown in figures 4.2 and 4.3 is the fact that there is only a very small correlation between the estimated fatality rate as a risk indicator and the evaluation of both the risk and its subjective acceptability. Risk-specific attitudes seem to play an additional and also a very important role here.

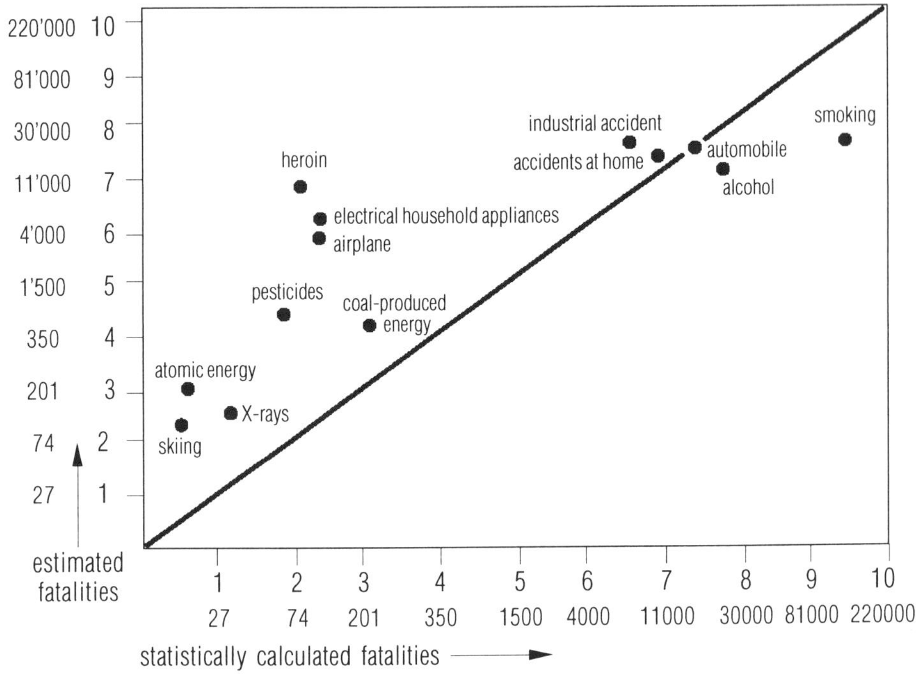

Fig. 4.2 Fatality rates for 13 risk sources, estimated by interviewees for an average year, compared to the statistically calculated rates (from Renn 1984, 148).

For, if we analyze risk assessments by lay persons, not only the robust heuristics in information processing prove to be relevant. At least equally important, if not more significant, are the contents of the associations linked in human minds to risks and to cases of damage for which these risks are indicators.

Even if these qualitative evaluation categories seem to be relatively homogeneous, at least for the populations of the U.S. and of Germany, we have to assume that for certain risk sources there are respectively specific associative mental patterns, which are activated when assessments are to be made and which strongly determine the result of an evaluation. Thus, for example, the factor „potential of catastrophes" seems to play a decisive role mainly for large-scale technological risks, particularly for the assessment of nuclear power or chemical production. Risk sources such as vaccinations or X-rays, on the other hand, are embedded in association nets that are tightly linked to less dramatic evaluation categories.

Important limitations of the subjective comparability of various risks seem to lie in the different significative value of the risk sources, the mental images linked to

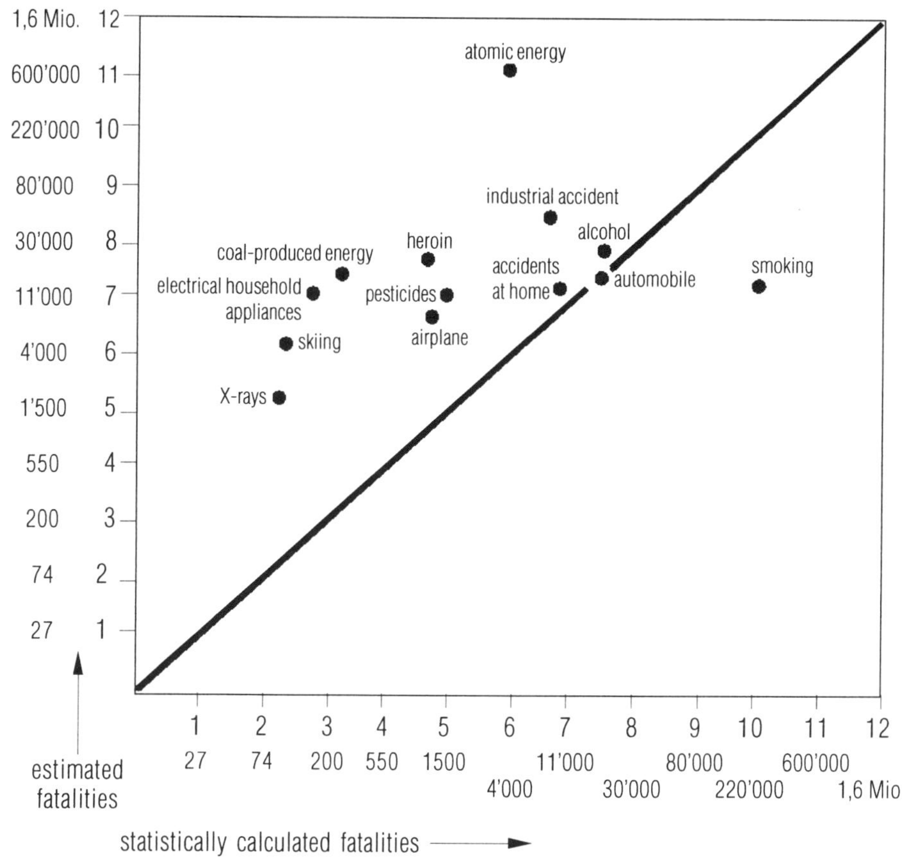

Fig. 4.3 Fatality rates for 13 risk sources, estimated by interviewees for a disastrous year, compared to the statistically calculated rates (from Renn 1984, 149).

them, and the fear-eliciting stimuli depending on those images. Although differential psychology might distinguish people according to their relatively general readiness to take risks, there is still the subjectively experienced dissimilarity of different risks. So, no matter how much it may make sense to describe damage probabilities in a relatively abstract, formalized and quantifiable way, making them thereby logically and statistically reciprocally comparable, this does neither correspond with our subjective experiences and the connected individual risk evaluations nor with the often bizarre profiles of acceptance or non-acceptance of certain hazards.

The latter aspect of risk evaluation already brings up the general question about individual differences in assessment formation. We have already pointed out certain correspondences as well as at serious differences between lay persons and experts. In general, it can be said that experts use a rather specific and well-defined notion of risk, whereas lay persons apply the concept of risk in a much wider, relatively vague and multidimensional sense. Experts take recourse to the probability of events and to the damage to be expected per event; lay persons concentrate mainly on the (material *and* emotional) consequences of assumed incidents of damage without adequately considering small probabilities. Of course, another difference between experts and lay persons is the fact that specialists base their assessment on more extensive and reliable information.

4.1.6 Relevance of General Value Orientations for Intuitive Risk Assessments

Apart from these cognitive differences between lay persons and experts, there are frequently also differences with respect to very general value orientations. Experts in the field of power supply are, almost without exception, people who rate the benefit of science and technology very high, who are convinced that the future well-being of society depends on the available energy, and for whom economic aspects therefore play an important role. Of course, not all experts believe in progress and orient themselves by an expedient rationality, but this does hold for a large number of them. On the other hand, it would be completely wrong to say that all lay persons were ecology-centered, skeptical about science and hostile to technology. Especially with lay persons, the differences in basic value orientations are extremely large. This statement remains a fact, even when we learn from the results of opinion research that in the German public the belief in progress is subsiding, the sensitivity towards environmental damage is growing, the confidence in expert judgments is dwindling and the average concern about hazardous technological developments that cannot be sufficiently controlled is rising (Weinert 1988).

Such general value orientations also influence current assessments of certain risks to a relatively large extent. On the basis of the existing empirical results, one could even ask whether proponents and opponents of nuclear power have come to their conclusions after and because they have cognitively assessed the risks and have found them acceptable or unacceptable, or if, conversely, they assess the risks differently because they are in favor of or against nuclear power due to their general value orientations. Thus, for example, it turns out that ecology-centered people in Germany, in particular those who classify themselves as belonging to the political 'alternative scene', expect a higher number of deaths to result from the use of nuclear power, they assume a greater potential of catastrophes and estimate the probability of an accident considerably higher than do people with a different value orientation. So, assessments of risks and the acceptance or non-acceptance of certain hazards also depend on the basic value attitudes of a person, which cannot be regarded as only an expression of specific preferences but which often have to be understood as an indicator for a certain way of life. Insofar, it is only natural that these values turn out to be individually stable and that a change of values takes place rather slowly. This has to do with the fact that information contradicting one's own value judgments is understood according to one's own convictions, interpreted differently or downgraded. For example, no matter how much the short-term attitudes towards nuclear energy changed after the Chernobyl accident, the more general value systems (e. g. the assessment of technology being a blessing or a curse) were not affected.

4.1.7 Social Influences on Individual Risk Evaluation

How such basic value orientations are acquired is currently a scientifically unanswered question and a matter of theoretical controversies. Apparently, we have to reckon with considerable individual differences in the genesis of relatively general systems of attitudes. After all, personal interests (e. g. linkage of professional objectives with the acceptance of certain risk sources), salient personal experiences (e. g. disasters as critical events in life), and especially the effect of significant reference groups have proved to be important. Whatever is quietly accepted, considered problematic, or attacked in a society altogether and in its subgroups has a strong effect on individual preferences, cognitive assessment processes and personal readiness to accept certain situations (Kasperson et al. 1988).

This social embedding of risk evaluations has led several times to theoretical and methodical criticism of an individualistic approach of research. On the one hand, such criticism pointed at the tendency to evaluate the suitability of subjective risk assessments only according to their deviation from rationally ascertained norms, a tendency which may lead to a discrimination of certain preferences of individuals and of social groups, as some science critics presume. On the other hand, the proneness of the cognition-psychological approach to artifacts has been emphasized. If individual assessment formation is rather strongly determined by general value orientations which are embedded in social reference systems, then – as some critics think – research on cognitive illusions could be a cognitive illusion itself. If, however, we regard the available results of the psychological research program, this criticism only seems to be justified insofar as it necessarily constrains the validity range of theories centered on individuals, and hence also the usefulness of empirical findings based on these theories for political planning and decision processes. Thus, these theories have to be supplemented with regard to social determinants and processes. This aspect will therefore be examined more closely in the next section.

4.2 SOCIAL PROCESSES AND CONFLICTS IN RISK EVALUATION

Section 4.1 focuses on how individuals subjectively deal with and evaluate risks and uncertainty. The following expositions will shift the discussion of actual risk evaluation from the individual to the social and institutional level. Just as a house is more than the sum of its bricks, to take up an example of the sociologist Simmel, the social evaluation of risks is more than the sum of all individual assessments: along with a higher aggregation level comes the transition to a new quality of the phenomenon considered.

The unintended side effects of social actions and their consequences are a good example of that (Schmid 1982, 22f). Description and analysis of social processes which transcend the actions of individual members of society often rest on abstract theories and models. Some authors have chosen system-theoretical models for that[4], others prefer the grid-group approach of cultural anthropology[5]; others again have brought the arena model from the political sciences into play for their analyses of the problem of risk in society.[6] An analysis of the problem of risk from a sociologic point of view which is not restricted to one specific approach can be found, among others, in Beck (1986, 1988) and in Perrow (1984). The following expositions largely represent a synthesis of these different approaches and attempt to introduce and discuss the most important results of research in the fields of sociology, anthropology and political sciences which impart a better understanding of a social dealing with risks.

The selective treatment of public issues in society and the forming of political arenas around social controversies (with arena-specific rules and groups) prevent us from speaking of risk evaluation in society. Risks are abstractions, allowing science, technology and politics to more easily classify phenomena according to systematic aspects and to reach a pragmatic consistency in initiating or evaluating actions. In daily life, however, notions with a medium range prevail, which interlink subjectively comparable situations without claiming to be universally valid. Such notions are valid for comparable situations and therefore depend on the context.

Thus, studies on individual risk perception have proved that most people have at least five different ideas of the term „risk", depending on the social context in which the term is used (Renn 1989a). Furthermore, social risk evaluation is not independent of the object which is the source of the respective risk. So, Lindell and Earle (1980), for example, report about an experiment in which test persons were to evaluate risks and benefits of different energy systems. Depending on whether or not the test participants were informed about which energy system was concerned in the evaluation, the risks, which were described with identical words (once with and once without naming the risk source), were evaluated quite differently. Even the numerical tradeoffs between dimensions of benefits and risks depended on the naming of the energy source (Lindell & Earle 1980). For these two reasons, it makes little sense to make generalized statements on the social perception of risk, since this perception depends on the context on the one hand and on the risk source on the other hand.

As a result, there is no comprehensive arena of „risk policy" in political reality; risk is rather an element of different social conflicts, which are mostly sparked off by concrete objects or activities. Radiation protection in the field of medicine, or exposure to natural radiation (for example by radon) are, despite their topical closeness to the problem of radiation protection with respect to nuclear power, two largely independent subjects with partly interconnected, but still autonomous political and social evaluation patterns. In a pluralistic society, the emergence of social conflicts and the respective definition of their topics cannot be regarded as a product of logically interconnected and topic-related affinities, but as the result of social processes which determine the contents and the scope of conflict topics (Coser 1965).

Even though we do not have to accept, from a normative point of view, such a historically grown separation into different political arenas, the analysis of the conflict field of „risk" has to start from the given social facts. For this reason, the structuring principle followed so far in this study, namely to regard radiation protection as a paradigm of risk management, has to be given up in the present section. Instead, it makes sense to choose a context in which social risk evaluation has exerted a significant influence on the social handling of conflicts and on political decision making. There is no doubt that the conflict about radiation protection has manifested itself with particular intensity in the debate on nuclear power. The knowledge that we can gain from the analysis of this debate clarifies the processes of social and political processing of risk information and, by means of a concrete political issue, shows the possibilities and limits of a risk policy orienting itself by scientific rationality.

4.2.1 Social Reception of the Concept of Risk

Although there have always been dangers to our health and to the environment by natural or technological events, risk has only lately become a public issue. The novelty of the „risk" issue can be put down to four factors:

(a) Humans have at all times taken precautions against hazards. However, the lack of anticipative knowledge made them regard negative events not so much as results of their own instrumental behavior but rather as „punishments from god" or as strokes of fate. With the increase in knowledge about cause-effect chains, society has instruments and institutions at its avail to foresee negative events and their consequences (anticipation) and to design or carry out corresponding counter-measures. Simultaneously, there is a growing moral demand to take risk precautions in order to rule out negative events or to limit them (Lowrance 1976; Renn 1984, 13ff).

(b) With the technological, medical and hygienic achievements, the relative portion of natural hazards (as, for example, infectious diseases) became smaller and the portion of man-made risks (due to technology, nutrition or leisure activities) grew larger. Traffic accidents, cancerous affections caused by smoking, an unhealthy lifestyle, and damage to the environment are regarded as dominant risk factors in modern industrial societies today. Such risks need explicit social or individual legitimation, because they are not natural but induced by human action (Hohenemser et al. 1983). In the case of natural risks, there is at most a moral obligation to minimize or prevent the negative consequences.

(c) Wide ranges of the evolution of technology are marked by a tendency to increase the potential of a catastrophe, while the probability of such an eventuation is reduced at the same time. The possibility of big disasters, no matter how small their probability may be, is consciously accepted in order to keep the individual probability of damage low on the one hand and on the other hand to benefit from economic advantages in the form of returns to scale (Perrow 1984; Renn 1984, 32ff). To travel by train is cheaper than to drive a private car, in terms of the economic use of resources, and it is safer with respect to the individual probability of an accident. Still, in case of a collision, the number of casualties is higher than normally occur in road traffic. This relation is far more dramatic in the field of nuclear power and large chemical plants. The increasing potential of a disaster with the simultaneously decreasing individual risk of damage requires collective decision processes (unlike the personal decision to take a risk) and therefore special attention to the distributive effects of risks (MacLean 1987, 572ff).

(d) Compared to the marginal utility of general health, a clean environment, and emotional well-being, the individual marginal utility of economic welfare has decreased against the background of economic prosperity and the great variety of consumables (Klages 1984, 107ff). It is all the more difficult, therefore, to justify risks for which the benefits are largely of an economic nature (Lynn 1986; Conrad 1981). The younger the persons interviewed and the more the economic situation of the individual is detached from the general economic situation (for example by state employment)[7], the more effective is this shifting of utility relations (Schröder et al. 1982, 248ff; Renn 1987, 81).

All four factors have contributed to the fact that risk has been recognized as a social problem and has gained political importance. With the improvement of forecasting techniques and the increasing moral self-commitment of modern society to limit risks, there is a growing demand of the public on social groups and most of all on political decision makers to actively shape the future and to react anticipatively to possible hazards caused by the natural and the technological environment. The increased variance of future development lines and the resulting instability of personal life plans intensify existential anxieties on the individual level and on the social level encourage political movements which have made it their task to limit the contingencies of social action (Lübbe 1987, Mitchell 1980). In Germany, safeguards against future hazards and foresighted risk management are therefore pivotal concerns of almost all population groups (Klages 1984, 82f).

Characteristic of the way society copes with the problem of risk is the selective spectrum of risks in which the discontent with modernity manifests itself (Berger et al. 1973). Social conflicts have general causes and entail consequences of general validity, but they are ignited by specific objects or issues. In the dispute about risks, it is, above all, nuclear energy and the connected accident and radiation risk on which, also symbolically, the conflicts about social risk management have vented themselves in Germany.

But why nuclear energy? First, an ideal interaction of the four factors mentioned above manifests itself here; second, the potential of a catastrophe was documented with particular intensity and concreteness by the dropping of the atom bombs over Japan; third, the intuitive perception patterns are especially sensitive to radiation risks, as was shown in 4.1; fourth, nuclear energy provides a good point to attack „big business" and the monopoly-like cooperation of state and economy; fifth, the symbolic position of nuclear energy for scientific progress and technological achievements was already ushered in by the „Atoms for Peace" program in 1954 and only had to be converted from technological optimism to pessimism;

and finally, sixth, local protests against nuclear plants acted as social intensifiers of the anti nuclear energy movement and gave it the desired publicity (Kasperson 1989; U.S. Office of Technology Assessment 1984, chapter 8; Paschen et al. 1987, 118ff; Renn 1984, 199ff). The fact that nuclear power became the pivotal point of the risk controversy can be explained on the basis of the historical development. Furthermore, this controversy is not confined to nuclear energy, but extends into other fields, such as chemical production methods, genetic engineering and the utilization of high-capacity computers.

The controversy about nuclear power also met with such a large response in society because, in this case, the canalization of political conflicts through the routinized mechanisms by which existing interest groups exert political influence, so typical for pluralistic societies, largely failed to happen. In Germany, at least during the first years of the intensification of the conflict, the debate on utilization and extended use of nuclear power went right across most political trends and could not be categorized in any of the existing political structures (Andritzky & Wahl-Terlinden 1978, 31ff; Renn 1985, 175). So, the issue of nuclear power exceeded the scope of the traditional absorption of issues by professional political groups, such as parties, for example. Neither the question, which social group would benefit from the expected additional profit resulting from the use of nuclear energy, nor the question of the power of management and control of nuclear installations was being discussed. Thus, the issue could hardly be positioned within the conventional German scheme of leftist and rightist political orientation.[8]

Traditional political conflicts are based on different opinions about plans concerning resources and their fair distribution, in other words, the allocation and sharing out of the socially produced wealth among the various parts of the population. On the macro-scale, a development of conflict issues can be identified which starts with the question of fair distribution of the socially produced wealth, leads from there to the conflict about distribution of economic power and social influence and finally results in the dispute about the distribution of risk consequences, i. e., about the question which risks may be reasonably imposed on different parts of the population. The nuclear energy conflict is a good indicator for this shift of conflict: neither the question who may participate in the wealth produced by nuclear energy, nor, apart from a few exceptions, the question on the right of disposal of nuclear power plants (keyword: democratization of steering functions) is being discussed. The crucial matter in dispute is rather the moral legitimation of a central type of technology with a high potential of catastrophes and an unjust distribution of risks in realation to social groups, regions and coming generations (Perrow 1984, 145; MacLean 1987; Beck 1986).

One consequence of this shift of the social conflict field was the fact that the existing channels of formation of political will were not adjusted to the new challenge. The lack of possibility to funnel the budding protest against nuclear energy into political processes and to shift it into representative institutions favored the development of spontaneous organizations in the early phase of the nuclear power conflict in West Germany. Mainly local and supraregional action groups and environmentalist organizations are to be named here. The dynamic power resulting from the combination of local protests with their publicity effect and of intellectual support by members of cultural elites stabilized the influence of the citizens' initiatives and extended their range of action towards other risk sources (Nelles 1984; Renn 1985).

The particularity of the debate on nuclear energy in Germany thus manifests itself in a new formation of social groups beyond the sphere of influence of political parties and interest groups. Since many people no longer found themselves represented sufficiently by parties or other organized interest groups with regard to the question of nuclear energy, new social movements developed which, on the basis of high symbolic integration (common value fund) and low role-specification (egalitarian steering structure), wanted to influence the result of a fighting out of the conflict (Raschke 1987, 21). The emergence of new and unconventionally operating groups increases the probability that the conflict might be fought out with more flexibility against the development of new rules and with more tolerance towards limited rule violations, due to the lack of role and norm traditions. Thus, the debate on nuclear energy proved to be a forerunner of new forms of expression and a new self-conception of the forming of political will within the political culture of Germany.

Modified political forms of expression lead to a qualitative change in the means and strategies that are used for carrying through political aims. According to polls among young people in Germany, unconventional methods of exerting political influence, which were introduced into German every-day political life by the movement of citizens' initiatives, are regarded as more attractive than conventional methods, such as elections, becoming member of a political party, or writing readers' letters (Sinus Institut 1983, 26ff).[9] The affinity between environment protection and life and health, the recruiting of the environment associations largely from members of the upper middle-class, who are able to express and justify their ideas better than others, and the political success of these methods made unconventional forms of passive resistance popular and at the same time legitimizable towards the „onlookers", who take part in the political events as media consumers (Kitschelt 1986, 286ff). We will later get on to the difficult question about the legitimacy and the legality of actions. For a description of the social reality, it suffices here to say that the environmentalist movement integrated the methods of protest and political exertion of influence which were developed and tested in the U.S. and in India into the political culture of Germany and made them a solid part of its strategic way of proceeding.

Such a diffusion of new forms of political articulation was only possible because the German governmental institutions of social and political control did not make full use of their power monopoly but tolerated partial infringements of the legal frame of political action, under public pressure and in order to avoid escalating conflicts. The question whether or not this was legally and politically justified is not the issue here. The decrease in loyalty to the government as a regulating body and the far-reaching social acceptance of demonstrative violations of rules helped weaken the rules of the political game, in the same way as the wide-spread perception of the government as a player bound by interests, who obviously promoted development and extended use of nuclear energy (Guggenberger & Offe 1984, 13ff).

The media and the general public in Germany, from the very beginning, took a lively interest in the new movement and the conflicts triggered by it. On the one hand, skepticism about nuclear energy was strong enough among the population to activate sympathy for the environmentalist movement. As surveys demonstrate, even supporters of nuclear energy were convinced of the importance and the necessity of a corrective for governmental environment policy.[10] On the other hand, the actions of the environmentalists were tuned completely to the selection criteria of the media: demonstrations, construction-site occupations, alternative congresses and similar events have a high actuality and news value for the media (Mazur 1984; Roth 1987, 77). Thus, an atmosphere of popularity and sympathy for the environmentalist movements developed. Apart from the undisputable objectionable environmental problems from which the population suffered and is still suffering, it was also the amplification by the media and the omnipresence of ecological problems in publicized opinion which made environment, risk and nuclear energy public issues in Germany which citizens had to deal with (Luhmann 1986).

Which conditions, then, mark the current state of the debate on nuclear energy? First of all, the conflict takes place in an extremely antagonistic way: on the one side, the environment associations, the „green party“, and a number of political and culturally active groups are fighting to stop any further use of nuclear energy; on the other side, industry, conservative circles and sections of the labor unions advocate a further use of this energy source.[11] The matter of dispute is not the question of „how" or „how much" but of „if at all". Here lies a main difference to the utilization of coal: forest decline and greenhouse effect prompted the environmentalist movement to demand stricter regulation of coal technology but by no means its abandonment.[12]

Finally, the arena of the nuclear power controversy has an international dimension: just as the development of the peaceful use of nuclear energy was laid out from the very beginning for international cooperation and control (IAEA) and led to international amalgamations on a scientific and an institutional level, the protest against nuclear energy was also an international phenomenon, which started in the U.S. and then spread to all countries that used or wanted to use nuclear energy (Mez 1979).

4.2.2 Actors in the Nuclear Power Conflict

The intensity of social conflicts depends largely on the numerical strength of the groups taking part in the conflict and the degree of their members' motivation to become active. Organizational factors, such as speed of reaction and media value, and receptive factors, such as affinity of the values and aims manifested in the conflict to generally accepted central ideas, also play a decisive role.

In the nuclear power controversy in Germany, these four factors have become effective in a special way. First of all, this issue has mobilized the German public stronger than any other issue before. Towards the end of the seventies, the national association of citizens' initiatives for environment protection (Bundesverband Bürgerinitiativen Umweltschutz BBU) had more members than the party strongest in number (Andritzky & Wahl-Terlinden 1978). As far as is known from surveys, the opinion spectrum concerning the nuclear-energy issue has become polarized more than is the case for comparative issues such as nature conservation, peace, or women's liberation. The number of indifferent persons is low, the numbers of firm opponents or staunch proponents of the anti nuclear energy movement are high and almost balance one another.[13] In a survey conducted by the Sinus polling institute among young people in Germany, it was again only nuclear power that led to a clear polarization in opponents and proponents, whereas other issues, such as the peace movement, met with widespread sympathy, or, such as squatting, with widespread rejection (Sinus Institut 1983). After the Chernobyl accident, the opinion spectrum shifted considerably in favor of the opponents of nuclear power, but recent survey data indicate a revival of supporting stands (Wiedemann & Jungermann 1989; Hohenemser & Renn 1988). The polarization effect is still clearly existent after Chernobyl.

The high mobilization potential has, on the one hand, psychological reasons, which were already discussed in detail in section 4.1; on the other hand, social factors which result from the special composition of the two antagonistic groups play an important role. From which social groups do the camps of proponents and opponents of nuclear power recruit their supporters?

The large number of opinion surveys on nuclear energy allows us to draw a typical profile – if only as a tendency – of proponents and opponents of nuclear power. The typical proponent of nuclear energy is middle-aged (between 40 and 55 years old), trained in a commercial or a technical profession and often works in private businesses (Becker et al. 1980; Frederichs et al. 1983; Heller 1980; Wiedemann & Jungermann 1989). An opposing attitude towards nuclear energy can often be found with people who have an academic degree, mainly in the humanities, with persons who work in cultural professions, with state employees and with people between 25 and 35 years of age (ibid. and Renn 1984). Persons who work in a technology-oriented profession tend to be in favor of nuclear energy, and all the more so, the higher their social level is (Wiedemann & Jungermann 1989; Wieland 1987). Conversely, it can be said for persons who work in the services sector that they tend to take an opposing stand towards nuclear energy. For persons occupied in non-technical professions and in jobs that have little to do with commerce, the opposing attitude also grows stronger with increasing social level. Finally, women are far more skeptical about nuclear energy than men.

Opponents of nuclear energy largely believe that the question of the use of nuclear energy is a political matter and should only be decided by political bodies (and not by technological or scientific institutes) (Institut für Demoskopie Allensbach 1984). They feel primarily obliged to environment protection and they are skeptical about the capability of technology and science to solve the problems we are faced with (Frederichs et al. 1983; Peters et al. 1987; Wiedemann & Jungermann 1989). Proponents of nuclear energy, on the other hand, express relatively strong confidence in the efficiency of technologies and in the competence of existing political institutions. They share rather conservative opinions on politics and society (Renn 1984; Heller 1980). Like the opponents of nuclear energy, they are convinced of the importance of environment protection; yet they regard nuclear energy as a risky, but still environmentally beneficial and economically necessary energy source (Renn 1984; Thomas et al. 1980). The classification of nuclear energy as a „risky“ option of energy supply is shared by both sides; the proponents, however, see economic and technological benefits, which, in their opinion, prove a positive cost-benefit balance, whereas most of the opponents neither believe in the safety, nor in the low cost, nor in the technological necessity of nuclear energy (Renn & Swaton 1984).

The puzzle of the survey data reveals a relatively homogeneous picture: social groups which take a rather opposing stand towards the economic and political institutions and are hardly convinced of their efficiency regard the fight against nuclear energy as an important political outlet to convince others of the necessity of a renewal of the institutions.[14]

These groups do not make use of the nuclear power conflict for their aims, as the popular explanation of the free-rider suggests; the fight against the nuclear energy policy practiced in Germany is rather an integral part of their political objective to change the political system in a revolutionary or evolutionary way. The position of nuclear energy as a symbol for the existing economic and political conditions makes it an ideal point of attack and a crystallization point integrating all groups which consider a renewal of the system necessary. It is not without good reason that all radical German political movements, rightists, leftists and greens, have integrated the fight against nuclear energy into their political programs. According to the representatives of these movements, the example of nuclear energy serves best to document to everyone the cynical abuse of power on the part of the existing decision-making authorities.

The recruiting of activists from politically radical movements has definitely increased the impact of the anti nuclear energy movement, but it is by no means the decisive factor for the popularity and the attractiveness of the new social movement. With increasing awareness of damages to the environment, with the higher popularity of postmaterialistic values (Fietkau et al. 1982) and growing uncertainty about the social controllability of complex large-scale technologies, the opinion climate has also changed profoundly among those who are content with the political system in Germany: an opposing attitude towards nuclear energy has rather become the social standard, and persons with proposing stands are now forced to justify themselves in every-day life (Lynn 1986; Institut für Demoskopie Allensbach 1984). So most of the opponents of nuclear energy are by no means political radicals or people out to change the system, but citizens feeling insecure and disappointed, who regard the decision of the political institutions in favor of nuclear energy as a „one-time mistake", but not as a general failure of the institutions. Even this majority of the opponents of nuclear energy, however, does not so much base their arguments on the instrumental disadvantages of the use of nuclear energy but is convinced that in the development of nuclear energy the historically grown congruence of interests between government, economy and science has annulled the necessary social control of an application of this new technology.

The proponents, on the other hand, are persons who regard just this alliance as a guarantee for a rational technology policy. The cooperation of these institutions to manage a complex technology seems to them to be a logical necessity to technically and organizationally limit the disaster potential of nuclear energy, which they, too, are aware of. Their confidence in the efficiency of the political-administrative system and their faith in the limitation of risk by means of further research and technological optimization make nuclear energy a paradigm of a life-world shaped by technology and rational organization. For them, the furthering of nuclear energy means a painful, but very rational compromise between economic prosperity, environment protection and security (Renn 1984, 300ff; Renn et al. 1985, 134ff). They, too, attach a symbolically high significance to nuclear energy and see in its use a political demonstration in favor of high-capacity technological systems and organizational efficiency.

This representative role of nuclear energy is the reason for the strong polarization effect: both sides link nuclear energy to different social concepts of a future structuring of the industrial society. Radical groups excluded (which are largely confined to the opposing side), it is not so much basic values and political convictions which divide opponents and proponents from each other but rather differences in the way these values are transferred into concrete action strategies for the future.[15] Due to their experiences with technology and their stronger attachment to the economic criteria of weighing goods against each other, proponents of nuclear energy see the values of environmental friendliness and minimization of risk ensured exactly in the application of this technology, whereas the opponents regard these values as endangered on the basis of their perception of lacking controllability and the oversize of this large-scale technology (Renn 1984, 302ff; Rayner & Cantor 1987). The fact that people with similar values and similar individual environments may come to diametrically opposed opinions on a question which both sides consider important amplifies the lack of mutual understanding and nurtures a stigmatization of one group or the other (Slovic 1987).

Cultural anthropologists have interpreted these different association patterns about nuclear energy as signs of two different cultures within modern industrial society (Douglas & Wildavsky 1982; Douglas 1982; Rayner 1987; Thompson 1980). The stance of the nuclear energy proponents is typical for the political group of „entrepreneurs“, who regard risks as necessary costs for further development and improvement of the quality of life. On the other hand, the „egalitarians“ are of the opinion that a society should provide an equal security level, the highest one possible, for everyone, even if this goes at the expense of wealth and economic development.[16] The first group regards technological innovation as a substantial guarantor of social progress, whereas the second group considers the uncertainty connected with innovations as socially intolerable, as long as consequences cannot be clearly estimated. At the same time, the „entrepreneurs" prefer to use analytical methods; as they believe that complex problems can be solved satisfactorily by splitting them up into simple problems which are aggregated afterwards. The „egalitarian" group, however, prefers holistic methods, in order to view complex problems in their entirety and to solve them, if possible. Under these circumstances, it is not surprising that complex technological systems constitute a personal and a social challenge for the group of „entrepreneurs", the successful managing of which brings them personal satisfaction and prestige in the social context, whereas, for the group of „egalitarians", the same systems lead to problems too complex to be understood clearly and to a perception of social control systems as being overtaxed.

The approach of cultural anthropology has met with much approval in the U.S. (Douglas 1982; Douglas & Wildavsky 1982; Covello & Johnson 1987). The intuitive plausibility of this explanation pattern must not, however, hide the fact that sufficient empirical proof for this theory could not yet be produced (Kasperson & Palmlund 1987). At the same time, critics hold against this approach that cultural differences between the two positions are not distributed exclusively to different social groups but are embedded in most individuals concurrently (at least in pluralistic societies), depending on their role and their social situation (Johnson 1987). Furthermore, the cultural-deterministic view of this approach has been criticized (Nelkin 1982).

Despite these reservations, the cultural-anthropological perspective can be regarded as a precious supplement to the results of the research on values and attitudes, since it creates a context of meaning between the different world views of individuals. So, positions on nuclear energy are not so much based on ideas about perceived instrumental advantages and disadvantages of this technology; they rather stem mainly from different opinions concerning the responsibility of social institutions to society and the desirability of options of social change. With respect to the discussion on environmental standards, the cultural-anthropological approach may offer an important explanation for the fact that social groups which can rather be assigned to the „egalitarian" side will, for fundamentally methodical aspects, reject results of balancing considerations on risks and benefits according to a means-ends rationality, no matter how plausible and clear the results may turn out to be (Renn 189b).[17] For this reason, it is important to take care that the different approaches to justify options of action are integrated in the process of decision making when establishing environmental standards.

4.2.3 Mobilization and Readiness to Take Action

Apart from the high mobilization effect, the intensity of commitment is another important characteristic of the conflict on nuclear energy. In the opinion of the German population, other issues, such as garbage disposal or water pollution are comparably as acute matters of concern as the use of nuclear energy (Fietkau et al. 1982). Nevertheless, there is not – apart from local protests – the intensity of commitment here which is so typical for the nuclear energy debate. The readiness to take political action in favor of one's own point of view is considerably more pronounced among opponents of nuclear energy than among comparable protest groups. Early in 1986 (i. e. before Chernobyl), two per cent of a representative sample of the West German population called themselves activists in the anti nuclear energy movement, 17% classified themselves as potential members and 20% as sympathizers (Watts 1987, 53). If we look at France for a comparison, the activation potential is considerably lower there: none of the representatively chosen 1003 French citizens classified themselves as anti nuclear energy activist, 5% called themselves potential members and 16% sympathizers.

Apparently, the mobilization potential in West Germany is large enough to motivate a critical mass of committed citizens to actively put their point of view into politically effective practice. The border between an uncritical mass, whose desires may largely be ignored by political decision makers without risk, and a critical mass, which requires a reaction of the decision makers, cannot be determined clearly. From empirical studies, it can be derived as a rule of thumb that approximately 5% of all citizens with a certain point of view about some problem are ready to actively support it, and approximately 1% actually do so in concrete situations (Wicker 1979). This percentage can turn out higher if the impression prevails that individual commitment can truly achieve something and, conversely, can decrease further if either other professional opinion elites are thought to sufficiently represent the cause or if individual commitment seems pointless. The wider-spread the support of a certain point of view in the population, the greater the numerical strength of the potential activists will be and the more likely it is that (ceteris paribus) it will come to an organized form of political action. With a population proportion of 2% activists and around 17% potential members, the conditions for a critical-mass basis of the anti nuclear energy movement are undoubtedly met in Germany.

Moreover, aggravating the conflict, the opinions of German citizens about nuclear energy are heavily polarized and the intensity of these opinions finds expression in a greater readiness for political activity. There is a positive correlation between the intensity of one's opinion and one's readiness to actively commit oneself to carrying through one's point by means of visible actions. In principle, this correlation exists both for proponents and opponents of nuclear energy, but it is far more pronounced among the opponents. In Germany, the personal readiness to take action is double as high among convinced opponents of nuclear energy as among convinced proponents (Renn 1984; Frederichs et al. 1983). This asymmetry has helped the opponents to get more publicity and has thus reinforced the impression in the general population that a considerably larger number of citizens has a critical opinion of nuclear energy than is in fact the case.[18]

How does it come to this asymmetry? German studies on opinions about nuclear energy suggest that the opponents' higher readiness to take action is based on four factors (Wieland 1987; Wiedemann & Jungermann 1989; Renn 1984): first, opponents of nuclear energy do not feel themselves sufficiently represented by other social groups and therefore see a more urgent necessity to become active than proponents, who feel themselves in accord with many social and political groups. In the course of the growing skepticism of many groups about nuclear energy, this motivation for action has recently become less relevant. Second, opponents hardly ever express doubts about the correctness and the moral superiority of their position in opinion polls, whereas proponents often deal with the inner conflict between perceived risk and perceived economic necessity and assert their position rather defensively. Third, the opponents believe that there are real alternatives to nuclear energy, which come off better in all dimensions of evaluation (profitability, safety, environmental friendliness). In their opinion, there is just a lack in political will to help these better alternatives to reach a breakthrough. Proponents of nuclear energy, on the other hand, do not regard the nuclear energy source as a superior form of energy supply compared to other options but, at best, as the lesser evil (Renn 1984, 246; Renn et al. 1985, 136). Accordingly, their motivation of action is less pronounced. Both groups, however, cling to the symbolic significance of the nuclear energy debate with great emotional intensity, and actively – if with different intensity – support their points of view. Finally, the number of those people among the opponents of nuclear energy who are politically action-oriented anyway, mainly young people with postmaterialistic values, is above average.

A high mobilization potential in the population and a high readiness to take action of the groups participating in the conflict are two significant preconditions for the emergence of social conflicts. The intensity of such conflicts and their temporal persistence moreover depend on three other factors: the degree of public support by social groups which are not directly involved in the conflict (and by the general public), the reception of the conflict by the media, and the role of governmental institutions in handling the conflict. In the following, these three factors shall be discussed in detail.

4.2.4 Roles of Government and Science in the Controversy about Nuclear Energy

The complexity of nuclear technology, the particular sensibility of the population for radiation risks, and the pivotal importance of an ensured energy supply for the national economy have favored a close cooperation between government and industry from the beginning of the commercial use of nuclear energy (Keck 1984). This cooperation was still intensified by the assessment of nuclear energy as symbolizing the international prestige of a country as a modern industrial nation. Seen from a historical point of view, there was the additional aspect that the development of nuclear technology in the U.S., originating from the field of governmental armament production, was transferred to governmental research and from there to private economic application (Deubner 1977).

These interlinked public and economic interests led to a dilemma: on the one hand, the government was the highest control body for regulation and supervision of nuclear facilities, on the other hand, it promoted these plants by means of governmental research and subsidies. In the U.S., this conflict became manifest with the dissolving of the Atomic Energy Commission (AEC). The control function of the AEC was assigned to the Nuclear Regulatory Commission (NRC); the function of promoting nuclear energy was allocated to several institutions and coordinated by the Department of Energy (Freudenburg & Baxter 1985). The crisis of public confidence in the governmental function of control and supervision has had after-effects on the trustworthiness of the NRC up to this day: according to opinion polls among the American population, the NRC is considered industrially friendly and immoderately generous when it comes to interpreting safety regulations (Farhar-Pilgrim & Freudenburg 1984).

The different functions of the government in handling the nuclear conflict gave rise to the impression of a conflict of interests between the roles of the government as licensing authority for nuclear installations and as an advocate and promoter of nuclear technology. The subjective perception of a monopoly-like bloc of government, governmental research and industry led to a loss of confidence in the objectivity of the government as a mediator between the interests of the nuclear industry and the environment associations. The environment associations were – virtually without any effort – able to use the sympathy and thus the solidarity of the public as a resource to „unmask" the government as a protagonist bound by interests. The government which, with a democratic constitution, depends after all on the solidarity of the members of society, reacted with an abundance of new regulations, laws and obligations, which all additionally complicated the licensing procedure but did not bring about the expected solidarity effect.

Industry saw itself placed in a dilemma by these actions: on the one hand, the opinion backed by all governments up to then that the use of nuclear energy was necessary for Germany's energy supply, helped it to legitimize its commitment to this matter and to maintain a limited amount of solidarity by the population; on the other hand, the new obligations were so expensive that the profitability of energy production by means of nuclear technology became more and more questionable. A fight against the regulations could easily have led to the loss of an ally considered important in the opinion of industry. In the U.S., this paradox situation led to a reversal of the ratio of profitability between nuclear energy and coal in many states. The investment expenses for all required safety conditions as well as the interdiction to pass the construction costs of nuclear power plants to the energy price during the construction period (together with a high interest level), made nuclear energy no longer competitive compared to local coal in most states (Cook 1985; U.S. Department of Energy 1983). Since 1978, no nuclear power plant has been built or commissioned in the U.S.

Although this threshold has not yet been reached in Germany (mainly because of the higher domestic coal prices), the average price of a nuclear power plant with constant capacity has quadrupled in the past 15 years.[19] This increase in price is certainly not only a result of the regulations and obligations alone, even if they doubtlessly contribute very much to this development. Nuclear industry and the energy-supply companies were understandably careful in using all their available resources against the flood of governmental regulations, because they did not want to lose one of their most important allies in the controversy on the acceptability of nuclear energy. Both allies, however, lost in this competition: the government, as the highest regulating authority, could not reach its aim of acquiring the citizens' solidarity with its aspired role as a fair but strict controller of nuclear technology; industry had to bear the financial consequences of the marriage of convenience without scoring any appreciable gain in prestige from its cooperation with the government.

Of course, it is problematic to speak of the government as an actor in this context. The German state governments have backed other opinions than the federal government. The parliaments react more sensitively to the new environment problem than do most of the administrations. Above all, however, courts play a role. Decisions on sites and licensing procedures are often delayed by administrative lawsuits, which raises the total costs of the respective plant considerably and at the same time signals to the conflicting parties the possibility of a course correction (Mutschler 1986, 1037ff).

Despite varied attempts to gain back confidence by means of public relations work and public self-portrayal, nuclear industry and the users of nuclear plants have not succeeded in mobilizing the proponents of nuclear energy more intensively for their own purposes and to meet with wide support in the population. There are many different reasons for this: the above-discussed symbolic status of nuclear energy as to large-scale technological risks, the perception of lack of governmental control, the alleged or real concealment of accidents, and the intuitive fear of radiation as a danger source may have played a part here. According to surveys in West Germany and the U.S., confidence in the nuclear industry is far weaker than in almost all other branches of industry (Renn & Levine 1988). A careless information policy in conjunction with exaggerated expectations in the beginning of the commercial use (for example concerning environmental friendliness and inexpensiveness) have contributed to this, just as have structural problems of internal control.

At the same time, however, the environment associations were not able to reach their goal of warding off nuclear plants in Germany either. They have succeeded in mobilizing a critical mass and in largely being respected by the population for their commitment, but none of this was enough to exert sufficient pressure on the government as the supreme instance of power and to push through an abandonment of nuclear energy (unlike in Austria and partly in Sweden, for example). The extension program in Germany has been slowed considerably and the costs for nuclear energy have increased significantly, but there is neither a moratorium for nuclear power plants, nor an official decision to withdraw from nuclear energy. The main reason for the relative failure of the anti nuclear energy movement in West Germany is the fact that voters do not regard the nuclear energy issue as an important element of their election decision (unlike in Sweden before the plebiscite, for example) and that at the same time, the convinced opponents of nuclear energy among the population are counterbalanced with a relatively large number of convinced proponents (Wiedemann & Jungermann 1989, 25).

In conflicts in which none of the parties involved are strong enough to decide the conflict in their favor and a strong polarization of opinions rules out the reaching of a compromise, the conflicting parties look with special intensity for allies, who offer their support on the basis of value congruences or common interests („If you scratch my back, I'll scratch yours"). Social groups or individuals who have high moral authority (e.g. churches), or enjoy a high social prestige (e.g. Nobel-prize winners), are particularly courted, since their support of the respective conflicting party means better chances in the political arena in which the conflict is carried out. In Germany, opponents of nuclear energy looked for and found support from many groups which, on their part, distrusted the governmental and economic institutions. At the same time, a wide-spread solidarity of culturally active or creative groups with the anti nuclear energy movement developed.[20]

On the other hand, the proponents of nuclear energy tried to mobilize for their purposes groups with a more economic orientation. The government, as the representative of nuclear research centers, mobilized the scientific world in the struggle for social support. Science still enjoys great social respect in the German population and is, in most cases, regarded as being unbiased and only committed to the cause (Infratest 1985). The expected legitimation effect of deploying scientists in information programs such as the „Bürgerdialog Kernenergie" (citizens' dialog on nuclear energy), however, did not take place (Weingart 1979). Instead, the research centers and the engineering sciences came under pressure of legitimation: the politicalization of scientific research destroyed the impression of an „objective" ascertainment of knowledge, and the so-called scientific controversies, in which professors of both sides accused each other of being unscientific and corruptible, mostly left the audience with confusion and anger (Rip 1985). The sacrificing of scientists in television-effective controversies neither served its purpose of providing governmental action with additional legitimation, nor did it improve the general knowledge level of the German population (Nowotny 1979).

The nuclear energy conflict in Germany is a prime example of the fact that none of the actors have reached their goals in their competition against each other, but nevertheless they have changed their own social role and the political atmosphere for good through often unintended side-effects of their actions. A new political party, the „Greens" has emerged, unconventional methods of the formation of political will have asserted themselves on a large scale, scientists, mainly those from the field of nuclear technology, have lost in public respect and, in public opinion, the government has lost in its authority as a regulating body.

Environment protection is firmly established in the canon of values of the German population now, and all parties are forced to consider it if they wish to retain the support of their electors. Maybe these changes would have come about even without the nuclear energy controversy. Yet this is a purely hypothetical speculation: nuclear energy and its risks have acted as forerunners for the described changes and have substantially influenced the political discussion on risk management in society.

4.2.5 The Role of the Media as Social Amplifiers of the Conflict

Conflicts in an open and pluralistic society are usually carried out on two levels: on the one side there is a political arena (a parliament or a commission, for example) in which the conflict parties deal with each other directly, on the other side the conflict parties avail themselves of the media in order to make their position known in the public and – if possible – to gain sympathy and support for their matters.[21] At the same time, the media report about the official arena and thus influence choice and relevance of publicly effective issues (gatekeeper function). By means of selection and amplification of the respective events, the media largely determine the priorities in the political arena (McCombs & Shaw 1972) and convey to those not involved in the conflict second-hand information, which always has a subjective touch, even if the respective journalists feel committed to objective news coverage (Sood et al. 1987, 30; Peltu 1985, 129f).

In contrast to reproaches often uttered by the protagonists in the German nuclear arena that journalists were either supporters of the „green party", as the one side says, or defenders of the status quo who are blinded (or even corrupted) by the Establishment, as the other side says, almost all empirical studies concerning this issue show that selection and processing of news are far more determined by professional standards of journalism than by ideological beliefs of the individual journalists.[22] These standards largely hold for all media, partly they are medium-specific (Battelle 1978; Peltu 1988; Mazur 1987; Rankin & Nealy 1979).

It is part of these standards that media normally take up current events and usually leave out continuing developments. The greenhouse effect as a global threat to mankind only then becomes news when cows die of thirst in the U.S. or when the German society of physics approaches the public with an insistent appeal. This general selection criterion favors the environment groups in the news coverage on nuclear energy, because, by their actions, they produce events which are worth reporting, while the continuing improvement of safety technology or the tightening of environmental standards have no news value. Closely linked with the quantity of news coverage on an issue is the idea that the topic has to be politically controversial, simply because of the abundance of information in the media, and that therefore special caution is required (Mazur 1981).

Just as important is the selection criterion of conflict and casting the blame. As American studies imply, the intensity of the news coverage on disasters is not so much determined by their physical impacts (such as the number of casualties or the loss of property), but rather by the intensity of the conflict about the necessary risk management and by the possibility of partially casting the blame for the event on someone (Adams 1986; Rubin 1987). Moreover, the local proximity to the disaster is, of course, relevant. Media reflect social events, not so much physical effects. So, if the participants in a conflict situation argue about the necessary forms of risk management, as this was the case after the Chernobyl accident, or if they blame each other for negative events, then these social events are important triggers and amplifiers for the news coverage of the media (Kasperson et al. 1988; Renn 1991). Whether the argument is only a mock fight or whether the reason for the argument is an „objectively" minimal risk hardly matters for the media-suitability of an event. Media react to the social construction of reality (Berger & Luckmann 1969; Seiderberg 1984) and not to „reality itself" or its scientific perception (Wilkins & Patterson 1987).

Applied to the nuclear energy debate, this selection mechanism means an amplification of the conflict nature of the controversy and a moralization of the actors in the arena. In their search for conflicts, journalists soon find in the nuclear energy issue what they are looking for. Yet they can neither judge the validity of positions in terms of scientific reasoning nor the representative nature of a diverging opinion (Peltu 1985; Sharlin 1987). Thus, the impression is created in the public that all statements concerning nuclear energy are controversial and that the effects of small radiation doses on health are assessed differently in the scientific world. From this perspective, the step towards moralization is only a small one: whoever backs a technology the effects of which are so controversial can only be driven by selfish motives.

The selection criteria described so far clearly favor the environment associations and other nuclear energy opponents, because they accord them legitimation as „proper" adversaries and moral authority. Other selection criteria, on the other hand, work the other way round: the search for extraordinary events draws the attention of the media to spectacular actions of chaotic and militant persons who only make up an insignificant section in the spectrum of the environmentalists. Their actions, however, are placed in the center of a report on a demonstration, with the result that many media-consumers are „appalled", turn away from the environmentalists and identify the entire movement with these riots (Guggenberger 1987, 330f). At the same time, empirical studies in West Germany have shown that many newspapers and magazines accept without further examination journalistically well-featured statements to the press on the part of industry and governmental bodies, but omit those on the part of environment associations or reassure themselves towards authorities or industry representatives first (Peters 1984; Peltu 1985).

Selection criteria are unwritten laws, which journalists acquire in the course of their training and which are in force across all ideological camps. Apart from that, political loyalties are, of course, also relevant, especially in Germany (Köcher 1986). These loyalties are shaped more by the preset direction of a newspaper or of an electronic medium than by the values and political preferences of individual journalists. Conservative or progressive newspapers select news according to their target group. Interestingly, with respect to the nuclear energy issue, it can be seen that the printed media take relatively balanced approving or rejecting stands to nuclear energy and show no biased favoring of one side (Battelle 1978). The idea which the public gets about nuclear energy is thus marked out more by general selection criteria than by political preferences of individual media or reporters.

4.2.6 Public Opinion in the Nuclear Arena

Mobilization potential and intensity of the conflict also depend on the reception of the respective issue by the politically non-active audience, which influences political decisions by showing sympathy or by silently tolerating political actions. Not least, opinion research has helped decision makers to know the barometer of the moods of the „silent" majority and to align themselves to it. So, in order to carry through political programs, not only the solidarity of important political groups is required but also the support of a large enough number of citizens. This does not have to be the majority of the citizens, but without a certain sympathy of a broad public for an intended step, even a parliamentary majority decision can hardly, or not at all, be carried through (Sahner 1986; Steffani 1986, 577ff). Thus, political conflicts are indirectly influenced by the uninvolved spectators, be it because the conflict parties want to be sure of the support of large parts of the population, be it because the legitimate decision makers take the public opinion ascertained by surveys into account.

What are now the aspects according to which the population grants support or even sympathy to individual conflict parties or positions in a controversy? Individuals in society seldom have the possibility to verify the correctness of assertions in the press on the basis of their own experiences, and they also lack the knowledge or the intuitive insight to evaluate the truth content of statements of the various actors in the nuclear energy debate (Lynn 1986; Nelkin & Fallows 1978; Nowotny 1976). This is all the more difficult as the effects of radiation doses can be of a stochastic nature, thus not allowing of conclusions about individual cases.

Due to the over-abundance of conveyed knowledge and the impossibility of checkung the truth content of some information, individuals have to develop their own selection criteria in order to accord or deny one or the other source trustworthiness in the case of contradicting information. Since the mid-sixties, it can be observed that the attribution of trustworthiness increasingly follows so-called peripheral signals.[23] This means that characteristics or heuristics which are not related to arguments are used to assess the credibility of the source or the argument. Examples of such characteristics are the length of the report, its position in a newspaper, the prestige of a medium, the form and the make-up of the message (glossy paper, for example), the personal charisma of the communicator in the case of telecommunication, or the way an actor treats his or her opponents (Eagly & Chaiken 1984). Particularly relevant to the nuclear arena are the following heuristics:

(a) Possibility of identification: The more receivers of news can identify themselves with a communicator, that is, recognize values, attitudes, gestures and modes of expression in him or her which concur with their own or correspond with their ideal of themselves, the more likely they are to believe this person or group of persons (Lee 1986; Eagly & Chaiken 1984). „Brave“ environmentalists who endanger themselves for the sake of preserving the environment (such as Greenpeace activists) are usually more likely to trigger identification processes than senior ministry officials or business managers who insist on pragmatic values and obviously speak for their own interests. On the other hand, business persons are often better able to represent common sense.

(b) Search for suspected motives: If the truth content of statements cannot be assessed, then at least one can speculate on why an actor obviously stands for one or the other opinion. The more interest bonds are suspected as motive of a communication, the lower the trustworthiness of a message is rated (Chaiken & Stangor 1987; Renn 1986). This mechanism, of course, leads necessarily to the fact that institutions which are obviously interest-bound (e.g. industry) have a worse starting position in the public contest of opinions and arguments than institutions which are only latently linked with interest groups or value groups (e.g. environmentalists).

(c) Consent with one's own interest: If the message is beneficial for the receiver, that is, if the receiver sees the possibility of a personal advantage in the statement, he or she is more likely to suppose that the source is trustworthy (McGuire 1985). In the nuclear arena, industry and government frequently have a better hand, because economic interests are often latently rated higher than social or moral commitment to the environment.

(d) Moralization of factual problems: The more factual knowledge a certain issue requires and the less comprehensible the context of the argumentation of rivaling actors is, the more difficult it becomes for individuals to investigate this subject-matter and to form their own opinions or attitudes. If they believe that one of the social groups involved represents their interests, they can cross off the issue and hope for an adequate delegation of their interests. As soon as citizens no longer feel represented by certain interest groups, however, and lack plausible signals for assessing the motives of the actors, or if these signals are inconsistent, moral assessment of the actors and of their positions plays an important role (Scheuch 1980; Renn 1986). Usually, nuclear energy opponents can more easily convince the audience of the moral integrity of their arguments.

It is exactly the moralization of factual problems which shows that the arena and its actors react to supposed or real sympathy effects of the population. For the actors, as well as the media, adapt themselves to the receivers' reduction of complex factual statements to moral evaluations. When certain technological, economic or political factual questions are elevated to moral questions, detailed knowledge is no longer relevant (and is mostly even disturbing). In a moralized controversy, points are scored in the public debate by appeals, the casting of blame, real or mock consternation and moral indignation. As mentioned, these are precisely the social events which the media react to and which they pass on to their consumers. The moralization of positions is thus the more intensified by arena conditions the less factual arguments can be examined for their correctness and the more uncertainty there is about the true scale of the risk (Renn 1986).

Moralization of factual problems and positions is employed by all contenders in the nuclear arena. As regards the effects on the receivers, however, environmentalists seem to be more successful here. Of course, nothing can be said in principle against a moral evaluation of technological effects. Yet, if moral argumentation is used as a substitute for technological expertise, interest conflicts can no longer be settled by consensus, because compromises between „good“ and „bad“ cannot and must not exist. Conflicts are replaced by religious wars.

4.2.7 Social Evaluation and Rational Finding of Standards

Events or social evaluations become public issues in a society only when important actors take the chance to make use of them as a means to improve their being equipped with social resources[24], such as power, prestige, knowledge etc. and when, at the same time, the media classify this process as worthy of becoming information. This is the more likely the case the more the opinions in a political controversy have hardened and polarized while at the same time public institutions can no longer solve the conflict by internal routines (Peters 1986). Analysis of the nuclear energy conflict has shown here that perceived instrumental advantages and disadvantages of nuclear energy are far less important for a social mobilization of groups and individuals than symbolic associations linked to this energy source. Positive attitudes towards nuclear energy are accompanied by a positive attitude towards political institutions of technological and social control and by a preference for analytical utility considerations. Conversely, negative positions towards nuclear energy are usually linked to a holistic way of looking at problems and to a general mistrust in the functionality of complex technologies and political organizations. These two contrary world views clash all the more as governmental regulation authorities and „established" social groups and parties have not been capable of channeling the conflict potential and of integrating it into the existing routines of political decision making.

From the analysis of the social and political processes which lead to the described conflict situation, a number of important conclusions can be drawn concerning the issue of rationality of environmental standards and of their establishment. First of all, it has probably become clear that the problem field of risk in terms of technological-scientific risk analyses played a subordinate role in the emergence of the conflict and its development. Even if all conflict parties reached a consensus regarding the risk (which they will probably not do, for tactical reasons), it would be highly unlikely that they would thus also change their positions to nuclear energy, since the symbolic assignment patterns would remain valid, even with modified risk perception. How can, under these conditions, rational procedures be initiated in order to establish the acceptability of risks and to limit potential dangers in a way which is adequate to the problem?

The simple solution, to leave the conflict about environmental norms to the play of political forces, may indeed increase the acceptance of political decisions but hardly increase their acceptability. Social acceptance is an important component of acceptability, but not the only one. Yet how can a procedure of setting standards be institutionalized which corresponds with the social and political requirements of politically reaching a consensus and which, at the same time, meets the criteria of normative acceptability?

First of all, it is obvious that in democratic societies a model of a „benevolent" dictator can just as little claim normative validity as can a confinement to science as decision-making institution (Fiorino 1989; Rushefsky 1984, 140ff). Value commitment, economic efficiency, social acknowledgement and legal execution of power are forms of social formation of consensus which have developed in the social evolution and can claim normative validity, provided that they are rationally integrated in a decision procedure. Therefore, a procedure for the social dealing with risks has to be found which does not annul or replace by authoritarian decision processes the effectiveness of the politically developed decision structures and of the political culture, but which reconciles the conditions of the political culture with the requirements of a rational decision process. Here, we refer less to the utopian and mostly frightening social technologies of decision making, but rather to the creation of social conditions which make a rational discourse on risk policy feasible (Dietz et al. 1989; Fiorino 1989).

In the field of nuclear technology, it is mainly two studies on the social compatibility of energy systems which attempted to provide answers to the question about a rational and democratic risk policy. The first study, conducted by Meyer-Abich and Schefold (1981 and 1986), set itself the task of identifying and classifying the objective conditions and consequences of the various energy systems with respect to the social order.[25] Criteria for this analysis were derived from the German constitution and from the generally accepted values in society. Accordingly, the properties of the different energy systems, especially of nuclear energy and solar energy, were objectively evaluated, yet unlike in technological risk analyses, the dimensions of the evaluation were oriented to social categories, such as flexibility of life-style, preservation of individual freedom, etc. (Bauerschmidt 1984).

The second study, which was conducted by the research center Jülich (KFA), was oriented more to the question of subjectively experienced violations and realizations of values (Renn et al. 1985; Renn et al. 1986). It was the aim of this study to examine the consequences of different energy scenarios as to their correspondence with the prevailing value conceptions and objectives, with the help of a random sample of citizens. For this, in a first step the evaluation criteria were obtained empirically, namely by interviewing interest groups, the second step was to ascertain the consequences of the energy scenarios as objectively as possible by interviewing experts, and the third step was to present the property profiles of the different scenarios to „citizen juries" for an evaluation. Despite the different approaches, both studies attained similar results. Both studies had in common that they regarded expert assessments alone as not sufficient to make risk-policy decisions. They rather considered a procedure to be necessary that couples rational decision making and participation.

The question about participation is one of the central questions of risk policy. Self-appointed speakers of the population can neither represent the people, nor can they adequately articulate the concerns of the persons affected. Opinion research is completely overtaxed with this task, because feedback requires information. Only if a person knows about the consequences of a measure, can he or she assess its desirability. Being informed is an absolutely imperative precondition for proper assessment of probabilistic consequences. Therefore, an instrument is required which combines information and representation.

It cannot be the purpose of this chapter to describe such instruments in detail. A suggestion for this will be formulated in chapter 6. Here, we can at best point to a number of conditions which have to be fulfilled to reach a rational risk dialog. First of all, it is obvious that a procedure for establishing standards which is based on decision-theoretical principles and scientific data will not necessarily be accepted by the actors in the conflict. The actors will ask themselves whether or not such a procedure will increase their chances of winning. If the protagonists have the impression that the acceptance of a procedure goes against their interest, they will be able to collect sufficient arguments to successfully mobilize social support against such an innovation. Under these conditions, the original conflict will be repeated on another level; it would most likely prove to not be useful.

In order to make the actors change the procedure and to obtain enough social support for such a change, the procedure has to be more attractive for all participants than the procedure followed so far. Although a dominant solution which satisfies everyone will remain an illusion, a number of criteria can be named which increase the attractiveness of procedures:

(a) Openness of the result: For both the factual formulation of questions and the interpretations and evaluations of the results, methods have to be negotiated in advance which are intuitively plausible and put nobody at a disadvantage from the start. This „legitimation by procedure" will then be accepted by the actors if they see no more chances to win in a conventional procedure, if, therefore, a new procedure which at least grants a prospect of winning puts them in a better position than before. The issue of procedures, however, is a central element of dissension in the nuclear energy debate, because one side regards analytical cost-benefit considerations as an adequate instrument of decision making, whereas the other side rejects this procedure as being „technocratic" and prefers holistic techniques of assessment formation. In order to find acceptance for a new procedure, it will therefore be necessary to admit both analytical and holistic methods as legitimate instruments of reaching and justifying decisions.

(b) Implementation based on consensus: Linked to the openness of the result are the discussion on and the elaboration of the individual procedural steps of the procedure in a body which includes as many conflict parties as possible. In direct communication, untenable arguments and irrational procedural suggestions are far more difficult to defend than on the political stage with its role coercions and the omnipresence of the media. The more heterogeneous the composition of such a body is (inquiry commissions, for example) the more likely we may reckon with social acknowledgement and trustworthiness. This is the result of a series of international studies on the trustworthiness of multiple information sources (Midden 1988; Chaiken & Stangor 1987).

(c) Participation in decision making: As soon as the procedures for setting standards are accepted, scientifically ascertained measurement values and dose limits have to be transferred into units utility and the different assessment criteria have to be translated into weightings. This task is of a political nature and therefore requires special legitimation. If, as is the case in the nuclear controversy, political bodies of the government have lost in trustworthiness, their decisions will no longer be regarded as collectively binding, and doubt will be cast upon their legitimacy through respective protest actions. Yet, if the conflict parties can integrate their value conceptions in the process of decision making and can argue out value conflicts among one another, the legitimation foundation will be expanded and new confidence in the system will develop.

(d) Transparency of the decision procedure: If the individual steps in the decision process are laid open and the value conflicts are described, it is easier to understand and comprehend the sense of standards. Similar to participation, sense is a functional equivalent to trustworthiness and confidence. The more a procedure makes sense when it is applied in practice, the more likely it is to meet with approval and support from the actors and the on-lookers. This step in the process, however, requires special communication efforts, particularly in the field of risk communication which is concerned here. This will be discussed in the final part of this chapter. Moreover, the special selection criteria of the media have to be considered in a communication program and, if possible, productively made use of. This is often very difficult, because agreements on procedures do not have a large sensation value.

(e) Compensation of negative distribution effects: If it is determined from the start that expected distribution effects of risks will be compensated collectively (not individually), then those actors, for whom social justice is one of the central values in their assessment of risks, are more likely to agree. Compensations can easily be mistaken as bribes, especially if the distribution effects are not regarded as inevitable consequences of a decision. As compensations in the field of environment, the restoration or creation of new biotopes might be taken into account; as compensations for possible health risks, prophylactic examinations or special medical precautions might be considered. Compensations in the form of money, however, should be avoided. They are too easily mistaken as bribes.

Naturally, all those measures cannot guarantee that a modified procedure of establishing standards will be accepted by the actors, but they do increase the probability of acceptance. In any case, it is not enough to demand procedures of establishing standards which are highly rational and adequate to the problem. In a democratic society, they have also to be acceptable. The model of a „mixed scanning", as elaborated by the American social scientist Etzioni, comes nearest to our suggestions (Etzioni 1967). Options are first to be designed and elaborated according to rational criteria and then handed over to the free play of political forces in a structured procedure of participation. Thus, the acceptance of different just, normatively acceptable solutions is tested. The better this process is structured in advance, the more likely a pacification between acceptance and acceptability is to be expected. How exactly such a procedure is to be structured will be explained in chapter 5.

4.3 GOVERNMENTAL EVALUATION OF THE RADIATION RISK BY LEGAL REGULATION

Beyond all social processes and evaluations, a network of governmental evaluations of the radiation risk in the form of binding juristic norms has existed for quite some time, which is constantly becoming more dense. Of course, these norms do not claim „eternal" validity, either. They undergo changes in the light of new scientific knowledge and due to changed political valuations.

4.3.1 Development of the German Law on Radiation Protection

The first regulation of the German radiation protection law was the ordinance of 12 May 1925 (RGBl. I, 69), in which illnesses caused by X-rays and radioactive material were acknowledged by the „Reichsversicherungsordnung" (German Reich insurance regulations) as occupational diseases.

Occupational and professional associations, as well as the German society of radiology, then developed protection and safety regulations within the framework of regulations for the prevention of accidents, which, however, could not acquire any legally binding effect.

The first ordinance on X-rays (ordinance on protection against X-rays and radioactive substances in non-medical businesses) of 7 Feb. 1941 (RGBl. I, 88 and 162) contained mainly legal regulations for protecting workers dealing with X-rays and radioactive substances, but left medical application of radioactive radiation completely unregulated. Only in 1953, by a modification of the trade regulations (BGBl. I, 1459), was the German federal government authorized to issue radiation protection supervision regulations, but this was not made use of.

After the enactment of a nation-wide law on atomic energy had failed at first in 1957, several states saw themselves obliged to make use of their legislative competence in the fields of safety of nuclear reactors and radiation protection and enacted their own radiation protection laws or decrees (e.g. Baden-Württemberg, Bavaria, Berlin, Hamburg, Hesse, North Rhine-Westphalia). These state decrees also had a strong influence on the first national ordinance on radiation protection of 24 June 1960, which was issued on the basis of the authorization by the law on atomic energy of 23 Dec. 1959. The third version of the ordinance on radiation protection of 13 Oct. 1976 (BGBl. I, 2905) (ordinance on protection against damages by ionizing rays), which annulled the first ordinance on radiation protection and the second ordinance on radiation protection of 18 July 1964, was prompted by the adjustment obligation of the German law on radiation protection to regulations of the European Atomic Energy Community. The law of the European Atomic Energy Community is also the basis for the second amendment of the ordinance on radiation protection) of 18 May 1989 (BGBl. I, 943), which came into force on 1 Nov. 1989.

The basic principle of the amendment is to develop the existing structures of the ordinance on radiation protection, as far as this has proved necessary, but to avoid a complete reformulation of the law on radiation protection like the comprehensive amendment of 1976, in order to further the certainty of nuclear law, as it says in the general part of the official grounds (BT-Drs. 594/88 and 149/88). Recently, the ordinance on radiation protection was again amended by Art. 1 of the ordinance to establish a radiation protection register of 3 April 1990 (BGBl. 1990 I, 607), first of all introducing a § 63a by means of which the radiation protection register was regulated more explicitly. The amendment came into force on 1 May 1990.

Other recent developments in the German law on radiation protection are mainly a reaction to the reactor catastrophe of Chernobyl of 26 April 1986 and its effects. It entailed a multitude of hurried political decisions, which are also of great legal importance: thus, especially

- the creation of the new Ministry for Environment, Nature Conservation and Nuclear Safety by the organizational edict of the German Chancellor of 5 June

1986 (BGBl. I, 864) was a result of the reactor accident. This meant, at the same time, handing over to the new Minister for Environment, Nature Conservation and Nuclear Safety the responsibility for radiation protection, which had been the competence of the Federal Minister of the Interior since 1972.
- the law on preventive radiation protection, which was introduced by the governing parties on 29 Sept. 1986, was passed on 11 Dec. 1986 (BGBl. I, 2610). Its aim is the removal of different values about health-damaging contaminations in different German states, which had caused confusion among the population after the Chernobyl accident.

The main regulations in the field of radiation protection in Germany today are:
- the law on peaceful use of nuclear energy and protection against its hazards;
- the law on preventive protection of the population against radiation exposure;
- the ordinance on protection against damages caused by ionizing rays (StrlSchV);
- the ordinance on protection against damages caused by X-rays.

Moreover, there are radiation protection regulations in numerous other special laws, regulating effect and application of radioactive substances and rays with respect to their special protection purpose, e.g.:
- in the law on revision and adjustment of legal regulations concerning the dealing with food, tobacco products, cosmetic material and other consumer goods, §13, Abs. 1 of the law (of 15 Aug. 1974; BGBl. I, 1945)
- in the law on dealing with pharmaceuticals of 24 Aug. 1976 (BGBl. I, 2445), § 7 of the law.

4.3.2 Influence of International and European Law on the German Law on Radiation Protection

The German law on radiation protection is influenced by and bound to international regulations. This is the result of the purpose determination of § 1 Nr. 4 AtG, where Gemany's compliance with international obligations in the field of nuclear energy and radiation protection is named as one of the purposes of the AtG.

Especially important in the field of international radiation protection are the recommendations of the ICRP (International Commission on Radiological Protection), which is one of several special committees of the International Society of Radiology. The recommendations are not legally binding, but because of the acknowledged competence of the ICRP, they do have a fundamental influence on the recommendations of other international organizations. This affects even the German law on radiation protection, which, in § 28 I StrlSchV, has incorporated, at least for the most part, the radiation protection principles of the ICRP (the so-called ALARA-principle (as low as reasonably achievable)). Especially important for the German law on radiation protection are the basic norms on protection of the health of the general population and of the labor force against the hazards of ionizing rays (so-called Euratom basic norms), which the European Communities are authorized to ordinance by article 30 of the Euratom contract. The establishing of basic norms in the form of a directive is binding for every member state (Art. 33 I EAG-V); the ways and the means of converting these norms into domestic laws, however, are left to the member states. The presently relevant directives are the directives 80/836, 84/466 and 84/467 of the Euratom.

Yet, it has to be emphasized here that the German ordinance on radiation protection makes considerably higher demands as to the exposure of the population in the vicinity of a nuclear plant than the Euratom basic norms do. Article 12, section 2 of the Euratom directive 80/836 of 15 July 1980 admits the dose limit of 5 mSv (= 0.5 rem) per year as effective dose for individuals of the general population, whereas § 45 Abs. 1 of the German ordinance on radiation protection determines the dose limit at 0.3 mSv, which corresponds with the mean variation range of natural radioactivity on the German territory.

As a result of the effects of transboundary radioactivity due to the reactor accident of Chernobyl, two new agreements on an international level were reached at a special conference of the IAEA (from 25 to 29 Sept. 1986 in Vienna). At this very meeting, Germany has signed and declared as binding for itself both the

- agreement on early information in the case of nuclear accidents, and the
- agreement on assistance in the case of nuclear accidents or radioactive emergencies, of 26 Sept. 1986.

Moreover, after the Chernobyl accident, West Germany reached additional agreements, beyond the existing bilateral agreements on operation and construction of nuclear facilities near borders and on mutual assistance in the case of disasters (among others with Switzerland, France, the Netherlands, Belgium, Luxembourg), with the former German Democratic Republic (BGBl. 1987 II, 598) on the exchange of information and experiences in the field of radiation protection, and with Switzerland (BGBl. 88 II, 519f) and Denmark, supplementing the already existing arrangements particularly with respect to radiation protection.[26] Thus, the concept of the BVerfG was also taken into account, which, already long before the Chernobyl accident, pointed at the transboundary risks due to nuclear facilities and nuclear power plants in the „Kalkar decision" (E 49, 89 (197)), and expressed a need for increased bilateral and international cooperation, specifically to protect against such hazards.

Other international organizations are only entitled to make recommendations to their member states, which are not binding for the latter, however. Furthermore, it can be observed that the IAEA, especially in the period of time following Chernobyl, has increasingly been acknowledged by the community of nations as a center of international efforts to protect against radioactive disasters and a center for mutual exchange of information, as is also shown by the special conference in September 1986 with its deliberation of the two above-mentioned agreements.

4.3.3 German Ordinance on Radiation Protection

Radiation protection means all devices and measures to protect against damages by ionizing rays of any kind. Respective regulations are laid down in the German law on atomic energy and in the ordinance on radiation protection, which was issued on the basis of the authorization of §§ 10–12, 54 AtG.

4.3.3.1 Purview of the Ordinance on Radiation Protection

The purview of the StrlSchV results from the authorization formulated in the law on atomic energy as well as from § 1 StrlSchV. According to § 10 AtG, exceptions can be made from the supervision regulations of §§ 3 to 7 and 9 AtG, which, among other things, regulate, by means of legal decrees, import and export (§ 3), transportation (§ 4), and keeping and possession (§ 5) of radioactive material. The StrlSchV has made use of this by requiring mere notification in certain cases (§§ 4 I, 12) or by exempting certain activities from any need of permission (§§ 4 II, 9, 13).

The more important purview of the StrlSchV results from the authorization of § 12 AtG, which, to achieve the purpose defined in § 1 AtG, states that protective measures for activities in connection with nuclear fuel and other radioactive substances may be established by legal decree. According to this, all material radiation protection regulations (apart from the licensing rules regulated in the AtG itself) are found in the StrlSchV, such as, for example, the establishing of principles of radiation protection and the determining of dose limits for the general population and for occupationally exposed persons. Therefore, the StrlSchV is, especially due to recent events (malfunctions and their concealment), of decisive significance for the law on atomic energy altogether and even beyond it.

4.3.3.2 Systematics of the German Ordinance on Radiation Protection

The first part, §§ 1 and 2, regulates the factual purview and serves for the necessary definitions of terms. The second part of the StrlSchV, §§ 3–27, contains those regulations which are supposed to serve governmental control insofar as it is not yet regulated in the law on atomic energy. Handling, transportation, import and export of radioactive substances as well as the construction and operation of nuclear plants mostly require permissions. This is not the case for substances which, due to their amount or their radioactivity, present no danger to the individual or the environment.

The third part of the StrlSchV contains the actual core of the radiation protection rules. These material radiation protection norms of §§ 28–80 basically hold for all activities in connection with radioactive material, hence also for activities which require no permission according to the German law on atomic energy. Particularly important here are the radiation protection principles as well as the determination of the dose limits. The fourth part of the ordinance contains regulations on administrative fines, where § 81 Abs. 2 Nr. 1–11 especially charges with fines violations on the part of the radiation protection responsible and the radiation protection officer.

4.3.3.3 Principles of Radiation Protection

In § 1 AtG, the purposes of the law on atomic energy in general are established. It contains both the basic decision to promote peaceful use of nuclear energy and, in Nr. 2, the commitment to ensure the protection of life, health and property from the hazards of nuclear energy and the damaging effects of ionizing rays.

According to the German constitutional order of values, the protection purpose of the AtG takes precedence over the promotion purpose of the law.[27] Thus, especially the radiation protection regulations and the principles contained therein represent the exposition of the purpose determination of § 1 Nr. 2 AtG.

The radiation protection principles laid down in § 28 Abs. 1 StrlVG demand, on the one hand, that any unnecessary radiation exposure is avoided (so-called radiation avoidance rule), on the other hand, that any radiation exposure or contamination, even if it is below the limits established in the radiation protection decree, is kept as low as possible, in consideration of the state of science and technology (so-called radiation minimization rule; see chapter 3.1).

4.3.3.4 Regulations to Ensure the Principles

(a) § 28 Abs. 1 Nr. 2 of the ordinance on radiation protection contains the rule to keep any unavoidable radiation exposure or contamination as low as possible. Since the application of radioactive substances in the fields of medicine, research

and technology is necessary for compelling reasons (so the official grounds to § 28), a complete avoidance of any radiation exposure cannot be demanded but only an optimum of radiation protection (Winters 1978). On the other hand, the „as low as reasonably achievable"-regulation is based on the safety-oriented hypothesis that no dose limit exists below which no radiation effect occurs at all. So it is presumed that any exposure to radiation, however low it may be, entails a certain if small radiation effect, which is correspondingly dangerous (stochastic effects).

The equipment of the radiation minimization principle with the characteristic formula „in consideration of the state of science and technology" does not relativize the concept of an optimal radiation protection. Radiation protection measures must be taken according to what is possible, given the respective state of science and technology. This autonomous protection regulation results in the obligation to always take those measures which are possible according to the latest developments in science and technology (see BVerfGE 49, 89 (136)).

§ 28 Abs. 1 Nr. 2 StrlSchV moreover demands that any radiation exposure or contamination, even if it is below the limits established by the decree, is kept as low as possible. So the radiation protection principles of § 28 are linked with the dose limits of the decree, i.e. the regulations cannot be regarded on their own. This connection is extremely important for an effective radiation protection.

By enacting the law on atomic energy and the StrlSchV, the legislator has imposed a certain radiation risk on the population (OVG Münster ET 1975, 220 (225)). This does not mean, however, that operators of a nuclear plant can simply exhaust the decreed limits and refrain from taking any radiation protection measure below the limits. The regulations about limits operate here as an orientation for the principle „as low as achievable" (OVG Münster ET 1975, 220 (224); Winters 1977, 331 (334)).

Still, the radiation minimization principle, unlike the radiation avoidance principle, does not hold without reservation, for § 28 Abs. 1 Nr. 2 is not a complete interdiction of any radiation exposure. The minimization principle is subject to the principle of proportionality of means and ends. This also relativizes the indirect statement of the StrlSchV that economic aspects are of no relevance to the implementation of radiation protection measures. This indirect statement is implied by the fact that the principle standardized in § 28 Abs. 1 Nr. 2 StrlSchV, unlike the ICRP recommendation, does not include any economic considerations. That means, on the one hand, that no measures can be demanded which, despite considerable expenses, would only result in an insignificant reduction of radiation exposure below the limits (Winters 1978, 67). On the other hand, it means that measures which lead to a considerable reduction of radiation exposure must be taken, notwithstanding high expenses.

The question whether the radiation minimization rule of § 28 Abs. 1 Nr. 2 is of a third-party protecting nature, i.e. if it grants a third party affected a subjective right to claim application and observation of the rule, has been denied by the German Supreme Administrative Court in the „Stade decision“ (E 61, 256 (267f)). (A contrary position taken before: OVG Lüneburg DVBl. 79, 686; DVBl. 78, 67 (69).) Two arguments are given in support of this decision: first, it is completely undetermined whether or not a person belongs to the circle of persons protected by the radiation minimization rule, and second, if a third-party protecting nature is assumed, dose limits will lose their function as limits of maximal admissible exposure to radiation, which, in turn, will relinquish legal certainty (BVwerGE 61, 256 (267 and 268)).

Yet, the lack of third-party protection only means that the citizen cannot refer to it directly. The rule remains fully effective for the licensing and supervising authorities, though. They must see to it when they grant the license that even below the limits established by the StrlSchV radiation exposure is to be kept as low as possible. They must also supervise and ensure this by means of the instruments of imposition and revocation while the plant is in operation.[28] Therefore, a radiation exposure below the dose limits (which again are considerably lower than the maximal admissible doses according to the Euratom basic norms) can be classed as belonging to the sphere of residual risk, which may be reasonably imposed on the population.[29]

(b) *Determination of dose limits (§ 45 StrlSchV).* §§ 44, 45, 49, 51 StrlSchV establish dose limits for different radiation protection fields. A distinction is made here between protection of the population (§§ 44–46) and protection of persons exposed to radiation (§§ 49ff). The dose limits to protect the population do not represent thresholds of risk or hazard; they are rather still within the variation range of natural exposure to radiation[30] and considerably below the maximal admissible doses according to the Euratom basic norms. Here, § 45 StrlSchV, with its concept of immission protection by establishing dose limits, represents the central regulation for the protection of the population (Kramer & Zerlett 1980, § 28 StrlSchV Anm. 1). It is supplemented by the regulation of emission limits of § 46 StrlSchV. The norm obliges the operator of a nuclear facility to design the plant in a way that outside the plant radiation exposure due to emission of radioactive material is as low as achievable, but in no case exceeds an effective dose of 0.3 mSv per year.

In the 1976 version, this limit was 30 mrem per year. Numerically, the present value of 0.3 mSv corresponds with this (see chapter 2). The new version of § 45 StrlSchV is therefore substantially not an abandonment of the so-called 30 mrem concept. Whereas up to now, however, only limits for the whole-body dose or for doses concerning parts of the body were established, now effective doses are taken into consideration for dose limits, in correspondence with the Euratom directives about basic norms (see chapter 2). The dose limits established in § 45 constitute the highest limit of exposure to artificial radiation which the legislator holds reasonable for individuals wherever they may be situated outside the control areas of nuclear power plants (so the „Stade decision", BVerwGE 61, 256 (264)).

According to the prevailing opinion, dose limits are of a third-party protecting nature, unlike the radiation minimization principle (BVerwGE 61, 256 (264); Haedrich 1986 – preceding annotation 12 d 120 with further documentation). This results from the concretization which §45 StrlSchV gives of the demand for damage precaution according to the state of science and technology, expressed in § 7 Abs. 2 Nr. 3 AtG.

Since § 7 Abs. 2 Nr. 3 AtG is of a third-party protecting nature, the concretizing regulation of § 45 StrlSchV, which determines the maximal admissible exposure of every individual, must also have a third-party protecting effect (Schattke 1982, 1083). The dose limit of § 45 StrlSchV, however, only holds for normal operation, i.e. for the proper operation of the plant (BVerwGE 61, 256 (264); OVG Lüneburg NVwZ 87, 75). Effects of malfunctions or accidents are thus not covered by the limit. The limits of § 45 StrlSchV are site limits, not plant limits. That means that the sum of all emissions from nuclear installations and other users of radioactive substances must not exceed the dose limits established in § 45 StrlSchV. For the procedure of licensing new plants, this means that exposures due to existing facilities must be allowed for and taken into account as well. Exposure to natural radiation, on the other hand, has not to be considered in the licensing procedure.

It is not clear whether existing exposures caused by foreign nuclear plants have to be considered here as well. The problem becomes more difficult when radiation exposures caused by domestic plants are increased to a point beyond the admissible limits by new additional foreign influences. It is sometimes held that foreign exposures have also to be considered in granting a license, since § 45 S. 3 StrlSchV does not distinguish between foreign and domestic exposures (e.g. Kramer & Zerlett 1980, § 45 StrlSchV Anm. 18). In contrast to that, recent legislation (OVG Lüneburg NVwZ 87, 75) holds the opinion that the factual purview of the StrlSchV only includes domestic plants, and that therefore the concept of the plant or facility in § 45 S. 3 StrlSchV, in default of other regulation, is to be understood in the same way, so that existing exposures from foreign plants do not have to be taken into account.

It is questionable whether this can be correct in that form. Thus, it has to be considered that the StrlSchV is supposed to ensure the most perfect possible protection of the population and of the environment by means of the radiation minimization rule and the determination of the dose limits. It would therefore contradict the idea and the purpose of the rule to consider only German pollution in planning and licensing German facilities. Moreover, § 45 StrlSchV formulates the site concept, i.e., the total immission occurring at the respective site must not exceed the normatively stipulated limit. But immissions occurring at some sites include also the immissions from foreign plants.

The value of 0.3 mSv or of 30 mrem must be calculated under the most unfavorable presumptions in order to thus ensure the best possible protection. This refers to the place where the distribution of radioactivity released into the environmental media is expected to cause the highest radiation exposure (see chapter 3.1).

§§ 49 and 51 StrlSchV in connection with enclosure X regulate the dose limits for persons occupationally exposed to radiation and for persons working in the control sectors of plants. Accordingly, persons who are occupationally exposed to radiation may be exposed to higher radiation doses (§ 49 iV, enclosure X) than the population in general. Depending on the category, the effective dose limit is 15 mSv or 50 mSv per year. This remains within the scope of Art. 8 Abs. 1 of the Euratom directive 80/836 of 15 July 1980. These limits, too, are not reference values, but here again the principle of radiation minimization must be considered. A novelty introduced in § 49 Abs. 1 sentence 3 by the amendment of 1989 is the dose limit for the entire occupational life of a person occupationally exposed to radiation. The sum of the effective doses ascertained in all calendar years must not exceed 400 mSv. If this limit is reached, the person must no longer work in areas which are exposed to radiation.

Another limit is established with the malfunction dose of § 28 Abs. 3 StrlSchV. Accordingly, power plants must be planned in a way that even in the most unfavorable case of malfunction against which the plant is designed, individuals in the vicinity of the plant must not be exposed to a higher amount of radiation than is admissible for occupationally exposed persons. The rule, however, refers only to the planning of the plant, but not to its later operation (Kramer & Zerlett 1980, Anm. 16; Winters 1977, 331 (338 first column)).

(c) *Consequences of the Chernobyl accident on § 45 StrlSchV.* As a result of the Chernobyl accident and its impacts on German territory, it had to be examined whether the radiation protection ordinance and especially the dose limit determinations are applicable to the radiation exposure caused by the accident. Particularly, it had to be clarified whether the additional radiation exposure should not lead to a decommissioning of domestic power plants or facilities if the dose limits of § 45 StrlSchV were exceeded.

In a statement in reply to a question of the „green party", the German federal government took the opinion that, according to the territoriality principle, the StrlSchV only holds for the sovereign territory of Germany and therefore is not applicable to the effects of the reactor catastrophe (BT-Drs. 10/8008, S. 1). This opinion, which the OVG Lüneburg held as questionable in its decision of 28 October 1986 (NVwZ 87, 75f), is simply wrong. According to the territoriality principle, a country's sovereign acts hold only for its own territory. International law requires that a country must not take sovereign measures on the territory of another country or permit such measures which affect the territory of another country without its approval.

This, however, is not what the present case is all about. International law does not forbid arranging matters which take place on one's own territory and do not affect other subjects of international law. The question of whether foreign pollution is of relevance to the continued operation of German nuclear power plants both concerns only German law and affects only German territory. Every country is allowed to take the protection measures which it considers necessary for the sake of its people.

A decisive fact regarding the nonapplicability of the StrlSchV is that radiation exposure caused by malfunctions and accidents of nuclear plants lies beyond the purview of this ordinance and is not covered by its regulations. The limits established in § 45 StrlSchV only refer to proper operation of the plants, hence radiation exposure caused by malfunctions or accidents must not affect licenses or continued operation (OVG Lüneburg, decision of 28 Oct. 1986 NVwZ 87, 75 and OVG Lüneburg, decision of 23 Sept. 1986 UPR 87, 153). Therefore, the emission rates permitted by the StrlSchV cannot have become illegal due to the impacts of the Chernobyl accident.

Even that pollution which must be considered according to § 45 S. 3 is only pollution occurring during normal operation. The regulation serves preventive radiation protection and requires a prognostic decision on possible pollution.

Accidents, however, like the one in Chernobyl, are unpredictable events and are thus beyond any prognosis (OVG Lüneburg UPR 87, 153, (154)).

Precautions against radiation released due to malfunctions and accidents can only be taken within the framework of § 7 Abs. 2 Nr. 3 AtG, which demands an optimal risk precaution and protection against dangers, so that damaging events become unlikely and reactor accidents are practically ruled out. The regulations for prevention of damages have their own purview and therefore are independent of the regulations which, according to the StrlSchV, hold for normal operation of nuclear plants (OVG Lüneburg NVwZ 86, 75 (76)). For reactor accidents, which, until quite recently, have been thought to be almost impossible, the StrlSchV has not set any dose limits. The legislator did not consider this problem when he issued the decree. Therefore the StrlSchV cannot be consulted in cases of emission of radiation as a result of accidents.

(d) *Institutionaliziation of responsibilities – the Radiation Protection Responsible and the Radiation Protection Officer.* The StrlSchV names two persons who are in charge of radiation protection, the Radiation Protection Responsible according to § 29 Abs. 1 and the Radiation Protection Officer according to § 29 Abs. 2. They are the main addressees of the radiation protection decree. Radiation Protection Responsible is the licensee, the company owner or the person obliged to notify, i.e. the industrialist who carries out an activity which is listed in the StrlSchV or who has such an activity carried out (official grounds to §§ 29, 30 in Kramer & Zerlett 1980, § 30 StrlSchV S. 162). Radiation Protection Officer is the person who has been nominated by the Radiation Protection Responsible to manage or to supervise the activity that is notifiable and requires a license (§ 29 Abs. 2). Preconditions for the nomination are reliability and expertise on radiological safety (§ 29 Abs. 4). The nomination must be done in writing and is notifiable to the authority in charge (§ 29 Abs. 2 and 3). Moreover, the sphere of competence of the Radiation Protection Officer within the company must be specified in writing. The nomination as such is subject to civil law and does not entrust the respective person with public tasks (Kramer & Zerlett 1980, § 29 StrlSchV Anm. 7; Bischof & Pelzer 1983, 150). It is the job of the Radiation Protection Officer to autonomously carry out the radiation protection tasks assigned to him or her, which he/she has also to guarantee to the radiation protection responsible (Winters 1978, 69).

According to § 31 Abs. 1, the Radiation Protection Responsible is obliged to respect the sphere of competence of the appointed officer. By making suitable rooms, protection facilities and safety equipment as well as sufficient staff available, he/she must ensure that radiation protection rules are observed.

Moreover, the Radiation Protection Responsible cannot shirk his/her responsibility for the observance of the protection regulations by appointing an officer; what changes by the nomination is only the sphere of his/her duties, which is transferred to the corresponding organization and supervision duties. The continuation of the responsibility is expressed in § 30 Abs. 1, which obliges the Radiation Protection Responsible to inform the Radiation Protection Officer without delay about all defects that lead to a reduction of radiation protection.

The StrlSchV demands a certain minimum sphere of competence for the Radiation Protection Officer within the company. Included, according to § 30 Abs. 4, is the authority to order immediate measures to ward off dangers. That also results in the authority to issue directives to all other employees in the company in such cases (BVwerG DVBl. 69, 586).

In order to achieve optimal radiation protection within a company, the Radiation Protection Responsible and the Radiation Protection Officer are to cooperate as conjointly as possible. If on some radiation protection measure suggested by the radation protection officer no agreement can be reached, the Radiation Protection Responsible must specify his/her reasons in writing and hand them in to the workers' council or to the authority in charge if he/she wants to reject the suggestion (§ 30 Abs. 1 sentence 3). § 81 Abs. 2 StrlSchV emphasizes the Radiation Protection Officer's and the Radiation Protection Responsible's obligation to respect the safety rules, by making violations punishable with penalties as administrative offenses.

A rule which deserves special attention because of recent events (malfunctions in nuclear power plants and their delayed publication) is § 36 StrlSchV. Sentence 2 of the rule obliges the Radiation Protection Responsible or the Radiation Protection Officer to notify any malfunction or accident or any other safety-relevant event to the nuclear law control board without delay, hence without culpable hesitation. Non-observance is an administrative offense according to § 81 Abs. 2 Nr. 5 StrlSchV. The obligation to notify such events is supposed to ensure that experiences about the safety of certain facilities and plants are analyzed and considered in other licensing procedures (official grounds to § 36 in Kramer & Zerlett 1980, § 36 StrlSchV).

4.3.4 Procedures of Licensing Plants under Nuclear Law

Whoever in Germany sets up, runs or otherwise owns a stationary plant for the production, the processing, or the fission of nuclear fuel or for the reprocessing of irradiated nuclear fuel, or who substantially modifies such an installation or its operation, needs a license (§ 7 I AtG). The same holds for the decommissioning of a plant as well as for the safe encasement of the definitely decommissioned plant or for the dismantling of the plant or of parts of the plant (§ 7 III AtG). The license must ensure the observance of the nuclear-specific and non-nuclear-specific requirements established in § 7 II AtG (§ 7 II Nr. 6 AtG) which result from the objective of the law expressed in § 1 AtG, with the protection purpose taking precedence over the promotion purpose. In particular, the license must only be granted if the necessary precautions against damages due to construction and operation of the plant have been taken, according to the state of science and technology (§ 7 II Nr. 3 AtG). Uncertainties beyond the „threshold of practical reason“ dependent on the limits of human cognitive capacity („residual risk“) may be disregarded. Rights of third parties result from the legal principle of precaution and protection, insofar as it intends to protect the individual from hazards and risks of nuclear energy – as it does by the dose limits of § 45 StrlSchV – but not insofar as it only refers to interests of the general public („population risk“) – as it does in form of the radiation minimization rule of §§ 28 I Nr. 2, 46 I Nr. 2 StrlSchV. In its decision on the license, the authority has a scope of due discretion – which is the exception with preventive interdictions with the reservation of granting permission – in order to be able to also take into account hitherto unpredictable conditions. That nuclear energy may be used for peaceful purposes at all, despite the risks, has been determined by the legislator in the law on atomic energy, and with constitutional approval (Art. 74 Nr. 11a, 87c GG).

Licensing authority is the respective supreme state authority in charge, which thus executes the law on atomic energy on behalf of the federal government (Art. 87c GG, § 24 II AtG). The licensing procedure is partly regulated by the AtG itself, and partly by the ordinance on procedure under nuclear law (AtVfV) in form of the promulgation of 31 March 1982 (BGBl I S. 412). The check on the part of the licensing authority does not only concern the license requirements of § 7 II AtG but also compliance with the other regulations of public law relevant to the project (§ 14 AtVfV). This does not, however, replace other licensing procedures. Everybody can raise objections during the licensing procedure. The participation of the public takes place in five phases regulated by the AtVfV: (1) public announcement of the project (§§ 4, 5 AtVfV); (2) laying the application and the documents open to public inspection (§ 6 AtVfV); (3) objections of third parties (§ 7 AtVfV); (4) discussion with the applicant and the objectors (§§ 8–13 AtVfV); (5) serving of the applicants and the objectors with the decision, together with instructions about legal remedies (§§ 15–17 AtVfV). In order to reach the objectives named in § 1 AtG, the license may be of limited range, and it may be subject to impositions, but not to conditions (§ 17 II 2 AtG). Later impositions are admissible, possibly only against compensation, insofar as they are necessary to ensure the purposes named in § 1 Nr. 2 and 3 AtG (§ 17 I 3, 18 III AtG).

As regards the preconditions for licensing a plant under nuclear law, above all two rules are relevant for environment protection, namely: according to the state of science and technology, the necessary precautions against damages due to construction and operation of the plant must be taken (§ 7 Abs. 2 Nr. 3 AtG), and substantial public interests, especially with respect to clean water, air and soil, must not conflict with the choice of the nuclear site (§ 7 Abs. 2 Nr. 6 AtG).

The first of these preconditions concerns the rules of indispensable defense against dangers and the preceding risk precaution, limited by the principle of proportionality. According to § 7 Abs. 2 Nr. 3 AtG, the licensing authority – regardless of all the problems of scientific and technological knowledge – must only check the concrete fulfillment of generally-abstract requirements and must not assign any environmental resources or pollution rights according to planned economic purposes. The same holds for the supplementary minimization rule of the radiation protection law (§§ 28 Abs. 1, 48 Abs. 1 Nr. 2 StrlSchV), in which the rule of risk precaution has found an expression. The minimization rule, unlike the the dose limits of the StrlSchV and the basic norm of § 7 Abs. 2 Nr. 3 AtG, lacks a neighbor-protecting effect.

The license for a plant which fulfills all the legal requirements of § 7 Abs. 2 AtG may be refused on the basis of the discretion of denial only in order to observe one of the protection purposes regulated in § 1 Nr. 2–4 AtG. Moreover, it would be an infringement of the promotion purpose of § 1 Nr. 1 AtG if the licensing authority generally and absolutely refused to grant a license for a plant although all the requirements of § 7 Abs. 2 AtG were fulfilled.

4.3.5 German Law on Preventive Radiation Protection (StrVG)

The law on preventive protection of the population against radiation exposure, enacted on 19 Dec. 1986 (BGBl. I 2610), was discussed and enacted in a great hurry, because of the organizational confusion in Germany dur to the impacts of the Chernobyl accident.[31] In this context, the federal government made use of its legislative competence according to Art. 74 Nr. 11a GG.

Article 3 of the law on the establishment of a Federal Office for Radiation Protection of 9 Oct. 1989 (BGBl. 1989 I, S. 1830), amends the law on preventive radiation protection in a way that former spheres of competence of other authorities are transferred to the Federal Office for Radiation Protection. These amendments have been in force since 1 Nov. 1989.

4.3.5.1 Objective of the Law

The StrVG is meant to create legal foundations for an effective and coordinated operation of the respective national and state authorities in their preventive protection of the population with regard to wide-spread contamination caused by radioactive accidents. The radiation protection law in force up to then did not provide any regulations about at which concentrations of radioactive substances action was to be taken to protect the health of the population. Moreover, there was no basis for an authorization to prohibit the selling of contaminated food, to prohibit the import of contaminated products from abroad or to issue behavior recommendations to the population.

The only regulation which established a limit, § 45 StrlSchV, refers merely to planning the normal operation of a domestic nuclear plant and therefore cannot be applied to the impacts of an accident. The only norm which regulated the consequences of transboundary effects of nuclear events was the liability regulation of § 38 Abs. 1 AtG. A consequence of the missing regulations was that after the Chernobyl accident and its impacts on the German territory every German state issued its own limits for contaminated food and every municipality issued its own behavior recommendations to the population.

An effective and coordinated action shall be ensured:
- by a nation-wide collection and evaluation of data about radioactive contamination of the environment, on the part of the federal and state governments,
- by a nation-wide uniform determination of measuring methods, and by ensuring the centralized collection and storage of all measurement data, as well as
- by creating the preconditions for establishing dose limits and contamination values after nuclear accidents and other events with significant radiological effects (BT-Drs. 10/6639).

4.3.5.2 Systematics of the Law and Measures Provided for

The first section of the law determines the purpose of the law, which is directed towards
- monitoring radioactivity in the environment
- in cases of events with possible significant radiological effects, keeping radiation exposure of humans and radioactive contamination of the environment as low as possible, by means of appropriate measures, in consideration of the state of science, and taking into account all circumstances (§ 1 StrlVG).

This determination of the purpose in the first section is more than simply a description of the objectives of the law, rather it is of decisive importance in view of the legal decrees which are to be issued according to the StrVG (see Czajka 1987, 556 (557)).

The second section regulates the monitoring of environmental radioactivity. §§ 2–5 StrVG standardize the assignment of tasks to the federal government and the state governments. Here, the task of the state governments is, first of all, to merely ascertain radioactivity. Of special significance is § 2 Nr. 4 StrVG, which grants the federal government an evaluation monopoly on whether a certain degree of contamination is to be classified as injurious to health or endangering health. According to § 4 StrVG, the Federal Minister for Environment, Nature Conservation and Nuclear Safety has the responsibility to sum up the ascertained data in an information system (he/she is also entitled to evaluate the data of the radioactivity permissions (§ 5 StrVG)). Insofar, in its second section, the law is a mere system of gathering and processing information, apart from its assigning the data evaluation competence to the Federal Minister for Environment, Nature Conservation and Nuclear Safety (Peine 1988, 115).

The third section includes the catalog of measures of preventive radiation protection. Above all, it also includes authorizations of individual federal ministers to issue legal decrees. Thus, according to § 6 I StrVG, the Federal Minister for Environment, Nature Conservation and Nuclear Safety can establish dose values and contamination values as well as calculation procedures by legal ordinance (for the question about the constitutionality of the authorization to issue decrees, see Peine 1988, 115ff; Rengeling 1987, 204f).

According to § 7 I and § 7 II StrVG, the Federal Minister for Youth, Family, Women and Health as well as the Federal Minister for Food, Agriculture and Forestry are entitled to issue decrees prohibiting and restricting the act of putting into circulation as well as the import and export of food, feed, pharmaceuticals and other substances. The respective legal decrees require the approval of the Bundesrat (the upper house of the German parliament); but the approval is not necessary if the matter is urgent (§§ 6 II, 7 IV). § 8 StrVG regulates the police competence concerning transboundary traffic. Similar to the police general clause, the authorities in charge of import and export control may take measures to ensure observance of the contamination values according to § 6 StrVG. According to § 9 StrVG, the Federal Minister for Environment, Nature Conservation and Nuclear Safety may address certain recommendations to the population in order to achieve the purposes named in § 1, but this is supposed to take place in agreement with the supreme state authority in charge. In case of locally confined events, the states may also issue behavior recommendations (§ 9 II).

Finally, the fourth section of the StrVG regulates the execution of decrees issued according to the law, in the form of administration by commission on the part of the states (§ 10), as well as the right of taking samples and, in connection with this, the right to enter (§ 12).

4.3.5.3 Criticism of the German Law on Preventive Radiation Protection (StrVG)

The StrVG only once provides for an exclusive state competence, in § 9 Abs. 2, and this only for recommendations to the population in case of events with exclusively regional effects. Surely, this rule is a result of the impacts of the Chernobyl accident. In future, contradicting recommendations of the federal and the state authorities shall be ruled out.

The regulation of § 9 appears to be problematic. First, it can hardly be imagined, when, of all events, a radiation accident should ever be an event with exclusively regional effects. According to this regulation, the states could not even make use of § 9 II StrVG if regionally varying radiation degrees occurred in different areas (just like after Chernobyl, when the wind direction determined the radiation degrees in individual areas). In this case, there would be a contradiction between the protection purposes of the law, namely to regulate all measures in a nation-wide coherent way on the one side and on the other side to ensure the best possible protection of the population. Moreover, this gives rise to the question; what should be done if the behavior recommendations of the Federal Minister for Environment should be delayed for some reason.[32]

Another criticism of the StrVG refers to the question whether the authorization to issue decrees is constitutionally sufficiently determined, especially with regard to the dose values and contamination values of § 6 I StrVG (Peine 1988, 115ff). Here, the reproach that they should have been determined by the law itself, is not tenable. The rule of definiteness of Art. 80 I 2 GG requires that the content, the purpose and the extent of the authorization must be determined in the law itself, but it is sufficient that this ensues from the context of the law. And if we regard the purpose of the law and the catalog of measures in their context, there can be no doubt that this is the case here. Since the „Kalkar decision" of the BVerfG (E 49, 89 (139f)), it is legally incontestable if such limits are determined by legal decrees and not by the law itself.

The fact that the law was enacted in a great hurry and without sufficient consultation of experts, gives rise to doubts concerning the law itself. Still, it remains to be seen whether the StrVG will stand the test in its present version or if a reconsideration of certain norms and, as a result, an improvement for the sake of the protection purpose of the law will be necessary.

4.3.6 German Ordinance on X-Rays (RöV)

§ 15 of the ordinance on X-rays establishes the principles of protection against damages due to X-rays, with the radiation avoidance rule being established in Nr. 1 and the radiation minimization rule being formulated in Nr. 2. §§ 31 and 32 RöV define the dose limits for persons occupationally exposed to radiation, for persons particularly in need of protection, and for other persons. The admissible values are given in milliSievert and correspond with the limits of the radiation protection decree.

There is no rule which corresponds with § 45 StrlSchV in the ordinance on X-rays, because the X-rays to which this ordinance applies cannot have effects beyond the control area or the monitoring area of the respective firm or institution, to which only persons with special permission have access (§ 22 Abs. 1 RöV).

Addressees of the ordinance on X-rays are, according to § 13 RöV, the Radiation Protection Responsible and the Radiation Protection Officer. §§ 23–28 RöV regulate the application of X-rays to human beings. Here, the application restrictions of § 24 RöV (application for medical reasons, Abs. 1) and the application principles of § 25 RöV (application only in case of a medical indication) must be observed.

Based on Art. 2 of the ordinance on a radiation protection register of 3 April 1990 (BGBl. I, 607), a § 35a has been inserted which regulates the radiation protection register more precisely.

4.3.7 Penal Provisions

§§ 311a and 311d StGB make the misuse and the inadmissible emission of ionizing rays liable to prosecution. The rule of § 311d, which traces back to the repealed § 47 AtG, is of special importance here with respect to radiation protection (§ 311d Abs. 1 Nr. 1). The offense must be committed „under violation of the offender's duties according to administrative law.“ This constituent fact covers all duties which result from the regulations about protection against ionizing rays, i.e., the radiation protection ordinance and the ordinance on X-rays.

The violation, however, must be a gross breach of duty (§ 311d Abs. 4), i.e., there must be a particularly heavy offense against regulations of radiation protection (see Steindorf 1988). The act has to be likely to harm life or limb of another person or to damage other people's property of significant value (injuries to health comprise especially burns, induction of cancer, malformations of unborn children, loss of fertility, as well as genetic injuries which affect only descendants). The act of emission means that the radioactive substances are left to take their course freely and uncontrollably. This act may also be committed by omitting necessary action if one is in the position of a guarantor (Lackner 1987, § 311d Anm. 3a). So any behavior is included which allows noxious radiation to get out of control (e.g., generation of rays without sufficient protection provisions or without the necessary safety measures).

Thus, the regulation concentrates mainly on observance of the protection regulations of the StrlSchV and the ordinance on X-rays by the Radiation Protection Responsible and his or her Radiation Protection Officer.

4.3.8 Conclusion Concerning Risk Assessment

4.3.8.1 Discretion of Denial because of the Special Hazards of Nuclear Energy

It has already been shown that the licensing procedure under nuclear law basically starts from the legal figure of a preventive prohibition with the reservation of granting permission. Usually this leads to the fact that the applicant may claim to be given the license if the preconditions required for a permission are fulfilled. In the case of licensing a nuclear plant, however, the authority may deny the permission applied for even if the requirements for the permission according to § 7 Abs. 2 AtG are met. As the law concedes an additional discretion of denial to the authority in charge, in view of this form of the license facts, a special justification is required, which the BVerfG sees in the high potential hazards of nuclear energy. In this context, the BVerfG expounds the following:

> „According to the constitution, the legislator was allowed to (...) assess the scientific, technological and industrial state of knowledge and experience with regard to these possible hazards and their controllability as insufficient (for the time being) and to subject these substances to a special regulation. This special position of nuclear law and the reasons for it make it seem justified, under certain circumstances, to deviate from principles which are acknowledged in other legal fields. Such an exceptional case is given here. It is true that the legislator is obliged by the principle of legal proviso, also in the field of nuclear law, to determine the general license requirements himself if he makes use of the instrument of preventive interdiction with

> the reservation of granting permission. Yet, if he has done so, as in § 7 Abs. 2 AtG, there can be no objection on the basis of the German constitution, in view of the high potential hazards of the plants which require a license according to § 7 Abs. 1 AtG, against his being especially cautious, by conceding an additional discretion of denial to the executive, thus allowing it to refuse a license which actually should be granted, if special and unpredictable circumstances should make it necessary. This holds all the more since the legislator – apart from the fact that there is no such thing as a completely free discretion (see BVerfGE 18, 353, 363) – has defined precisely enough the scope of discretion which the licensing authority is entitled to by § 1 AtG, especially by the standardized protection purposes" (BVerfGE 49, 89, 146f).

This statement of the BVerfG convincingly defines the preconditions for the exertion of the discretion of denial. First of all, it must be safety considerations – beyond the license requirements – which could, as an exception, justify the denial. Although the provision for waste disposal of § 9a AtG is no precondition for a license in the sense of § 7 Abs. 2 AtG, according to the prevailing opinion, it has to be taken into account in the authorities' exertion of discretion.

4.3.8.2 Ensuring the Required Precaution against Damages According to the State of Science and Technology: § 7 Abs. 2 Nr. 3 AtG

(a) *The inherent dynamic protection of constitutional rights.* The material license requirement which is most important for the fulfillment of the protection purpose is § 7 Abs. 2 Nr. 3 AtG. According to this regulation – apart from the other material requirements named in Abs. 2 – „the necessary precautions against damages due to the construction and the operation of the plant, according to the state of science and technology" must be taken. The BVerfG points out, in the context of its investigation on whether § 7 Abs. 2 Nr. 3 AtG meets the constitutional principle of definiteness, that especially in the field of technological safety law, reference to the „state of science and technology" serves the dynamic protection of constitutional rights, because it helps to realize the protection purpose of § 1 Abs. 2 AtG in the respective best possible way. It should be mentioned that, on the other hand, a legal determination of a certain safety standard by means of setting up rigid regulations would, if it were possible at all, inhibit rather than promote technological progress and the respective appropriate ensuring of constitutional rights. With reference to such a determination, the court speaks of a step back at the expense of safety. Nothing has hereby been said against a normative determination in the form of legal decrees.

The executive may concretize the matter by legal decrees on the basis of legal authorizations; the federal government has made use of this authorization by issuing the radiation protection ordinance on the basis of § 12 Abs. 1 Nr. 2 AtG. Concretization by legal ordinance is unobjectionable from the point of view of constitutional law, also in the field of nuclear law, as the BVerfG confirms (BVerfGE 49, 89, 138).

Apart from the radiation protection ordinance – and the decrees on provisions for coverage and on X-rays, which are not relevant in this context – no such decrees concretizing the license requirements of § 7 Abs. 3 AtG have been issued. In particular, no use has yet been made of the possibility of § 12 Abs. 1 Nr. 1 AtG to normatively determine certain requirements as to the safety-technological design of nuclear plants. There are, however, criteria and guidelines on the part of the executive as well as recommendations and codes of technical rules for the design of nuclear power plants. Among them are the safety criteria issued by the Federal Minister of the Interior.

(b) *Concretization by the radiation protection decree, and the problem of the residual risk.* The required damage precautions according to the state of science and technology must be taken for each operational state of the plant. It is usual to distinguish between normal operation, malfunction and accident (hypothetic malfunction).

Normal operation is the proper operation of the plant. The admissible discharge of radioactive substances from the plant via waste air or waste water during normal operation is limited in § 45 StrlSchV by the demand that at the location with the most unfavorable exposure and under consideration of all relevant exposure pathways including the food chain, the radiation exposure of humans is to be kept as low as possible and must amount to, at most, 0.3 mSv or 30 mrem per year (effective dose). The dose limits established here must not be exceeded and are of a third-party protective nature; a potentially affected third party has the right to claim their observance.

The BVerfG points out that, by means of the determination of these dose limits, the required damage precaution in consideration of the state of science and technology is concretized according to § 7 Abs. 2 Nr. 3 AtG, and confirms with the following detailed reflections that the scope of the authorization norm is adhered to:

> „The dose limits according to the '30-millirem-concept' are far below the maximal admissible doses according to the Euratom basic norms (see Art. 11 of the directive of the council of 1 June 1976 (76/579/Euratom, Abl. Nr.

L 187/1), which again are based on the internationally acknowledged recommendations of the International Commission on Radiological Protection (ICRP); according to these recommendations, the whole-body dose limit is 500 millirem per year. The directive of the council of 15 July 1980 (80/836/Euratom, Abl. Nr. L 246/1) also retains the individual dose of 500 millirem per year for exposure of the whole body (Art. 12 Abs. 2), where it is irrelevant for the legal controversy concerned here that the calculation method has been changed and calculations are now carried out according to the concept of the 'effective dose'. After all, the dose limits of § 45 StrlSchV still lie within the regional variation range of natural radiation exposure (see the information in the official grounds to the StrlSchV, BR-Drucks. 375/76, 12); in practice, they are calculated on the basis of pessimistic assumptions, which thus include additional security reserves. The hypothesis of a linear dose response relation on which they are founded can hardly lead to an overestimation of the risk linked to them (see recommendations of the International Commission on Radiological Protection, number 26 (1978), 11; Rausch, Fourth German symposium on nuclear law (1976), 277 (281)); this risk is lower than the risk linked to the natural radiation which every individual is inescapably exposed to from the beginning of his or her life, and it is by several orders of magnitude lower than other civilization and life risks (see the statement „Gefährdung durch Kernkraftwerke" (endangerment by nuclear power plants) by the scientific advisory council of the German federal medical board in: Deutsches Ärzteblatt 1975, 2821 (2824)). According to the standards of practical reason, it had therefore not to be taken into account." (BVerfGE, DÖV 1981, 295).

The radiation minimization rule for normal operation in the StrlSchV makes additional demands on planning, design and operation of the plant. As an expression of precaution thinking, which has to include the radiation risk for the entire population in its considerations, this radiation minimization rule, when put into practice in the licensing procedure, regularly results in discharge rates for waste air and waste water being determined by the license which ensure an immission below the dose limits of the radiation protection ordinance for normal operation. However, the individual has no right to claim adherence to this radiation minimization rule; third-party protection cannot be legally claimed here.

In comparison to normal operation of the plant, a *malfunction* is a course of events which, if occurring, require that the operation of the plant must no longer continue for safety reasons. These are events which are unlikely according to the state of science and technology, but which cannot be ruled out and for which the plant must be designed. The plant must therefore, according to § 28 Abs. 3 StrlSchV, be designed in a way that if such events take place, no inadmissible emission of radioactivity can occur. As to such malfunctions, the so-called „malfunction limits“ according to § 28 Abs. 3 of the StrlSchV (effective dose 50 mSv) must be observed. Even in the case of a malfunction – and this is to be taken into account when the plant is being designed – the rule holds that radiation exposure is to be kept as low as possible below the dose limits established in the radiation protection decree. The dose limits mentioned here refer to the protection of persons outside the so-called internal and external company control areas. For persons inside these control areas, the StrlSchV determines other dose limits.

With regard to the other category, the *accident* or the hypothetic malfunction, § 7 Abs. 2 Nr. 3 AtG demands that the plant is designed on the basis of all safety systems, which are often redundant, in a way that such an event can practically be ruled out. The scientific-deterministic proof of such an impossibility is the goal to be aimed at here; in support of this endeavor, the negligibility of the residual risk is to be proved on the basis of probabilistic risk considerations, for which a high degree of experience in the field of reactor technology is already available.

The check of whether the requirements of § 7 Abs. 2 Nr. 3 AtG are met focuses primarily on the question of whether a certain event could actually not lead to an inadmissible release of radioactivity from the plant, in other words, whether this is indeed only a „hypothetic“ malfunction. The question about the safety standards is decisive: when is additional precaution needed, when does the acceptable residual risk begin?

Since the already mentioned „Kalkar decision“ of the BVerfG of 8 August 1978 (E 40, 89), jurisdiction has oriented itself by the following criteria formulated in the decision:

> „As far as the damages to life, health and property are concerned, the legislator has set up a standard by the principles of best possible danger defense and risk precaution established in § 1 Nr. 2 and in § 7 Abs. 2 AtG, which allows to grant a license only if it appears to be practically impossible according to the state of science and technology that such damaging events will take place (see Breuer, DVBl. 1978, 829ff, 835f). Uncertainties beyond this threshold of practical reason have their cause in the boundaries of human cognitive capacity; they are inescapable and insofar have to be born as socially adequate loads by all citizens. With the present form of the nuclear law, a violation of protection duties on the part of the legislator can thus not be found.“

Hence it follows that:
- best possible defense of dangers and risk precaution is imperative, on the basis of § 7 Abs. 2 Nr. 3 AtG.
- a dangerous situation, in which the eventuation of a certain course of events with damaging results is sufficiently likely must not occur in any case. Corresponding precautions must be taken against such a situation until, according to the standard of practical reason and in consideration of the state of science and technology, the course of events with damaging results is ruled out.
- an absolute exclusion of damage is not even demanded in the field of danger defense, but the principle must be respected that the greater the possible extent of a damage, the lower the probability of its eventuation must be.
- risks – in other words, possibilities of the eventuation of a damage – which have their cause in the boundaries of human cognitive capacity are to be accepted as socially adequate; they cannot be ruled out with any technology and are therefore „inescapable".
- the standard of practical reason distinguishes risks with danger quality from risks without danger quality.
- in the field of risks without danger quality – „below the threshold of danger" – further precaution measures are to be taken, yet subject to technological feasibility as well as to the proportionality of costs and benefits.

§ 7 Abs. 2 Nr. 3 AtG does not mention the concepts of danger and risk; the law merely uses the concept of necessary damage precaution. Yet, in doing so it does not abandon the usual legal evaluation of safety-technological phenomena, as the BVerfG appropriately expounds; the concept of danger and the concept of risk, which mean graded demands on the plant, are rather inherent in the wording of the law.

With the safety standard designed in this way, the BVerfG refers to that formulation of the conventional safety-technological concept of danger which establishes a connection between the extent of damage and the probability of its eventuation; new, however, is the difference between danger defense – which has strictly to be undertaken – and risk precaution – which has to be taken below the threshold of danger: the greater the possible damage is, the higher demands must be made on the degree of its unlikeliness.

4.4 ORGANIZATIONAL AND STRUCTURAL FACTORS IN ESTABLISHING ENVIRONMENTAL STANDARDS

Apart from the psychological perception of risks, the social evaluation of behavior relevant to the environment, and the legal regulations, the organizational structure of the administration establishing or implementing the standards also plays an important role in the development of environmental norms. This section focuses on an analysis of the structural factors which either necessarily result from the conditions set by an administration based on the division of labor or which have developed in the concrete reality of German administration.

Yet, it is not the method of decision making, i.e. the rules according to which decisions are made in organizations, that will be our main concern here (for this, see chapters 5 and 6). The often analyzed gap between decisions and implementation is also left aside (Mayntz et al. 1978; Knoepfel & Weidner 1982). The point here will rather be the interdependence between organizational structure (allocation of spheres of competence, division of labor, career, composition of the staff etc.) and establishment of standards under various external conditions. Since chapter 4 focuses on the deficiencies of the way society deals with environmental risks, here the main emphasis will be on the problems of an organizational realization of environmental policy. First of all, however, it should be clear that the environmental policy developed and executed in the German federal and state ministries already largely complies with our ideas of a rational procedure in the establishing of environmental standards.

Organizations are social institutions which execute a function for a specific purpose, on the basis of a division of labor and according to certain rules (Scheuch & Kutsch 1972, 170; Hartfiel 1972, 488; Mayntz 1972, 14f). Organizations are marked by five characteristics (Mayntz 1972, 40ff; Luhmann 1981, 354ff):
- by a structure of responsibilities and spheres of competence (formal hierarchy),
- by a delimitation from other social units which either provide information for the organization or which have to bear its decisions (interaction with the institutional environment),
- by an allocation of tasks within the organization according to certain rules and routines (division of labor),
- by a sequential splitting up of labor into different partial steps (process of decision making),
- by a dualism of the behavior of their members, on the one hand being task-oriented and on the other hand maintaining the organization (internal dynamics).

With growing differentiation of the social division of labor and the sociation of social functions, organizations are becoming necessary elements of a social order (Weber 1973, 136). The complexity of the collective performance of tasks demands a competent cooperation between individuals and social groups, which orients by an expedient rationality and maintains continuity. This holds mainly for administrations; here, in the past decades, a shift of emphasis from the order and service function to a political structuring function has gained ground (Mayntz 1973, 99). Structuring means to act and not only to react; structuring is a future-oriented and complex activity.

As necessary and indispensable the organization of collective tasks is in a differentiated society, as much do the choice or growth of organizational structures entail disadvantages for an accomplishment of tasks which orients by an expedient rationality. These disadvantages are often linked to the specific organizational form. Nevertheless, a number of general problems of organizations entrusted with collective tasks can be identified:

(a) Division of labor and sequential solving of problems often lead to a fragmentation of proceeding and to a lack of cooperation between working units with related functions (Scharpf 1975, 85; Müller 1986, 6). The classic example of this is the systematic shifting of pollution from one medium to another (e.g. from air to water and from there to solid waste).

(b) Since not all conceivable functions can be covered and an organization is always forced to reduce complexity (i.e., it is less complex than the matter which it is regulating), gaps develop within the system, which often remain unnoticed until an actual crisis occurs (Luhmann 1981, 367). Gaps in the system are especially problematic when organizations are not flexible enough to react with structural adjustments to changed general conditions or to external expectations.

(c) Since tasks are transformed into routines and procedures are bound to regulations, interdependences of functions are often overlooked and consequences of interventions are not considered (Scharpf 1973b). Moreover, when career and department-internal prestige are linked to the observance of the regulations, and an external control of the decision results either does not take place or has no internal influence on positive or negative sanctions, it often comes to innovation inhibitions or immobility (Mayntz 1973).

(d) Organizations develop their own interests and routines which transcend the tasks which they handle. Most organizations strive for more autonomy, independence, the maintaining of their structure and the scattering of formal responsibility for institutional decisions (Müller 1986, 477). These self-interests might collide with the functional demands. The striving for institutional autonomy is often incompatible with the necessity of coordinating tasks with other institutions. The scattering of responsibility often leads to inconsistencies and incoherent problem solutions, because nobody ultimately feels responsible for the entire solution package. Obviously, there is neither the ideal organization that avoids these and other problems, nor are there magic formulas to eliminate the structural problems of organizations. Still, there are basic suggestions for structural innovations and reforms, which aim at improving the efficiency and transparency of organizational decision procedures (Mayntz & Scharpf 1973, 140ff). Such suggestions have to start from two central ideas: on the one hand there must be enough competence in organizations to elaborate proper and cost-efficient solutions to complex problems; on the other hand, organizations must be able to react flexibly to external changes and new demands (Luhmann 1970, 163f; Grottian & Murswieck 1974, 29). In the first case, the capacity to solve problems is concerned, in the second case, the basic openness towards new tasks.

4.4.1 Structure of the Ministerial Administration in Germany

Ministries are political organizations which actively prepare decisions and carry through political resolutions on the basis of legally or politically set targets. Like every other large organization, they develop their own dynamics and sometimes determine the outcome of social conflicts about political priorities. They are neither neutral agents and bodies executing political programs, nor are they an extension of social interest groups (Müller 1986, 12). Rather, they channel social demands, adjust politically set targets to their routine operations, and are characterized by own interests which are independent of their tasks (Weber 1973, 153ff). What are the most important characteristics of the ministerial structure in Germany and what strengths and weaknesses result from this structure?

(a) *Congruence of the allocation of tasks and the structure of the organization:* The fundamental unit in the organizational structure of every ministry is the departmental section: these sections are the working level of the departments. Every task within the authority of a department must be assigned to a departmental section (Müller 1986, 23ff). Hence every working unit can only handle the tasks for which it is responsible, with the capacities at its avail. If it wishes to influence the performance of tasks of other units, it can only do so on the basis of participation rights which arise from the factual connection of its own tasks to other fields of work.

(b) *Hierarchical structure of responsibilities:* On the lowest level rank the departmental sections, which are structured according to specific issues and can independently propose initiatives concerning their respective specific issue. At the same time, it is their task to react to politically set targets, to external pressure, or to proposals from related sections. Ideas or reactive statements are then passed on to the next higher units, the subdivisions. In mutual agreement between departmental section and subdivision, the proposals are gone over again and then passed on to the next higher level of the hierarchy: first to the heads of division and finally to the undersecretaries of state or to the minister (Mayntz 1973, 102). This formal structure of responsibilities is, however, loosened by informal contacts and coordination talks inside and outside the ministry and is thus handled less rigidly in practice.

(c) *Internal orientation by consensus*: The departmental section in charge is obliged to make all sections participate which are possibly concerned, according to the established allocation of duties or the nature of the matter (Müller 1986, 26ff). This participation mainly takes place in the form of a „negative coordination“ (Wahl 1973, 144). This means that other ministries involved may reject a proposal, but are not obliged to introduce a new constructive suggestion or to help elaborate one. The consequence is that proposals will put those interests at a disadvantage which have no advocate in one of the participating sections and that in most cases, the range of program suggestions is reduced and „coordinated down“ (Scharpf 1975, 86).

(d) *Parallel solution of conflicts:* Conflicts among departments are first handled on the same level within the hierarchy. Superiors are only included if no agreement can be reached on the parallel level. If necessary, the federal cabinet has to decide. Conflicts within a department are settled by the respective superior (Müller 1986, 30). Apart from special cases, there is no possibility to „appeal“ in this case.

(e) *Reflexive allocation of tasks:* Fields of work of departmental sections can overlap in two ways. First, there are basic sections that deal with interdepartmental issues and point at their consequences in the different fields. Then there are so-called reflexive sections which deal with the tasks of another department from the viewpoint of their own special tasks (Müller 1986, 17). Basic sections are found in all ministries (sometimes subdivided according to general issues); reflexive sections are fundamental elements of the Federal Chancellery and the Federal Ministry of Finance.

(f) *Anticipation of the interests of clients:* The respective departments are shaped by the central idea of „public welfare“ and identify themselves with the role of the advocate of the citizens affected by their field of work (clients are, e.g., businesses for the ministry of trade and commerce, or employees for the ministry of labor). Although external groups may exert direct influence by hearings and other forms of participation, the heads of the sections regard themselves not so much as mediators between different interest groups (as is the case with the adversarial style in the U.S.), but rather as administrators of social problems or tasks which are to be dealt with, according to expedient rationality, for the benefit of the clients who are affected by these tasks (Scharpf 1973a; Bauman & Renn 1989, 35)

(g) *Political weight of the departments:* Carrying through administrative proposals also depends crucially on the relative weight of the department and the political position of the respective head of department (Müller 1986, 23). The stronger the internal power position of the respective department is, the more likely it is that there will be new initiatives and innovative solution suggestions. Thus, the separation of environmental policy from the powerful Ministry of the Interior and the creation of a separate department might have reduced the political significance of environmental policy.[33]

Aside from these general structural characteristics of the ministerial bureaucracy in Germany, a number of more informal behavioral patterns can be identified which have developed from the administrative practice, referring mainly to choice and motivation of the staff. Apart from the top leaders of a ministry, all permanent employees are employed for life, in accordance with the traditional role of the civil servant as an expert administrator of public welfare. This ideal of the independent expert, however, collides with the prevailing employment methods, according to which new members of staff are often only employed if they sympathize with the respective political orientation of the minister (i.e., if at least they are members of the same political party) (Müller 1986, 26ff). This has led to an increasing politicalization of the ministerial civil servants. This may have three negative consequences: the continuity with which tasks are carried out, which is typical for bureaucracies, is interrupted or at least disturbed, the energy for structural reforms is used up by these political adjustments, and a problem management orienting by expedient rationality is easily replaced by the political conviction test of party-membership or by opportunism. As a counter-tendency to this politicalization, the ministries have developed counterweights by internal regulations and career channels.

While the representatives of the staff (staff councils for civil servants) want to grant prospects of promotion according to the principle of seniority (and thus favor inertia and persistence), many superiors in the departments are interested in promoting those employees which incrementally perform their duties in faithful observance of the regulations – and without taking risks (Scharpf 1973a, 89; Mayntz 1982, 190; Müller 1986, 20ff). Such civil servants, who are often called indispensable experts, run less of a risk of being transferred or ignored when it comes to political shifts. If career chances are bound up with incremental working methods, however, sensible innovations are blocked and individual initiatives are throttled.

The ignoring of problems or the insisting on traditional routines is to be expected primarily when the pressure from outside is great and internal solidarity seems necessary to ward off external attacks (Müller 1986, 468). Problems which demand innovative solutions are only taken up if the heads of sections are sure that they have the full support of their superiors, primarily of the ministers. If such support is doubtful, they will rather turn down an initiative or suggest a solution which is routine-oriented. Further-reaching solutions may also fail because other departments have rights of participation and use their possibilities of influence to reject the suggestion. This is to be expected all the more in the case of innovative suggestions, since they are mostly connected with significant risks and moreover have often unknown effects on other fields of work. Since the system is ruled by the veto principle, support helps less than opposition harms (Scharpf 1973a, 178).

Yet, incremental methods of action and routine orientations also have positive consequences. Since politics, the media and the public mostly react to crises and often make too high demands on selective solutions of problems, the sections responsible for certain tasks approach solutions more systematically and less emotionally. Thus, they can maintain their relative autonomy towards the policy of the day. This has, for example, led to a largely nation-wide and comprehensive environmental regulation in Germany, whereas in the U.S., the Environmental Protection Agency (EPA) has achieved only partial regulatory success because of stronger politicalization and the connected necessity to manage conflicts (Uppenbrink 1974). Thus, effective and nation-wide initiatives can, to a far extent, be conceived in the sections (without public pressure) and then be introduced into public debate. This capacity of defining problems is a major power factor of bureaucracies (Andritzky & Wahl-Terlinden 1978, 31; Müller 1986, 469f).

Routine handling of problems results in proposals, which are presented to the social forces while internal conflicts are still carried out, or afterwards. Then, these external forces can only react and get changes accepted by means of public pressure. Conversely, initiatives of external groups to introduce their own suggestions into the bureaucracy have failed in most cases up to now. For even under great political pressure, the ministerial bureaucracy is able to successfully refuse to deal with the respective task.

The power of defining the problem at the section level has the advantage that, independent of the ecological damage that is referred to as crisis at the moment, solutions to problems can systematically and consistently be worked out. But at the same time, it has the disadvantage that legitimate initiatives from outside are regularly blocked, even if they point at necessary corrections. The blocking of external demands most often takes place by means of two strategic tricks: postponement and packaging (Müller 1986, 470ff). In the first case the „hot potato" is sent through the official channels, hoping that it takes less time for the issue to cool down than for the the matter to be officially dealt with. In this case, the externally demanded setting of priorities is signalized by granting additional research funds and employing additional staff, without changes being necessary in the policy sector. In the second case, a proposal which is already being worked on is packaged in a way that it appears to be an answer to the presently prevailing ecological crisis. Illustrative material for this case is provided by almost every answer to a parliamentary interpellation.

The necessity to reach agreements on proposals across the borders of departments reinforces the mechanism of blocking external demands as well as the incremental way of proceeding in the choice of ecological measures (Mayntz & Scharpf 1973, 126). This will, above all, be the case when zero-sum games are concerned, i. e., when there will be winners and losers among the departments (Müller 1986, 38). The potential losers will spare no effort to make the proposal fail by means of negative coordination, i.e. by making use of their veto power. The measure introduced by a department either has to be moderated to an extent that the losses of the zero-sum game are equally distributed or the proposal is declared a matter of conflict by the superiors in charge, and then it depends on the respective power of the departments and the political circumstances in determining which side will win.

Normally, sections will avoid the way of fighting out conflicts, except if they are absolutely sure that they have internal and external political backing. Innovative and integrative solutions are only to be expected if either the distribution mass can be changed by means of positive coordination and all sections involved may gain by cooperation, or if the internal integration mechanism of the department is so well elaborated (e.g. by reflexive sections in other departments and basic sections in the own department) that interdepartmental issues can be tackled and integrative solutions can be negotiated among conflicting pluralistic interests (Mayntz & Scharpf 1973, 136). It has been empirically proved that such attempts depend on two conditions: a relatively strong power position of the respective department compared with other departments (perhaps even direct backing by the Chancellor) and congruence between internal objectives of the department and external demands (Müller 1986, 40). This was, e.g., the situation in the first offensive phase of environmental policy in the early seventies (Müller 1986, 465ff).

4.4.2 Integration of Environmental Policy in Public Administration

As a reaction to the reactor accident in Chernobyl, the German government decided in 1986 to separate the fields environment, reactor safety and nature conservation from the ministries which were responsible for them up to then and to integrate them into a new ministry. If and to what extent new cooperation structures have been developed in this ministry has not yet been systematically examined. The available literature on environmental policy therefore refers to the situation before 1986, when the ecological tasks were distributed among different ministries under the overall control of the Federal Ministry of the Interior. But the new situation which was caused by the creation of the Ministry for the Environment probably differs only little from the old structure. Although the field of nature conservation was excluded from the Federal Ministry of Agriculture and integrated into the new Ministry for the Environment, other important ecological tasks remain in the responsibility of the Ministry of Transportation, of the Federal Ministry for Youth, Family, Women and Health as well as of the Ministry for Regional Planning, Building and Urban Development, among others. For this reason, the ministerial structure in its new form since 1986 keeps following the basic idea of dividing ecological responsibilities up among different ministries under the overall charge of one department.

At first glance, a concentration of ecological responsibilities in one ministry seems to be beneficial, because the possibility of negative coordination is cut back and the probability of zerosum games is reduced. But this holds only to a certain extent:

(a) When responsibilities are distributed among different ministries, often allies can be made across ministerial borders. To stand up jointly for something helps to effectively overcome opposition from others and to decide the conflict in one's favor (Müller 1986, 40ff).

(b) If several ministries are included, there is a greater possibility to „appeal“ if the superior in one ministry wants to block the proposal. Since, at first, all conflicts are always fought out on the parallel level among the ministries, the veto of a superior in one department can be weakened by the support of an equal-ranking civil servant in another ministry; in this case the matter has to be passed on to the next higher level of the hierarchy (Müller 1986, 474ff).

(c) If environmental policy is not concentrated in one single department, this affects the way the ministry in charge of coordinating environmental policy handles problems internally. The transfer of tasks to other departments relieves the department in overall charge from the necessity to anticipatively introduce the various interests of the other departments into the integrative process, since this balancing act is already performed by the sections transferred to other departments. That leads to a more homogeneous environmental policy within the department in overall charge and to more solidarity among its civil servants, which results in a high internal positive coordination. From the point of view of the superordinate political goal, subordinate tasks in such departments therefore also have the nature of an outpost or a commissioner, which can favorably influence the solution of the overall problem. This holds, however, only for constellations of problems and tasks which bear the possibility of non-zero-sum games (Müller 1986, 484ff).

(d) A dividing up of ecological tasks among different departments can also be of help if the respective department in overall charge has a weak political position within the ministerial bureaucracy or if it is put under extraordinary external pressure. Under these circumstances, the environment sections in other departments can at least partly compensate for the weakness and stand up as allies in the struggle against external pressure. This, again, holds essentially only for non-zero-sum games, because otherwise the other ministries involved, weighing the conflicting interests of environmental matters and their own actual tasks, will give preference to the latter. Thus, both options, the concentration of all environmental responsibilities in a single department and the allocation of tasks to different departments, have their advantages and their disadvantages. For conceiving an integrative environmental policy and solving zero-sum games, the concentration strategy turns out to be more suitable; for cross-section measures, i.e. for the evaluation of the environmental friendliness of department-specific measures, the outpost strategy is more suitable (Müller 1986, 486). A combination of both is ideal.

4.4.3 External Pressure and Effects on the Outside World

So far, this analysis has largely referred to the effects of the organizational form on the accomplishment of environment-political tasks. Organizations, however, are not autonomous structures, but partly orient themselves by external influences and at the same time cause changes in their external environment. This environment can be subdivided into two components: the functionally preceding or succeeding institutions, and the third parties which influence the organizations in the course of the communication process. As to German environmental policy, functionally preceding organizations are of little significance. While in the U.S., for example, the legislative, the presidential executive, and social interest groups often have the power to define problems and actually make full use of this in practice, in Germany, it turns out, problem definitions neither on the part of the political formation of will, nor on the part of social interest groups have found their way into the ministerial administration. If anybody at all, the state research institutions and semiofficial associations, such as the AGU or the VDI, have taken up new problems and introduced them into the administrative environmental policy.

Politicians and interest groups try to pressure the departments into dealing with the respective current problem definitions in society, but the above-mentioned possibilities of refusal and of the postponement or packaging strategy hinder the success of such attempts to exert influence. Problems are defined in the departmental sections, where the topicality of subjects evidently plays an important indirect role in the selection process. The division of functions among the individual sections guarantees, however, that issues which are not topical are also dealt with. The more comprehensive the field of work of the individual sections is, the more likely a comprehensive environmental policy is to be expected.

Yet, on the output side, things look totally different. The committees of the upper and the lower house of the German parliament, the working group of the state environment ministers and the committee of their heads of divisions, the respective state ministries right down to the regional control authorities (e.g. the trade supervision authorities) have the possibility to change the decisions or recommendations of the department or are free to interpret them differently. Many times, proposals are rejected or modified during the deliberations of the „environment committee“ in the upper or the lower house of the parliament. However, the ministerial bureaucracy has reacted to the diversity of possibilities of influence of external forces with an anticipative orientation policy (Müller 1986, 26). Before proposals are presented to succeeding organizations, it is checked in advance whether opposition has to be expected. So, modifications are made before they are forced by external organizations.

Equally, the attempt is made to limit the discretion scope of the regional control and implementation authorities by means of detailed regulations. Here again, the comparison between the U.S. and Germany is helpful: the EPA is a decision and execution authority at the same time. Decisions of the EPA, once they are made, can only be appealed against by means of legal action: neither the legislature nor the states, nor even the president have a say in this matter. Therefore it takes longer for the EPA to make a decision. But once the decisions are made, they are implemented within relatively short time in a uniform and consistent way (if they are not appealed against in court). The EPA has regional offices whose responsibility it is to monitor the local environmental situation and to continuously control the observation of environmental standards. In contrast to that, the environmental authorities in Germany can react rapidly and effectively to current problems (such as forest decline); yet, the enforcement of the measures taken can often be delayed or diluted by long-winded negotiations with the legislative and the functionally succeeding instances, which change the substance of the measures.

The same holds for groups which are not functionally involved, but are indirectly participating in the decision process by public pressure and political exertion of influence. Each department feels obliged to a certain group of clients, but at the same time depends on compromises because of the participation rights of the other departments. At the same time, external demands of clients are modified and limited by the filter of the respective prevailing political programs (Mayntz & Scharpf 1973, 117). Participation and compatibility with the superordinate political line often cause an alienation between the clients and the respective department. This has become clear primarily in environmental policy: while in the beginnings of environmental policy, administrations and external environment associations were allies for an environmentally beneficial policy, an increasing polarization between environment administration and the environment associations was, above all, observed during the defensive phase of the seventies. The rather positive stance of the Ministry of the Interior towards nuclear energy might have contributed to this; but probably this alienation can be explained by the necessary obligation of coordination between the departments.

Usually, the administration is interested in a participation of groups concerned in the establishment of standards, because an early involvement of interest groups in the preliminary stages of socially fighting out a conflict helps to overcome unnecessary opposition. This coordination mostly takes place „behind closed doors" in Germany. Suggestions for environmental standards are presented to the organized social groups to have their opinion. In exploratory talks between the interest groups and the representatives of the environment ministries (and the experts of the Federal Environmental Protection Agency), the scope for compromises is successively explored (Bauman & Renn 1989, 31ff). The talks with the interest groups here often serve as a means to overcome a negative coordination with the ministries which regard the respective interest groups as their clients, by explicitly involving these clients.

In the political sciences, this negotiation procedure is called (neo)-corporatistic regulation style (Alemann & Heintze 1979; O'Riordan & Wynne 1987, Renn 1989c). As opposed to the American adversarial style, where opposing interests fight in a public political arena, the corporatistic style offers a number of advantages, but also several problems:

(a) The exclusion of the public from these negotiation talks has the advantage that interest groups do not have to pay attention to their public image so much and are rather open to factual arguments. Above all, the scientists who take part in the talks have an important integration function (Coppock 1985, 389f).

(b) The routine work of the departmental sections is not unnecessarily hampered by public pressure. The various teams can systematically continue their coordination work, even if the issue has long been banned from public discussion (Mayntz & Scharpf 1973, 132).

(c) If there is clear knowledge about an environmental damage, which no group can deny in the presence of experts, there will be rapid and effective solutions. This was the case, for example, with the ordinance on large-scale firing plants, when the damages to the forest due to acid rain started to become apparent (Knoepfel & Weidner 1982, 85f).

(d) But the same example of establishing standards in the case of large-scale firing plants also points at a significant disadvantage of the corporatistic style: the possibility of interest groups to protract solutions by endless negotiations when there is a high degree of uncertainty about the dose response relation and a lack of political pressure (Müller 1986, 18).

(e) In most cases, the organized environment associations remain excluded from the negotiations between the environment department and external interest groups (Bauman & Renn 1989, 33ff). This situation, which seems paradoxical at first sight, is explained by the simple fact that these associations are represented in the ministries by nobody else than by the environment section itself. Thus, the possibility of a negative coordination through participation of other ministries is inapplicable for these groups. Moreover, the environment associations were and still are only to a certain degree willing to adhere to the rules of the corporatistic style. While industry, the unions and other groups often benefit from the fact that the public is excluded from the negotiations, the environment associations deprive themselves of their most important political resource, namely public support, when they let themselves in for negotiations „behind closed doors" (Rucht 1982, 289ff).

(f) The de facto exclusion of the environment associations from the negotiations, negotiations behind closed doors, and the possibility to drag out negotiations until a current crisis arises, have led to the impression in the public that in environmental policy nothing occurs without political pressure and that the ministries, in a fiddling policy with the interest groups, are all too ready to sacrifice principles of environmental protection for economic necessities (Böhret 1982, 145; Fietkau et al. 1982). Compared with the environmental policy of other countries, this impression is at least one-sided, if not even wrong. No doubt, there are implementation deficits in environmental policy, and an integrative, all-media environmental jurisdiction is largely lacking. Still, in international comparison, Germany comes up with a comprehensive and effective environmental policy (Bauman & Renn 1989, 48ff; Coppock 1986, 25ff; Brickman & Ilgen 1982, 3ff). So, the impression of a lack of initiative in the field of environmental protection on the part of the government authorities is not so much due to a lack of effectiveness of environmental policy, but rather due to the perception of the decision process, where the non-transparency of the decision making, the exclusion of many social groups from the process of defining the problem, as well as the necessity to reach compromises in the preliminary stages of decision making almost inevitably give rise to the impression that ecology is treated like the stepchild of public politics.

4.4.4 Structural Suggestions for a More Effective Environmental Policy

The organizational structure of environmental policy is an integral part of the administrative structure of Germany. A change in this structure is therefore only possible insofar as it can be brought into line with the principles of the German political administrative culture and its structures, which have developed in the course of time (Dunn & Swierczek 1977, 142ff; Scharpf 1975, 120ff). All attempts to bring about a lasting change in the administrative structure have remained unsuccessful up to now (Böhret 1982, 138ff). Still, modification possibilities can be identified within the range of the presently existing ministerial structure, which are in congruence with the administrative culture and at the same time help to remove the inadequacies of environmental policy diagnosed above. If we consult the criteria for evaluating organizations suggested by Luhmann (Luhmann 1970, 163f), namely proved competence and basic openness towards social demands, the environmental policy in Germany comes off relatively well with regard to the first criterion, whereas it shows deficiencies with regard to the second one. The most important deficiencies are:

(a) There is a lack of integrative environmental policy due to the splitting up of fields of competence in sections with fixed duties.

(b) There is a lack of creative and innovative suggestions, due to the rights of participation of other departments and the negative coordination that is often linked to it (approval helps less than opposition harms).

(c) Incremental way of proceeding and adjustments to internal career criteria, which often have nothing to do with the achievements in the field of environmental policy, are promoted by internal regulations.

(d) Internal objectives of the organization, such as maintaining power, gaining a positive image in public, or avoiding too much work, are often substituted for objectives of environmental policy.

(e) By the strategies of packaging and postponement as well as by other measures of denial, important environment-political measures can be sabotaged or delayed.

(f) The monopoly of the power of defining problems constrains the participation of other organizations and interest groups in the phase of problem definition. They can either merely react or exert pressure on the political mandataries, provided they have enough public support.

(g) The corporatistic style of participation of external groups in the process of decision making has two serious disadvantages: first, the decisions can be put off until a current crisis arises, second, the public gains the impression that nothing is done in the field of environmental policy.

Since ministerial environmental policy comes off quite well with regard to the first criterion of effectivity, a rectification of the deficiencies analyzed here should not jeopardize the obviously positive aspects of environmental policy, especially the relative independence of current environment-political measures from topical environmental events. If we assume that, in future, environmental policy will increasingly have to be cross-section policy, i.e. has to check measures of the other departments for their environment-friendliness, reforms appear sensible and necessary. They have to be compatible with the organizational structure which has developed in the course of time and must not jeopardize the effectiveness of environmental policy, though. In the relevant literature, the following improvements of the system have been suggested:

- in order to strengthen positive coordination and non-zero-sum solutions, every ministry should establish its own environment section, which checks all measures that are suggested by the respective department (and that may influence environmental quality) for their environment-friendliness (Müller 1986, 524ff).
- the environment ministry should, in analogy to the ministry of finance, support these outpost sections in other departments by corresponding reflexive sections in its own department. The double appointment of section and reflexive section offers the advantage that zero-sum games are avoided within the specific ministry, but support of another ministry in cases of conflict is made possible (Müller 1986, 528ff).

- In the environment ministry, the basic sections should accelerate the progress of an integrative environmental policy, which helps to avoid particularism in establishing environmental standards and which prevents the so far common shifting of ecological damage from one medium to the other. Most of all, the cross-section function of environmental policy has to be emphasized by directing the individual fields of work in the sections in an integrative way, and by identifying interdependences between the development of specific programs and their consequences for the environment (Mayntz & Scharpf 1973, 135f).
- In preliminary negotiations with the interest groups, the environment associations should not be excluded, even if that means that the negotiations become more difficult and the possibility of a negative coordination with the other departments becomes more likely (Mayntz & Scharpf 1973, 123). In the long term, such an exclusion will lead to more legitimation deficits, which might paralyze a successful environmental policy.
- Social groups should be conceded more rights in an ordered procedure as to the problem definition. First, important information is lost if the power of defining problems is monopolized, second, here again there is an increasing pressure on the administration to legitimize its actions. Finally, a participation in problem definition also means shared responsibility for the ecological consequences. This way, the scapegoat function of politics can effectively be taken back (Bauman & Renn 1989).
- The fields of work of the functionally succeeding organizations should be streamlined as far as possible and the accomplishing of tasks should focus on contents. With the establishing of an environment committee in the German national parliament, the first step towards a successful direction control has already been made (Müller 1986). Coordination between the federal government, the state governments and the municipalities is basically also given, but needs further intensification. A superordinate authority should continuously evaluate the respective ecological measures, from their conception to their implementation, in order to verify that the intended results take place, and to identify the unintended side-effects (Mayntz 1973, 101ff). Only with this feedback mechanism can the process of environmental policy as a cross-section measure continuously be improved.
- The transparency of decision making has to be improved. The style of the negotiations should become more open and more public, without running the risk of degenerating into a spectacle for the public. For example, selected representatives of the public could be permitted to take part in the negotiations (Scharpf 1973b, 181) or a record transmitted to the press could lay open the arguments and the decision process based on them.

All of these suggestions aim at structural changes within the ministerial administration. Analogous ideas can, of course, also be developed for the states or the municipalities. Beyond organizational changes, however, it is also necessary to critically investigate the decision process within the ministerial bureaucracy. Above all, the necessity of transparency of the decision process and of rational justification of political measures imply the necessity of a decision-making process which orients by decision analytic procedures.

ANNOTATIONS

1 Compare, however, the critical objections to this theoretical position by Gigerenzer 1993. The bounded rationality of probabilistic mental models. In. K. I. Manktelowe, D. E. Over (eds.), Rationality, Psychological and Philosophical perspectives (pp. 284–313). London: Routledge.

2 See, for example, Kahnemann et al. 1982.

3 See, for example, Covello et al. 1985; Slovic 1987; Vlek & Stallen 1981.

4 For example, Luhmann 1986.

5 E.g., Douglas 1985 and Douglas & Wildavsky 1982.

6 See Kitschelt 1980, 18ff.

7 Huber, for example, proves in an empirical study on images of technology in the German population that 71% of the engineers, 38% of the civil servants, 25% of the artists and 15% of the people working in social fields have a utopian (intentionally positive) conception of technology (quoted from: Wiedemann & Jungermann 1989, enclosure 1, table 102).

8 Until the late seventies, the number of the opponents of nuclear energy among the electorate of the three political parties then represented in the German national parliament was almost identical (Renn 1984, 278f); ever since the coalition of CDU/FDP took over governmental power, a clearer polarization of the positions towards nuclear energy has developed between supporters of the CDU and the SPD (Schmitt et al. 1981, 516ff). According to a poll of the year 1983, 42% of the SPD voters, but only 26% of the CDU voters considered nuclear energy to be not environmentally friendly. More than 38% of the SPD voters believed that politicians should not accept nuclear power as an option of energy supply. Only 26% of the CDU voters thought alike (Wiedemann & Jungermann 1989, table 82). 32.5% of the CDU voters, but only 19% of the SPD voters called themselves staunch proponents of nuclear energy (ibid., table 50). After Chernobyl, however, the percentage of the proponents of nuclear energy has decreased significantly among the CDU electorate as well (Peters et al. 1987).

9 In a survey conducted by the German Sinus polling institute, 11% of the 15 to 30 year-olds expressed that they felt to be members of the peace movement, 10% regarded themselves as members of the environmentalist movement and 8% as members of the anti nuclear energy movement, whereas traditional German political junior organizations met with little response (4% for the young labor unionists, 2 % for the 'Junge Union', 1% for the 'Jungsozialisten', and 1% for the 'Jungdemokraten'. It is also interesting that the number of young people who classify themselves as being squatters is just as high as the number of those who belong to the two junior organizations of the SPD or the FDP. The numbers of sympathizers, too, lie relatively close together (18% for the 'Jungsozialisten', 16% for the 'Jungdemokraten'and 14% for the squatters). As Watts proves in his study, the preference for social movements is no longer merely a phenomenon among young people but has met with great response in the population. In 1986, 2% of the people interviewed in a representative sample survey saw themselves as active members of the anti nuclear energy movement, 17% as potential members and 20% as sympathizers. Similar results also turn up for other movements, such as the environmentalist movement or the feminist movement (Watts 1987, 52f).

[10] In a survey conducted by Renn in 1980 in five German sites of nuclear power plants (percentage of nuclear energy proponents was 44% there), only 10% expressed the viewpoint that citizens' initiatives against nuclear energy only stood for their own interests; 63% agreed to the opinion that citizens' initiatives were indicators of the fact that something was wrong, and 27% had no reservations about letting citizens' initiatives take part in political decision making (Renn 1984, 291). In 1986, 42% of the West German population called themselves activists, potential members or sympathizers of the environmentalist movement; the corresponding percentage for the anti nuclear energy movement is 39%, and for the peace movement 62% (Watts 1987, 52f).

[11] Asked whether the decision of the politicians in Germany to use nuclear energy should be accepted, 44% of the white-collar workers, 52% of high-school graduates, and 75% of interviewees with so-called postmaterialistic values (environment and social justice as central values), answered 'No', whereas the respective percentage for the entire sample was only 34% (Wiedemann & Jungermann 1988, table 78). In an international study conducted by the Institute for Environment and Society of the Scientific Center of Social Research in Berlin (WZB), social scientists ascertained a particularly strong connection between political ideology (conservative versus progressive) and the support of the anti nuclear energy movement. The correlation coefficient (Pearson's r) amounted to 0.45 in 1982 and to 0.40 in 1986 (Watts, 1987, 58).

[12] In a survey conducted in the U.S. in 1988, 46% of those interviewed were in favor of an increased use of coal for the energy supply, despite problems with acid rain and global warming. About 15% voted for maintaining the present share of coal, and about 25% wanted a reduction (Bisconti 1989). Only 26% of the interviewees would be willing to accept higher electricity prices in order to make coal less polluting. At the same time, however, 53% of the persons interviewed wanted stricter rules and regulations to reduce the pollution caused by coal. Surveys conducted in Germany in the early eighties show similar results concerning the issue of coal utilization and acid rain (Wiedemann & Jungermann 1989, 10ff).

[13] According to a survey of the WZB from 1986 in Germany, 2% classified themselves as activists in the anti nuclear energy movement, 17% as potential members, 20% as sympathizers, 8% as indifferent, 27% as critics and 27% as opponents of this movement (Watts 1987, 53). The number of the indifferent persons is lower than it is with regard to other comparable issues, whereas the polarization in two antagonistic camps is stronger.

[14] The study of the WZB proves for 1986 that leftist political opinions have a positive correlation coefficient of 0.40 with the mobilizing potential of the anti nuclear energy movement, and dissatisfaction with democracy has a positive correlation coefficient of 0.19 (Watts 1987). These correlations are considerably stronger for activists of the movement. In the period between 1982 and 1986, however, correlations with leftist ideology and the necessity of a social change have gone back strongly (ideology from 0.45 to 0.40; necessity of a social change from 0.33 to 0.19). This can be interpreted as an indication of the progressing integration of „average citizens" in the anti nuclear energy movement, mentioned below.

[15] As to this point, however, social scientists have different opinions: on the basis of his empirical studies, Renn comes to the conclusion that opponents and proponents of nuclear energy hardly differ in their basic values (Renn 1984, 261ff), whereas Bechmann et al. find considerable value differences between the two groups in the survey results of the studies conducted by the nuclear research center Karlsruhe, Germany (Bechmann et al. 1981).

16 While Douglas and Wildavsky presume that there are two antagonistic groups, the traditional approach of the cultural-anthropological risk concept is based on a division into four groups, each with different reaction patterns to environment and technology conflicts. The four groups are: „competitive/individualist", „bureaucratic/hierarchical", „egalitarian" and „stratified individualism" (Rayner 1987, 13).

17 In the course of the past ten years, the readiness to accept a compromise between environment protection and economic costs has decreased considerably. In an American survey, the percentage of persons who wanted more environment protection, no matter how much it would cost, rose from 45% in 1981 to 66% in 1986 (Dunlap 1987). The Allensbach polling institute obtained similar results in a time-series analysis from 1979 to 1984 (Institut für Demoskopie Allensbach 1984).

18 In almost all German surveys concerning nuclear energy, the persons interviewed overestimated the numerical strength of the opponents of nuclear energy. The observation that minorities can create the impression of a majority opinion by means of high commitment and politically effective actions, is one of the core elements of the concept of the spiral of silence developed by Noelle-Neumann (1980).

19 According to the RWE, the final price of a light water nuclear power plant of 1300 megawatt has increased by approximately 195% from 1975 to 1985 (11% per year). However, the costs for hard-coal power stations have also increased heavily, due to desulfurization and denitration installations (Uffmann 1986, 493).

20 Although there is sufficient anecdotal evidence for this thesis, and most cultural groups tend to express negative opinions and to support opponents of nuclear power if, at all, they take a stand concerning nuclear energy, there is still no empirically verified proof for it. Indirect instances can be named here at best. Keeney and others have proved in a study on the preferences of engineers versus persons with jobs in the cultural sector that energy scenarios including nuclear energy were far more popular with the engineers than with the group active in the cultural sector (Keeney et al. 1986). In a study conducted by Frederichs and others in Germany, 65.1% of the persons interviewed approved of the following statement: „The majority of the politicians elected by us have decided in favor of nuclear energy. We must accept that." Among the white-collar interviewees, only 50.4% approved of this statement, and among persons with a postmaterialistic value attitude, only 23.5% did so (Frederichs et al. 1983).

21 Interestingly, Peters observed that conflict parties increasingly use the media in order to get in touch with other conflict parties or to strengthen their negotiation position. To do so, they use „open letters" or „public demands to ...". The addressed conflict partners are then indirectly forced to react to the demand publicly or to signal their readiness to negotiate (Peters 1984 and 1986).

22 Still, comparative studies about the self-image of journalists in different countries show that the interviewed journalists in Germany, more than their British colleagues, tend to regard their news coverage as a political job and not so much as a distanced observation of events (Köcher 1986). Nevertheless, deliberate manipulation of opinion in German media is the exception rather than the rule (Otway et al. 1987; Lichtenberg & MacLean 1988).

23 Socio-psychological research distinguishes between a central and a peripheral route to change opinions or attitudes. With the central route, the shift of opinion takes place on the basis of an intensive analysis of the arguments of the respective adversaries; with the peripheral route, issue-independent signals, such as the public prestige of the information

source, the featuring of the information, sympathy for the informers, and other aspects are consulted for an assessment of trustworthiness and as a cause for a change of opinion (Petty & Cacioppo 1986; Renn & Levine 1988).

24 What is to be called a resource in a society is not undisputed in sociological literature. The approach represented here primarily goes back to the conflict theory of Coser, who regards conflicts as results of the competition for scarce resources. According to his classification, there are three original resources: power, wealth and prestige (Coser 1967, 289ff). A similar classification is found in Merton (1968). Other authors prefer the concept of generalized media of exchange. Parsons distinguishes four generalized media which are assigned to the social fields of „economy", „politics", „social community" and „culture". These are: money, power, social commitment and influence (Parsons 1963, 399, and 1976, 302). In a critical statement on the media of exchange suggested by Parsons, Luhmann suggests another media scheme related to communication. He refers to the four generalized media: truth, love, money (or belongings, assets) and power (Luhmann 1975, 175). A mixture between Parsons and Luhmann can be found in Münch: „Money serves for exchanging it against individually desired goods and services, political power serves to carry through collectively binding decisions, arguments serve to convince others of truth, validity and correctness of symbolic constructions, and commitments serve to oblige others to the norms of the society" (Münch 1982, 103).

25 See also Jungermann 1986, 101.

26 See Bischof 1982, 585; Bruha 1984, 39f.

27 Since BVerwG DVBl. 1972, 678 (680), see also BVerfG 53, 30 (58).

28 See Kramer & Zerlett 1980, § 28 StrlSchV Anm. 11; Schattke 1982, 1083 (1085).

29 See BVerwG 61, 256 (267); BVerfG 49, 89 (143).

30 See official grounds in BT-Drs. 385/76, 12.

31 For a criticism of the development of the law, see: Peine 1988, 115; Czajka 1987, 556.

32 § 9 should be considered, see also Czajka 1987, 556f.

33 Concerning this point in general: Mayntz & Scharpf 1973, 142f; specifically referring to environmental policy: Müller 1986, 535ff

LITERATURE

Adams, W.C. (1986): Whose Lives Count?: TV Coverage of Natural Disasters, Communication 36, Nr. 2, 113–122.

Alemann, U. von, Heintze, J. (1979): Neo-Korporatismus. Zur Neuen Diskussion eines alten Begriffes, Zeitschrift für Parlamentsfragen 10, 469–482.

Andritzky, E.A., Wahl-Terlinden, U. (1978): Mitwirkung von Bürgerinitiativen an der Umweltpolitik. Forschungsbericht 77, Institut für Zukunftsforschung, Berlin, Umweltbundesamt.

Bandura, A. (1986): Social Foundations of Thought and Action: A Social Cognitive Theory, Englewood Cliffs, NJ., Prentice Hall.

Battelle Institute (1978): Die Kernenergie-Kontroverse im Spiegel der Tageszeitungen – Inhaltsanalytische Auswertung eines exemplarischen Teils der Informationsmedien. Bericht für das Bundesministerium des Inneren, Bonn.

Beck, U. (1988): Gegengifte – Die organisierte Unverantwortlichkeit, Frankfurt a.M., Suhrkamp.

Beck, U. (1990): Vom Überleben in der Risikogesellschaft, in: Risiko und Wagnis: Die Herausforderung der Industriellen Welt, M. Schüz (ed.), Pfullingen, Neske, 12–31.

Becker, G., Berg, J.V., Coenen, R. (1980): Überblick über empirische Ergebnisse zur Akzeptanzproblematik der Kernenergie, Karlsruhe, Kernforschungszentrum Karlsruhe.

Berger, P., Berger, B., Kellner, H. (1973): Das Unbehagen in der Modernität, Frankfurt a.M., Campus.

Berger, P., Luckmann, T. (1969): Die gesellschaftliche Konstruktion der Wirklichkeit. Eine Theorie der Wissenssoziologie, Stuttgart, Teubner.

Bischof, W. (1982): Zwischenstaatliche Übereinkommen über die gegenseitige Hilfe in radiologischen Notfällen – Überblick über die geltenden völkerrechtlichen multilateralen und bilateralen Vereinbarungen, Energiewirtschaftliche Tagesfragen 32, 585–590.

Bischof, W., Pelzer, N. (1983): Das Strahlenschutzrecht in den Mitgliedstaaten der Europäischen Gemeinschaften, vol. II Bundesrepublik Deutschland, Baden-Baden, Nomos Verlagsgesellschaft.

Bisconti, A.S. (1989): Polling an Inattentive Public: Energy and U.S. Public Opinion, in: Energy and the Public, Welt-Energiekonferenz, Subcommittee on Energy and the Public, Vol. 2, Country Reports, London, WEC.

Böbret, C. (1982): Reformfähigkeit und Anpassungsflexibilität der öffentlichen Verwaltung, in: Politikwissenschaft und Verwaltungswissenschaft, J.J. Hesse (ed.), Opladen, Westdeutscher Verlag, 134–150.

Beuwe, E. (1978): Gefahrenabwehr und Risikovorsorge im Atomrecht – Zugleich ein Beitrag zum Streit um die Berstsicherung für Druckwasserreaktoren, Deutsches Verwaltungsblatt 93, 829–839.

Brickman, R., Ilgen, T. (1982): Controlling Chemicals: The Politics of Regulation in Europe and the U.S., Ithaca, Cornell University Press.

Bruha, T. (1984): Internationale Regelungen zum Schutz vor technisch-industriellen Umweltnotfällen, Zeitschrift für ausländisches öffentliches Recht und Völkerrecht 44, 1–63.

Bundesministerium des Inneren (1984): Kernenergie oder Sonnenenergie?, München, Beck.

Chaiken, S., Stangor, C. (1987): Attitudes and Attitude Change, Annual Review of Psychology 38, 575–630.

Chase, W.G., Simon, H.A. (1973): Perception in Chess, Cognitive Psychology, 4, 55–81.

Conrad, J. (1981): Changing Lifestyles: Economic Stagnation, Postmaterialistic Values and the Role of the Energy Debate, in: Beyond the Energy Crisis – Opportunity and Challenge, R.A. Fazzolare & C.B. Smith. vol. 4, Oxford, Pcrgamon, 2127–2134.

Cook, J. (1985): Nuclear Follies, Forbes 2, 82–100.

Coppock, R. (1985): Interactions between Scientists and Public Officials: A Comparison of the Use of Science in Regularatory Programs in the United States and West Germany, Policy Sciences 18, 371–390.

Coppock, R. (1986): Regulating Chemical Hazards in Japan, West Germany, France, the United Kingdom, and the European Community: A Comparative Examiniation, Washington D.C., National Academy Press.

Coser, L.A. (1965): Theorie sozialer Konflikte, Neuwied, Luchterhand.

Coser, L.A. (1967): Sozialer Konflikt und sozialer Wandel, in: Sozialer Wandel, H.P. Dreitzel, Neuwied, Luchterhand, 278–294.

Covello, V.T., Johnson, B.B. (1987): The Social and Cultural Construction of Risk: Issues, Methods and Case Studies, in: The Social and Cultural Construction of Risk, V. T. Covello & B.B. Johnson (eds.), Dordrecht, Reidel, VII-XV.

Covello, V.T., Mumpower, J.L., Stallen, P.J.M., Uppuluri, V.R. (eds.) (1985): Environmental Impact Assessment, Technology Assessment, and Risk Analysis, Berlin, Springer.

Czajka, D. (1987): Das Strahlenschutzvorsorgegesetz, Neue Zeitschrift für Verwaltungsrecht 6, 556–559.

Daamen, D.D.L., Verplanken, B., Midden C.J.H. (1986): Accuracy and Consistency of Lay Estimates of Annual Fatality Rates, in: New Directions in: Research on Decision Making. B. Brehmer, H. Jungermann, P. Lourens & G. Sevón (eds.), North Holland, Elsevier, 231–243.

Deubner, C. (1977): Die Atompolitik der westdeutschen Industrie und die Gründung von EURATOM, Frankfurt a.M., Campus.

Dietz, T., Stern, P.C., Rycroft, R.W. (1989): Definitions of Conflict and the Legitimation of Resources: The Case of Environmental Risk, Sociological Forum 4, 47–69.

Douglas, M. (ed.) (1982): Essays in the Sociology of Perception, London, Routledge and Kegan.

Douglas, M. (1985): Risk Acceptability According to the Social Sciences, New York, Russell Sage Foundation.

Douglas, M., Wildavsky, A. (1982): Risk and Culture, Berkeley, University of California Press.

Dunlap, R.E. (1987): Public Opinion on the Environment in the Reagan Era, Environment 29, Nr. 6, 7–11, 32–37.

Dunn, W.N., Swierczek, F.W. (1977): Planned Organizational Change: Toward Grounded Theory, Journal of Applied Behavioral Science 13, 135–157.

Eagly, A.H., Chaiken, S. (1984): Cognitive Theories of Persuasion, Advances in Experimental Social Psychology 17, 268–359.

Etzioni, A. (1967): Mixed Scanning. A Third Approach to Decision Making, Public Administration Review 27, 385–392.

Farhar-Pilgrim, B., Freudenburg, W.R. (1984): Nuclear Energy in Perspective: A Comparative Assessment of the Public View, in: Public Reactions to Nuclear Power: Are there Critical Masses?, W.R. Freudenburg & E.A. Rosa (eds.), Boulder, Westview Press, 183–203.

Jungermann, H. (1982): Zur Wahrnehmung und Akzeptierung des Risikos von Großtechnologien, Psychologische Rundschau 33, 217–238.

Jungermann, H. (1986): Die öffentliche Diskussion technologischer Mega-Themen: Eine Herausforderung für Experten und Bürger, in: Die Analyse der Sozialverträglichkeit für Technologiepolitik, H. Jungermann, W. Pfaffenberger, G.F. Schäfer & W. Wild (eds.), 1992, HTV-Edition Technik und sozialer Wandel, München, High Tech, 92–101.

Jungermann, H., Slovic, P.: Die Psychologie der Kognition und Evaluation von Risiko, in: Risiko und Gesellschaft, G. Bechmann (ed.), Opladen, Westdeutscher Verlag.

Kahnemann, D., Slovic, P., Tversky, A. (eds.) (1982): Judgement under Uncertainty: Heuristics and Biasis. Cambridge, Mass., Cambridge Univ. Press.

Kasperson, R.E. (1989): Nuclear Power in International Comparison: An Outlook to the Year 2000 and Beyond. Unveröffentlichtes Manuskript, Worcester, Clark University.

Kasperson, R.E., Palmlund, I. (1987): Evaluating Risk Communication, Unveröffentlichtes Manuskript, CENTED, Worcester, MA und Stockholm, Clark University und Universität von Stockholm.

Kasperson, R.E., Renn, O., Slovic, P., Brown, H.S., Emel, J., Goble, R., Kasperson, J.X., Ratick, S. (1988): The Social Amplification of Risk: A Conceptual Framework, Risk Analysis 8, 177–187.

Keck, O. (1984): Der schnelle Brüter. Eine Fallstudie über Entscheidungsprozesse in der Großtechnik, Frankfurt a.M., Campus.

Keeney, R., Winterfeldt, D. von, Wetter, T. (1986): Preferences and Values among Engineers and Artists with Respect to Energy Scenarios, University of Southern California, Los Angeles, Kepplinger.

Kitschelt, H. (1980): Kernenergiepolitik-Arena eines gesellschaftlichen Konflikts, Frankfurt a.M., Campus.

Kitschelt, H. (1986): New Social Movements in West-Germany and the United States, Political Power and Social Theory 5, 286–324.

Klages, H. (1984): Wertorientierungen im Wandel, Frankfurt a.M., Campus.

Knoepfel, P., Weidner, H. (1982): Formulation and Implementation of Air Quality Control Programmes: Patterns of Interest Consideration, Policy and Politics 10, 85–109.

Köcher, R. (1986): Bloodhounds or Missionaries: Role Definitions of German and British Journalists, European Journal of Communication 1, 43–64.

Kramer, R., Zerlett, G. (1980): Strahlenschutzverordnung, Kommentar, 2. ed. Köln, W. Kohlhammer und Deutscher Gemeindeverlag.

Lackner, K. (1987): Strafgesetzbuch mit Erläuterungen, 17. ed., München, C.H. Beck.

Lee, T.R. (1986): Effective Communication of Information about Chemical Hazards. The Science of the Total Environment, 51, 149–183.

Levi, D.J., Holder, E.E.: Psychological Factors in the Nuclear Power Controversy, Manuscript Dept. of Psychology and Human Development, San Luis, Obisco, Cal., Cal. Polytechnic State Univ.

Lichtenberg, J., MacLean, D. (1988): The Role of the Media in Risk Communication, in: Risk Communication, H. Jungermann, R.E. Kasperson & P.M. Wiedemann, Jülich, Kernforschungsanlage, 33–48.

Lindbloom, C. (1959): The Science of Muddling Through, Public Administration Review 19, 79–99.

Lindell, M.K., Earle, T.C. (1980): Public Attitudes toward Risk Tradeoffs in Energy Policy Choices. Abschlußbericht, Seattle, Battelle Human Affairs Research Centers.

Lowrance, W.W. (1976): Of Acceptable Risk: Science and the Determination of Safety, Los Altos, William Kaufman.

Lübbe, H. (1987): Technischer Wandel und die individuelle Lebenskultur, in: Fortschritt der Technik – gesellschaftliche und ökonomische Auswirkungen,

H. Lübbe (ed.), Heidelberg, Decker, 49–63.

Luhmann, N. (1970): Soziologie des politischen Systems, in: N. Luhmann, Soziologische Aufklärung, vol. 1, 3. ed., Opladen, Westdeutscher Verlag, 154–177.

Luhmann, N. (1975): Soziologische Aufklärung, vol. 2, 2. ed., 1982, Opladen, Westdeutscher Verlag.

Luhmann, N. (1981): Organisationstheorie, in: N. Luhmann, Soziologische Aufklärung, vol. 3, Opladen, Westdeutscher Verlag, 335–413.

Luhmann, N. (1986): Ökologische Kommunikation, Opladen, Westdeutscher Verlag.

Lynn, EM. (1986): The Interplay of Science and Values in Assessing and Regulating Environmental Risks, Science, Technology, and Human Values 11, Nr. 5, 40–50.

Machjlis, G. (1989): Environmental Policies and Social Science: Lessons of Reevaluation, Unpublished Manuscript, Hartfield, University of Idaho.

MacLean, D. (1987): Understanding the Nuclear Power Controversy, in: Scientific Controversies: Case Studies in the Resolution and Closure of Disputes in Science and Technology, H.T. Engelhardt, Jr. & A. L. Caplan (eds.), Cambridge, Cambridge University Press, 567–582.

Mayntz, R. (1972): Soziologie der Organisation, 6. ed., Reinbek, Rowohlt.

Mayntz, R. (1973): Probleme der inneren Kontrolle in der planenden Verwaltung, in: Planungsorganisation. Die Diskussion um die Reform von Regierung und Verwaltung des Bundes, R. Mayntz und F. Scharpf (ed.), München, Piper, 98–106.

Mayntz, R. (1982): Soziologie der Öffentlichen Verwaltung. 2. ed., Heidelberg, C.F. Müller.

Mayntz, R., Derlin, H.-U., Bohne, E., Hesse, B., Hucke, J., Müller, A. (1978): Vollzugsprobleme der Umweltpolitik. Empirische Untersuchung der Implementation von Gesetzen im Bereich der Luftreinhaltung und des Gewässerschutzes, in: Materialien zur Umweltforschung, Rat der Sachverständigen für Umweltfragen (ed.), vol. 4, Stuttgart, Kohlhammer.

Mayntz, R., Scharpf, F.W. (1973): Kriterien, Voraussetzungen und Einschränkungen aktiver Politik, in: Planungsorganisation. Die Diskussion um die Reform von Regierung und Verwaltung des Bundes, R. Mayntz & F.W. Scharpf (eds.), München, Piper, 115–145.

Mazur, A. (1981): The Dynamics of Technical Controversy, Washington, D.C., Communication Press.

Mazur, A. (1984): The Journalist and Technology: Reporting about Love Canal Seveso and Three Mile Island, Minerva 22, 45–66. Berlin.

Mazur, A. (1987): Mass Media Effects on Public Opinion About Nuclear Power. Office of Plants, Unveröffentlichtes Manuskript, Syracuse, Universität von Syracuse. Age of McCombs, M.E. & Shaw, D.L. (1972): The Agenda-Setting Function of Mass Media, Public Opinion Quarterly 36, 176–187.

McGuire, W.J. (1985): Attitude and Attitude Change, in: Handbook of Social Psychology, G. Lindzey & E. Aronson (eds.), 3. ed., Vol. 2, New York, Random House, 223–346.

Merton, R.K. (1968): Sozialstruktur und Anomie, in: Kriminalsoziologie, F. Sack & R. König (ed.), Frankfurt a.M., Akademische Verlagsanstalt, 283–313.

Meyer-Abich, K., Schefold, B. (1981): Wie möchten wir in Zukunft leben?, München, C.H. Beck.

Meyer-Abich, K., Schefold, B. (1986): Die Grenzen der Atomwirtschaft, München, C.H. Beck.

Mez, L. (1979): Atomindustrie, Atompolitik und Anti-Atombewegung im internationalen Vergleich, Berlin, Pahl-Rugenstein.

Midden, C. (1988): Credibility and Risk Communication, Manuskript für den International Workshop on Risk Communication, 17.–20. Oktober 1988, Jülich, Kernforschungsanlage.

Mitchell, R.C. (1980): How ‚Soft', ‚Deep', versus ‚Left'? Present Constituencies in the Environmental Movement for Certain World Views, Natural Resources 20, 345–358.

Müller, E. (1986): Innenwelt der Umweltpolitik, Opladen, Westdeutscher Verlag.

Münch, R. (1982): Basale Soziologie: Soziologie der Politik, Opladen, Westdeutscher Verlag.

Mutschler, U. (1986): Genehmigungsverfahren, in: Handbuch der Kernenergie, H. Michaelis (ed.), vol. 2, Düsseldorf, Econ, 1004–1039.

Neidhard, F. (1985): Einige Ideen zu einer allgemeinen Theorie sozialer Bewegungen, in: Sozialstruktur im Umbruch, S. Hradil (ed.), Opladen, Westdeutscher Verlag, 193–204.

Nelkin, D., Fallows, S. (1978): The Evolution of the Nuclear Debate: The Role of Public Opinion, Annual Review of Energy 3, 275–312.

Nelkin, D. (1982): Blunders in the Business of Risk, Nature 298, 775–776.

Nelles, W. (1984): Kollektive Identität und politisches Handeln in neuen sozialen Bewegungen, Politische Vierteljahreszeitschrift 4, 425–440.

Noelle-Neumann, E. (1980): Die Schweigespirale, München, Piper.

Nowotny, H. (1976): Social Aspects of the Nuclear Power Controversy, Laxenburg, Internationales Institut für Angewandte Systemanalyse.

Nowotny, H. (1979): Kernenergie: Gefahr oder Notwendigkeit? Anatomie eines Konfliktes, Frankfurt a.M., Suhrkamp.

O'Riordan, T., Wynne, B. (1987): Regulating Environmental Risks: A Comparative Perspective, in: Insuring and Managing Hazardous Risks: From Seveso to Bhopal and Beyond, P.R. Kleindorfer & H.C. Kunreuther (eds.), Berlin, Springer, 389–410.

Office of Technology Assessment, US Congress (1984): Nuclear Power in the Age of Uncertainty, Washington D.C., OTA.

Otway, H., Haastrup, P., Cannell, W., Gianitsopoulos, G., Paruccini, M. (1987): An Analysis of the Print Media in Europe Following the Chernobyl Accident, Bericht des Gemeinsamen Forschungszentrums der Europäischen Gemeinschaft Ispra, Ispra, EC.

Parsons, T.E. (1963): On the Concept of Political Power, Proceedings of the American Philosophical Society 17, 352–403.

Parsons, T.E. (1976): Zur Theorie sozialer Systeme, hrsg. und kommentiert von S. Jensen, Opladen, Westdeutscher Verlag.

Paschen, H., Bechmann, G., Frederichs, G., Gloede, F., Heuser, F.-W., Hörtner, H. (1987): The German Risk Study for Nuclear Power Plants, in: Nuclear Risk Analysis in Comparative Perspective, R.E. Kasperson & J.X. Kasperson (eds.), London, Allen & Unwin, 86–125.

Peine, F.-J. (1988): Verfassungsprobleme des Strahlenschutzvorsorgegesetzes III Natur und Recht 10, 115–121.

Peltu, M. (1985): The Role of Communications Media, in: Regulating Industrial Risks, H. Otway & M. Pettu (eds.), London, Butterworth, 128–148.

Peltu, M. (1988): Media Reporting of Risk Information: Uncertainties and the Future, in: Risk Communication, H. Jungermann, R.E. Kasperson, & P.M. Wiedemann (eds.), Jülich, Kernforschungsanlage, 11–32.

Perrow, C. (1984): Normal Accidents, New York, Basic Books. Peters, H.P. (1984): Entstehung, Verarbeitung und Verbreitung von Wissenschaftsnachrichten am Beispiel von 20 Forschungseinrichtungen, Bericht der Kernforschungsanlage, Jülich.

Peters, H.P. (1986): Public Opinion as a Channel of Communication between Science and Other Parts of Society, Vortragsmanuskript für den 11. Soziologischen Weltkongress, New Delhi: August 18–22, Jülich.

Peters, H.P., Albrecht, G., Hennen, L., Stegelmann, H.U. (1987): Die Reaktionen der Bevölkerung auf die Ereignisse in Tschernobyl. Ergebnisse einer Befragung, Bericht der Kernforschungsanlage, Jülich, Kernforschungsanlage.

Petty, R.E., Cacioppo, E. (1986): The Elaboration Likelihood Model of Persuasion, Advances in Experimental Social Psychology 19, 123–205.

Rankin, W., Nealy, S. (1979): A Comparative Analysis of Network News Coverage of Nuclear Power, Coal, and Solar Stories, Seattle, Battelle Research Center.

Raschke, J. (1987): Zum Begriff der sozialen Bewegung, in: Neue soziale Bewegungen in der Bundesrepublik Deutschland, R. Roth & D. Rucht (eds.), Frankfurt a.M., Campus, 19–29.

Rausch, L. (1976): Die medizinisch-biologische Rechtfertigung von Dosisgrenzwerten, in: Viertes Deutsches Atomrechts-Symposium, Institut für Völkerrecht der Universität Göttingen & Bundesminister des Innern (eds.), Schriftenreihe Recht-Technik-Wirtschaft vol. 8, Köln, Carl Heymanns Verlag, 277–299.

Raymond, C.A. (1985): Risk in the Press: Conflicting Journalistic Ideologies, in: The Language of Risk, D. Nelkin (eds.), Beverly Hills, Sage, 97–133.

Rayner, S. (1987): Risk and Relativism in Science for Policy, in: The Social and Cultural Construction of Risk, V.T. Covello & B.B. Johnson (eds.), Dordrecht Reidel, 5–23.

Rayner, S., Cantor, R. (1987): How Fair is Safe Enough? The Cultural Approach to Societal Technology Choice, Risk Analysis 7, 3–13.

Rempel, J.K., Holmes, J.G. (1986): How do I trust thee? Psychology Today 2, Rengeling, H.-W. (1987): Das neue deutsche Strahlenschutzvorsorgegesetz, Deutsches Verwaltungsblatt 102, 204–207.

Renn, O. (1984): Risikowahrnehmung der Kernenergie, Frankfurt a.M., Campus.

Renn, O. (1985): Die Alternative Bewegung: Eine historisch-soziologische Analyse des Protestes gegen die Industriegesellschaft, Zeitschrift für Politik 32, 153–194.

Renn, O., (1986): Akzeptanzforschung: Technik in der gesellschaftlichen Auseinandersetzung, Chemie in unserer Zeit 2, 44–52.

Renn, O. (1987): Eine kulturhistorische Betrachtung des technischen Fortschritts, in: Fortschritt der Technik – Gesellschaftliche und Ökonomische Auswirkungen, H. Lübbe (ed.), Heidelberg, Decker, 65–100.

Renn, O. (1989a): Risikowahrnehmung: psychologische Determinanten bei der intuitiven Erfassung und Bewertung von Risiken, in: Risiko der Industriegesellschaft: Analyse, Vorsorge, Akzeptanz, H. Lübbe (ed.), Erlangen, Erlanger Universitätsverlag.

Renn, O. (1989b): Risk Analysis: A Need to Communicate, Forum for Applied Research and Public Policy 4, Nr. 2, 86–92.

Renn, O. (1989): Risk Communication in the Community: European Lessons from the Seveso Directive, Journal of the Air Pollution Control Association 40, No. 9, 1301–1308.

Renn, O. (1991): Risk Communication and the Social Amplification of Risk, in: Risk Communication, R.E. Kasperson & P.J. Stallen (eds.), Dordrecht, Reidel.

Renn, O., Albrecht, G., Kotte, U., Peters, H.P., Stegelmann, H.U. (1985): sozialverträgliche Energiepolitik. Ein Gutachten für die Bundesregierung München, HTV.

Renn, O., Albrecht, G., Kotte, U., Peters, H.P., Stegelmann, H.U. (1986): Das Projekt ‚Sozialverträglichkeit von Energiesystemen', in: Die Analyse der Sozialverträglichkeit für Technologiepolitik, H. Jungermann, W. Pfaffenberger, G.F. Schäfer & W. Wild (eds.), HTV Edition „Technik und sozialer Wandel", München, High Tech, 15–27.

Renn, O., Levine, D. (1988): Trust and Credibility in Risk Communication, in: Risk Communication, H. Jungermann, R.E. Kasperson & P.M. Wiedemann (ed.), Jülich, Kernforschungsanlage, 51–81.

Renn, O., Swaton, E. (1984): Psychological and Sociological Approaches to Study Risk Perception, Environment International 10, 557–575.

Rip, A. (1985): Experts in Public Arenas, in: Regulating Industrial Risks, H. Otway & M. Peltu (eds.), London, Butterworth, 94–110.

Roth, R. (1987): Kommunikationsstrukturen und Vernetzungen in neuen sozialen Bewegungen, in: Neue soziale Bewegungen in der Bundesrepublik Deutschland, R. Roth & D. Rucht (ed.), Frankfurt a.M., Campus, 68–88.

Rubin, D.M. (1987): How the News Media Reported on Three Mile Island and Chernobyl, Communication 37, Nr. 3, 42–57.

Rucht, D. (1982): Neue soziale Bewegungen oder: Die Grenzen bürokratischer Modernisierung, in: Politikwissenschaft und Verwaltungswissenschaft, J.J. Hesse (ed.), Opladen, Westdeutscher Verlag, 272–292.

Rushefsky, M. (1984): Institutional Mechanisms for Resolving Risk Controversies, in: Risk Analysis, Institutions, and Public Policy, S.G. Hadden (ed.), Port Washington, Associated Faculty Press, 133–148.

Sahuer, H. (1986): Sozialer Wandel und Konsens. Zur Legitimationsproblematik des Mehrheitsprinzips, in: Mehrheitsprinzip, Konsens und Verfassung, H. Hattenhauer & W. Kaltefleiter (ed.), Heidelberg, C.F. Müller, Jurist. Vlg, 93–113.

Scharpf, F.W. (1973a) Komplexität als Schranke der politischen Planung, in: Planung als politischer Prozess. Aufsätze zur Theorie der planenden Demokratie, F.W. Scharpf (ed.), Frankfurt a.M., Suhrkamp.

Scharpf, F.W. (1973b): Planung als politischer Prozess, in: Gesellschaftliche Planung. Materialien zur Planungsdiskussion in der BRD, B. Schäfers (ed.), Stuttgart, Enke, 169–201.

Scharpf, F.W. (1975): Politische Durchsetzbarkeit von Reformen, Göttingen, Otto Schwarz.

Schattke, H. (1982): Grenzen des Strahlenschutzes in kerntechnischen Anlagen, Energiewirtschaftliche Tagesfragen 32, 1083–1088.

Scheuch, E.K. (1980): Kontroverse um Energie – ein echter oder ein Stellvertreterstreit, in: Existenzfrage Energie, H. Michaelis (ed.), Düsseldorf, Econ, 279–293.

Scheuch, E.K., Kutsch, T. (1972): Grundbegriffe der Soziologie, vol. 1, Stuttgart, Teuhner.

Schmid, M. (1982): Theorie sozialen Wandels, Opladen, Westdeutscher Verlag. Schmitt, H., Niedermaier, O., Menke, K. (1981): Etablierte und Grüne, Zeitschrift für Parlamentsfragen 4, 516–540.

Schröder, D., Bülow, H., Wolff, H., Hofer, P. (1982): Politik, Wertwandel, Technologie – Ansatzpunkte für eine Theorie der sozialen Entwicklung. Schriftenreihe Technologie und Beschäftigung, vol. 6, Düsseldorf, Econ.

Schwarz, M., Thompson, M. (1990): Divided We Stand: Redefining Politics, Technology, and Social Choice, Philadelphia, University of Pennsylvania Press.

Seiderberg, P.C. (1984): The Politics of Meaning: Power and Explanation in the Construction of Social Reality, Tuscon, University of Arizona Press.

Sharlin, H.I. (1987): Macro-Risks, Micro-Risks and the Media: The EDB Case, in: The Social and Cultural Construction of Risk, V.T. Covello & B.B. Johnson (eds.), Dordrecht, Reidel, 183–197.

Singer, E., Endremy, P. (1987): Reporting Hazards: Their Benefits and Costs, Communication 37, Nr. 3, 10–26.

Sinus Institut (1983): Die verunsicherte Generation. Jugend und Wertwandel, Opladen, Sinus.

Slovic, P., Fischhoff, B., Lichtenstein, S. (1982): Why Study Risk Perception? Risk Analysis 2, 83–93.

Slovic, P. (1987): Perception of risk, Science 236, 280–29.

Sood, R., Stockdale, G., Rogers E.M. (1987): How the News Media Operate in Natural Disasters, Communication 37, Nr. 3, 27–41.

Steffani, W. (1986): Mehrheitsentscheidung und Minderheiten in der pluralistischen Verfassungsdemokratie, Zeitschrift für Parlamentsfragen 17, 569–586.

Steindorf, J. (1988): Kommentierung des § 311 d. StGB, in: Leipziger Kommentar zum Strafgesetzbuch, H.-H. Jescheck, W. Ruß & G. Willms (eds.), vol. 7, 10. ed., Berlin, Walter de Gruyter & Co..

Thomas, K., Maurer, D., Fishbein, M., Otway, H.J., Hinkle, R., Simpson, D.A. (1980): Comparative Study of Public Beliefs About Five Energy Systems. Wilkins, Laxenburg, Internationales Institut für Angewandte Systemanalyse.

Thompson, M. (1980): An Outline of the Cultural Theory of Risk, Laxenburg, Winters, Internationales Institut für Angewandte Systemanalyse.

Tversky, A., Kahnemann, D. (1985): The Framing of Decisions and the Psychology of Choice, in: Environmental Impact Assessment, Technology Assessment and Risk Analysis, V.T. Covello, J.L. Mumpower, P.J.M. Stallen & V.R. Uppuluri (eds.), Berlin, Springer, 107–129.

Uffmann, D. (1986): Stromerzeugungskosten und Wettbewerbsfähigkeit der Kernenergie, in: Handbuch der Kernenergie, H. Michaelis (ed.), Band 2, Düsseldorf, Econ, 477–515.

Uppenbrink, M. (1974): Organisation der Umweltplanung in den USA. Beiträge zur Umweltgestaltung, Berlin, Bundesumweltamt.

U.S. Department of Energy (1983): 1983 Survey of Nuclear Power Plant Construction Costs, Washington, D.C., Energy Information Administration.

U.S. Office of Technology Assessment (1984): Nuclear Power in an Age of Uncertainty, Washington, D.C., OTA.

Van der Pligt, J. (1985): Public Attitudes to Nuclear Energy: Salience and Anxiety, Journal of Environmental Psychology 5, 87–97.

Vlek, C.A., Stallen, P.J.M. (1981): Judging Risks and Benefits in the Small and the Large, Organizational Behavior and Human Performance 28, 235–271.

Wahl, R. (1973): Politische Planung in der differenzierten Gesellschaft. Notwendigkeiten, Grenzen und verfassungspolitische Probleme, in: Gesellschaftliche Planung, Materialien zur Planungsdiskussion in der BRD, B. Schäfers (ed.), Stuttgart, Enke, 127–157.

Watts, N.S.J. (1987): Mobilisierungspotential und gesellschaftspolitische Bedeutung der neuen sozialen Bewegungen, in: Neue soziale Bewegungen in der Bundesrepublik Deutschland, R. Roth & D. Rucht (ed.), Frankfurt a.M., Campus, 47–67.

Weber, M. (1922): Wirtschaft und Gesellschaft, 3. ed. 1947, Tübingen, Mohr. Weber, M. (1973): Soziologie, in: Universalgeschichtliche Analysen, J. Winkelmann (ed.), 5. ed., Stuttgart, Kröner.

Weinert, FE. (1988): Der Laie als „Chemie-Experte"? – Das Bild einer Wissenschaft im Lichte alltäglichen Wissens und Wertens. 125 Jahre Hoechst -Wissenschaftliches Symposium, Frankfurt a.M., Hoechst, 15–29.

Weingart, P. (1979): Das ‚Harrisburg-Syndrom' oder die De-Professionalisierung der Experten, Vorwort zu H. Nowotny, Kernenergie: Gefahr oder Notwendigkeit, Frankfurt a.M., Suhrkamp, 9–17.

Wicker, A.W. (1979): Attitudes vs Actions: The Relationship of Verbal and Behavioral Responses to Attitude Objects, Journal of Social Issues 22, 41–78.

Wiedemann, P., Jungermann, H. (1989): Energy and the Public: Country Report FRG, in: Energy and the Public, Welt-Energiekonferenz, Subcommittee on Energy and the Public, Vol. 2, Country Reports, London, WEC.

Wieland, B. (1987): Empirische Untersuchungen zur Technikakzeptanz, Bericht Nr.7/87, Forschungsstelle für gesellschaftliche Entwicklungen an der Universität Mannheim, Mannheim, FGE.

Wilkins, L., Patterson, P. (1987): Risk Analysis and the Construction of News, Communication 37, 80–92.

Winters, K.-P. (1977): Zur Novellierung des Strahlenschutzrechts, Deutsches Verwaltungsblatt 92, 331–340.

Winters, K.-P. (1978): Atom- und Strahlenschutzrecht, München, C.H. Beck.

5

FORMAL CRITERIA AND INSTRUMENTS FOR THE DECISION PROCESS

Environmental standards establish what risk is collectively regarded as acceptable after the utilities have been weighed up. As has been expounded in chapter 1, the question about procedures and criteria according to which the acceptable risk can be determined in a socially binding way cannot be answered on the basis of scientific knowledge alone, such as, for example, knowledge about the dose response relation (Mayntz 1990, 141). This chapter focuses on working out rational evaluation procedures and developing instruments that can become effective in the social decision process. Its reflections orient by the question how far the limits of a scientific way of proceeding can be extended in the face of the social decision processes described in chapter 4.

Establishing criteria depends on explicitly denoting the objectives to be reached by means of an environmental standard as well as on defining a process of balancing interests which allows of a systematic evaluation of costs and benefits under the condition of pragmatic consistency. In order to do so, three steps should be distinguished (Hansmeyer & Rürup 1975, 93; National Academy of Sciences 1983, 196ff; Hoerger 1987, 651):

- determining objectives which both serve the protection of life, health and environment and allow of a rational allocation of social resources (section 5.1);
- analyzing the effects that may result from the implementation of these objectives (section 5.2);
- weighing the social expenditure or damage, where the concept of damage has to include the loss of benefit that will be faced if these objectives are realized (section 5.3).

Rationality in establishing environmental standards does not refer to the objectives of society but to the selection of means and strategies with the help of which the objectives can be realized (Schürmann 1978, 39ff). From this selection, methods for a rational way of action are derived. When establishing environmental standards, the following aspects have to be taken into account, according to the three above-mentioned steps:

(a) In order to elaborate a profile of protection-worthiness, scientific research can draw the attention to new, still unknown dangers or ascertain the possible effects of a new technology or of other social activities (Dalkey & Helmer 1963). The limitedness of our methodical knowledge for assessing such effects will later be discussed in detail.

(b) Establishing measures to realize the objectives and to evaluate their possible effects first of all requires the decision maker to consider as comprehensively as possible the relevant scientific knowledge with the help of which he or she will be able to assess the consequences for health and environment. Moreover, he/she has to know the technological possibilities and the costs of reducing the risk (Watson 1982; Hyman & Stiftel 1988, 64ff).

(c) Both types of information are necessary, but they are not sufficient with regard to decisions; they say nothing about the standard which might be regarded as acceptable. The acceptability of the standard ensues from the relative desirability of the objectives and from balancing the expenditure for reaching the objective against the resulting damage.

The weighing of costs and benefits has also to consider the distribution effect on various groups of the society (Külp 1976, 84). There are rational models for this, with the help of which formal criteria of consistency, coherence and adequacy can be put into practice (Keeney & Raiffa 1976, 50ff; Krewski & Birkwood 1987, 404ff). Because of the fundamental importance of rational decision procedures for the establishing of environmental standards, a number of basic methodical realizations of decision-theoretical models are illustrated in greater detail in section 5.4. The results of these models are to be regarded as „evidence“ in the social decision process, but they are not a substitute for the process itself (Merkhofer 1984, 183ff; Fritzsche 1986, 506ff).

5.1 SOCIAL FORMATION OF OBJECTIVES

The first step in determining environmental standards is defining protection objectives. In the terminology of decision analysis, protection objectives are the guides which help the decision maker to determine the desirability of options relative to one another (Keeney & Raiffa 1976, 33). Unlike in our everyday linguistic usage, objectives are not to be regarded as deterministic borders between achievement and non-achievement, but as an open-ended measuring staff for evaluating options. This way, options can be evaluated in a differentiated way, even if they all lie above or below a threshold value which has been defined as an objective.

Science can make an important contribution to discovering and structuring protection objectives. The most important medium for determining objectives is the codified law (Fritzsche 1986, 87ff). If we presume that laws reflect the social consensus of legalized constraints on action and offers for action, then they can serve as a basis for finding objectives.

The existing legal regulations are, however, not the only medium for determining objectives; they are embedded in the political and social process of opinion formation. From the plurality of the social process of opinion formation it follows, at least for countries with a democratic constitution, that all sectors of the social formation of will, directly or by representatives, participate in specification and interpretation of objectives, or at least they should (Fiorino 1989). Formal scientific procedures can contribute to structuring the purposes and ordering them hierarchically. Advantages of a formalization of the process of definition of objectives are (Keeney et al. 1984, 30; Fritzsche 1986, 506):

- a check of coherence between purposes and the fundamental principles of political institutions (especially of nations);
- a check of coherence between the general protection objectives and the assigned purposes and criteria;
- comparability of patterns of protection objectives among various social groups.

Scientific procedures for early recognition and impact assessment help in finding objectives, since physical consequences are often not evident. Moreover, science can point at new possible dangers or consequences of „doing nothing“ or of new activities. Apart from structuring and systematizing objectives, this is done by methods of early recognition and impact assessment. Scientific assessment of effects saves society the often woeful way of trial and error (Häfele et al. 1990, 376f).

In the first step of setting standards, rationality means including into the formulation and structuring of set targets a scientific assessment of effects as well as a secured and procedurally reliable analysis of the objectives and the criteria which are to be aimed at by means of the respective standards.

5.2 POSSIBLE CONSEQUENCES OF MEASURES TO REACH OBJECTIVES

5.2.1 Basic Principles and Meaningfulness of Risk Studies

The second step in establishing standards is to study the effects of objectives and measures by means of risk studies (Fischhoff et al. 1984, 123ff; Krewski & Birkwood 1987, 339ff). These studies require that:

- the effects of the respective substances and their dependence on the dose are known;
- the technological and financial expenditure for reducing the risk can be assessed;
- the distribution of the effects of risks on various groups of the population, various regions, or future generations has been or is being studied.

The knowledge about the dose response relation, the costs of risk reduction, and the distribution pattern of who is affected can basically be acquired by scientific methods in an intersubjectively valid way (Morgan 1990, 22ff).

However, as has been stated several times, this knowledge is not enough to actually determine the standard. In risk ascertainment, it is mainly the following aspects which have to be taken into account for an interpretation of the results (National Academy of Sciences 1975; Rowe 1977, 23ff; Crouch & Wilson 1982, 20f; Fischhoff et al. 1984, 123ff; Schrader-Frechette 1985; Renn & Kals 1990, 67ff):

- the definition of what is called a damage and what is included in the ascertainment of risk;
- the selection and perhaps the aggregation of various damage categories (e.g. cancerous diseases, development anomalies, genetic mutations);
- the choice of the degree of certainty (significance level) from which onward deviations from the mean are no longer regarded as a matter of chance;
- the choice of the methodological tools to ascertain damage extent and failure probabilities (fault tree analysis, event tree analysis, reliability analysis etc.);

- the choice of the probability concept for objective (statistical, logical) or subjective probability;
- the treatment and calculation of uncertainty (statistical confidence intervals, expert assessments etc.);
- the choice of the reference quantity (e.g. expected damage per unit time or per driven kilometer or per unit of the gross national product etc.);
- the choice between individual risk versus collective risk;
- the linkage of probabilities and extent of damage (usually by multiplication of both elements, which averages damage over time).

The necessity of making such assumptions when identifying risks has led to a confusion in the public about the reliability of risk analyses. The choice in making such assumptions is not an arbitrary act in which researchers use the available scope of discretion according to their political conviction or their personal likings. This choice is rather amenable to rational judgment (Fritzsche 1986, 423).

This may be illustrated by a simple example. One of the necessary points to be established in the framework of risk analyses is the choice of the reference quantity (Merkhofer 1987, 595ff). As to passenger services with airplanes and with cars, e.g., the risk to become involved in a fatal accident in the U.S. is greater for cars if we relate the damage to the number of kilometers covered. If, however, we relate the damage to the amount of time spent in one of the two means of transportation, then flying is riskier (Evans et al. 1989). Thus, depending on the choice of the reference quantity, we reach different results.

So, which reference quantity is the correct one? That depends on the context. Usually, the idea of transportation is to get a person from A to B, i.e., the purpose of a trip is to cover a certain distance. Therefore, only the reference quantity „frequency of accidents per covered kilometer“ makes sense here. However, if, for example, somebody makes a decision between a forty-hour job as a pilot or as a truck-driver (a more realistic example: an insurance agent who could either take over a near or a far away district, spending in both cases the same amount of time, either in a car or in a plane), then the reference quantity „accidents per unit time“ is more appropriate when comparing both occupational options. So, it depends on the context (in this case on the purpose of the comparison), how the discretion scope is used. The choice of a reference quantity in our example can be made independently of the respective person's preference for flying or for driving a car.[1]

The realization that aspects like the ones mentioned can be rationally judged is of decisive importance for the reaching of standards. No matter if one likes nuclear power plants or not, the criteria remain the same according to which a rational decision can be made about which aspects are appropriate for measuring the impacts of nuclear power plants and which ones are not. And it may turn out that different presumptions are appropriate to the same degree.

On the basis of these reflections on the importance of risk studies as aids for an analysis of the possible effects of implementing objectives, the following basic principles for the translation of risk studies into the regulation process can be formulated – without laying claim to completeness (Renn & Kals 1990, 78ff):

(a) Risk studies provide a discretion scope for methodical execution and scientific interpretation. The choice of what risk studies are taken as a basis for establishing standards has to be made according to the target definition inherent in the problem. As opposed to a popular prejudice, such intersubjective criteria can be determined and operationally applied on the basis of a scientific discussion, even if the usual variation range of scientific statements remains (Hattis & Kennedey 1990, 161; Morgan 1990, 22f).

(b) Scientists are, like all other members of society, exposed to the political forces and influences of their surroundings and their time. Therefore, the danger that they will take their own preferences as a standard for judging risks is present (Linnerooth 1983). At the same time, scientific expertise and experience are the preconditions for an intersubjective judgment of presumptions, i.e., the necessary critical control can only take place within the system of science. It is necessary here to create organizational structures which ensure such an intra-scientific control (Rushefsky 1984, 138ff).

(c) If risk studies have been elaborated which meet the above-mentioned criteria of being scientific, they can be regarded as „state of the art" in the respective problem field. In most cases, there will be a range of results which reflect different aspects of knowledge formation and of dealing with the always remaining uncertainty. Which results are chosen as guidelines for political activity depends then on the purpose of the decision and on the respective context. Under certain conditions, optimistic and pessimistic assessments may simultaneously gain entry into political norms, as upper and lower limits (National Academy of Sciences 1983; Krewski & Birkwood 1987; Hattis & Kennedey 1990).

(d) The variation range of possible effects should determine the range of activities, especially of disaster control and disaster prevention measures. Insofar, many countries have adjusted their disaster regulations to reactor accidents that go beyond the basic design accident. While it makes sense to put more emphasis on the expected value in decisions about options, it appears expedient to allow for very rare events in the prevention of disasters (Lagadec 1987, 42ff).

5.2.2 Determining Costs

In order to enable a rational allocation of social resources, i.e., to organize and combine the socially provided production factors (labor, capital, raw material and know-how) in a way that the given objectives of a society are reached as far as possible, expenditure and returns of environmental standards have to be compared with one another. Such a comparison is necessary, since money spent for the reduction of risks is no longer available for other purposes. The comparison between expenditure and returns does not yet say anything about the acceptability of environmental standards; for this purpose, the individual target dimensions (health, environment, distribution justice and costs) have to be made commensurable. The procedures for this are described in sections 5.3 and 5.4.

How can the costs of environmental standards or of equivalent regulations (e.g. environmental taxes) be determined? In most cases, a tightening of environmental standards implies resulting costs in the following fields (DeFina 1977; Just et al. 1982):

(a) *Costs for technological retention measures or for organizational changes which arise in the respective company.* These costs can be and are usually passed on to the prices of the respective goods. For the economy as a whole, it is not relevant who eventually bears these costs, but this does affect the possible distribution effects of the cost increase (e.g. lowering the profit margin versus lowering the effective income of the employees).

(b) *Costs caused by the loss of domestic production due to improved cost-effectiveness of imported goods and worsened cost-effectiveness of exported goods.* In this case, the respective environmental standard reduces the international competitiveness of a production due to increased costs, and leads to a transfer of the production abroad. The amount of these costs is difficult to estimate: on the one hand, potential consumers benefit from the inexpensiveness of imports, but on the other hand, domestic producers lose competitive advantages, which negatively affects the domestic product. Moreover, social costs have to be considered, such as the social expenditure for unemployment benefits, or the loss of prestige of domestic products. Only by means of an analysis of the individual factors can the cost shares be determined somewhat reliably.

(c) *Costs due to the price effect of the tightening of standards.* According to common economic theory, a price increase of a good results in an adjustment of the demand, the extent of which mainly depends on the elasticity of the demand function. If the price increase caused by the environmental standard reflects exactly the value which is necessary to compensate for the remaining ecological damage, then the price fulfills its allocation-optimizing function. If the price increase goes beyond the amount of money needed for compensation, or if it is far lower than this amount, the result is a misallocation and thus an increase in social costs. How exactly these social costs are to be determined is controversial.

(d) *Transaction costs for enforcing and controlling environmental standards.* Regulating environmental standards, controlling the limits and tracing and penalizing polluters requires a significant amount of money for material and staff. Often the expenditures for implementation and control bear no relation to the saving of costs resulting from the thus achieved observance of limits. Such reflections, however, must not only take the cost-cost comparison as a standard for effectiveness, but have also to consider the symbolic signal effect of strict control. The perceived strictness of governmental control decisively influences the perception of the seriousness of environmental policy or the motivation to improve one's own ecological behavior.

(e) *Conflict costs.* Regulating and appeasing social conflicts about the acceptability of environmental standards causes social costs (often to a considerable extent). Delays in the licensing of plants, the securing of plants against protesting groups of the population, consequential effects of social protests etc. may cause costs to the operators of plants and the state. While in categories a) and b) costs rise with the tightening of environmental standards, conflict costs sink. Therefore, conflict costs can be optimized. Moreover, organizational measures (as are presented in chapter 6) may help limit conflict costs, so that actual denial of acceptance will mean a parallel shift downwards of the cost curve.

The existence of at least five partly contrary cost categories shows that the cost side also includes several individual dimensions and is connected with ambiguities and uncertainties, as is the risk side. An advantage of the determination of costs is, however, that here all dimensions concerned can be evaluated with the same clearing unit. As to risk, on the other hand, the various effects on health and on the environment have to be made commensurable by means of an equivalence rule (Schrader-Frechette 1984, 280ff). The intangible disadvantages of an environmental standard, however, remain ignored in the determination of costs (Schürmann 1978, 272ff).

Despite the contrary tendency of conflict costs and production costs or consequential costs, the prevailing economic doctrine assumes rising marginal costs due to environmental standards (Morgan 1990, 20). It can thus be concluded that with a continuous tightening of environmental standards, the connected costs do not only increase proportionally but at an even higher rate. If we start from this assumption, which has been empirically proved in many individual cases, the logical conclusion is that a reduction to a zero-risk will necessarily lead to a misallocation, i.e. a waste of economic resources, in almost all cases (Derby & Keeney 1981, 220f) and, moreover, will contribute to other long-term risks as possible consequences of the resulting innovation refusal (Wildavsky 1979, 32f). The immense funds required for a reduction from almost zero to zero could be used more efficiently in any case where another risk or several other risks could be brought down from relatively high values to lower ones (Merkhofer 1984, 183ff; Fritzsche 1986, 481). Only when all risks have reached a level of almost zero would a further reduction be worthwhile. But even in this case, the question remains whether the required means should not be used much more efficiently for other social objectives (e.g. education or prophylaxis).

While the expenses for environmental standards are to be booked on the debit side, the existence of environmental standards causes at the same time financial gains by preventing ecological damage which would have occurred without these standards. Quantifying this is a difficult task and is usually successfully done only by means of indirect indicators, such as changes in land prices, costs for restoring persons to health, or the raw material value of the destroyed environmental resources (Schürmann 1978, 272ff; Smith 1986, 16; Wicke 1990, 225ff).

In the context of a cost analysis, an attempt has to be made to offset the costs due to environmental standards with the cost savings due to the improved quality of the environment (Fischhoff et al. 1985; Morgan 1990, 20). Apart from the absolute cost level, the distribution of costs, i.e. the reasonableness of imposing costs on consumers, producers, social groups or public authorities, also plays an important role. If, e.g., a production sector is transferred to another country, the former domestic employees will have to take over the consequential costs for an improved environmental quality, which is of benefit to everybody. Distribution effects of environmental standards are therefore a further dimension which has to be included in their evaluation.

5.2.3 Distributive Justice

Like all legal norms in force, environmental standards must be binding for all citizens. Observing them, however, does not guarantee that citizens are protected against risks. Even if environmental standards are fully observed, there is a residual risk which is imposed on every individual. It is highly unlikely that all citizens are exposed to such residual risks to the same extent – maybe relativized with respect to their individual sensitivity. In fact, risk exposures in a society are distributed very unequally. Inequalities for which a justification is not recognized, are experienced as cases of injustice (Fischhoff et al. 1985; MacLean 1986). Therefore, prior to establishing a standard, its effects on distribution justice will have to be examined.

It is a matter of ethics, namely the theory of justice, to reconstruct principles of such justification procedures. Especially in connection with the risk issue, the philosophical theories of justice, mainly in the fields of economy and jurisprudence, have lately been heavily discussed again. Theories of justice provide the foundations for the justification of actual inequalities.[2]

Obviously, the formulation of principles of justice depends on the conception of ethics on the basis of which they are explicated. Elitist and conflict-oriented ethical conceptions are not considered here, since the point is exactly to ensure social generalizability for the unequal distribution of risks, i.e. to create risk justice. *Utilitarian and deontological* ethics diverge in their opinions about which conditions have to be given for generalizability to be achieved (Frankena 1963). The idea of *generalizability*, however, is common to both approaches. A conception of distributive justice for technologically induced risks is therefore only interesting within the framework of a universalistic, egalitarian and consensus-oriented conception of ethics (MacLean 1986). For a further treatment of the problem, the following tasks have to be distinguished:

(a) A distinction has to be made between dimensions of actual inequalities, e.g., regional, temporal and social differences. A by no means trivial task consists of measuring differences in these various dimensions and relating them to one another (e.g. for determining compensatory payments). Furthermore, it is not sufficient for a discussion of risk justice to know inequalities of risk distribution; these inequalities will, in any case, have to be related to the unequal benefits that the members of a society gain from accepting risks (Fritzsche 1986, 46).

(b) The normative aim of reflections on justice cannot always be to reach equality. There must also be reasons that justify inequality. Rawls suggests the following criterion: an inequality is justified if its removal would cause higher costs than to grant a compensation for the inequalities to those at a disadvantage (Rawls 1971).

(c) If an equal distribution is not possible or if it is not desirable for reasons of overriding importance, there is still the possibility of *compensation*. Compensations are nothing but the balancing of utility losses against utility gains. It is often felt that this is cynical and unnatural. In fact, the compensation concept is indispensable in a highly differentiated society (Raiffa 1982, 310ff; Diederichsen 1990, 155ff). Extra payment for „dirty work" may be regarded as an example of this: an unpleasant job which might involve health risks is compensated for by a monetary advantage. It is interesting here that the amount of money is not derived from a mysterious risk inherent in the „dirt", but is determined according to market mechanisms: the adequate amount of money is the minimum sum for which somebody is prepared to take over this „dirty" work. This example assumes, however, that, first, it is a matter of an individual decision (the risk is not imposed on anyone by force) and that, second, the person taking the risk has completely understood the risk for himself/herself and that, third, he/she is absolutely free from economic and social pressure in his/her decision. In the case of collective risks, compensation is more difficult, since every member of a collective links different utility profiles with the utility loss caused by the risk and with the utility gain caused by the compensation, so that an aggregation of utilities is necessary. For this, however, criteria are missing in most cases.

(d) The lacking transparence of the utility loss in case of individual risks and the problem of aggregation in case of collective risks de facto result in the necessity to establish general social boundaries for cases in which, due to complexity for different reasons, persons affected do not take over the role of the „ideal consumer" or cannot reach a consensus about a commonly accepted compensation. Therefore, it is the task of the government to lay down special standards for especially affected groups of the population (e.g. children, pregnant women, ill persons, groups of persons who are occupationally particularly exposed). For collective risks, collectively effective compensations may also be worked out by representative bodies.

5.3 BALANCING EXPENDITURE AND SUCCESS ON LOGIC AND MEANINGFULNESS OF STANDARDS OF COMPARISON

The third and decisive step for judging the implementation of objectives that are to be realized by environmental standards is the comparison of costs and benefits or of risks and chances. The above considerations have shown that normative stipulations are part of the balancing process in multiple respects. Rationality in the procedure which integrates factual knowledge with normative stipulations can only be granted if the latter are already an effective element of our reality. Commonly held convictions are therefore a precondition for the implementation of objectives which may claim general recognition.

Among the formal balancing procedures which serve as tools in the balancing process, the *risk-risk comparisons* (r-r comparisons) stand out insofar as they rely on critically tested actual convictions as a standard for the reasonableness of risks. Therefore, in the following, we will first start from a heavily idealized model of the risk-risk comparison, which will later be modified in view of the advantages of other procedures.

5.3.1 Risk-Risk Comparisons

As discussed in chapter 1, the demand for pragmatic consistency within life-forms (sectorial pragmatic consistency) forms the basis for a rational judgment of the reasonableness of risks. For justifying environmental standards and hence for judging their acceptability, risk-risk comparisons thus play a decisive role. The consistency postulate is here related to both individuals and collectives: risk-risk comparisons are based on the demand that already accepted risks are to be consulted in order to find a standard for judging the reasonableness of other risks. If a society has already classified certain risks as acceptable, then it is consistent to demand that options of action linked to smaller risks than the already accepted ones may also be accepted as long as the benefits and the risk distribution are the same (Wilson & Crouch 1987, 267ff; Merkhofer 1987, 581ff). For the issue of environmental standards, this approach has been elaborated in three variations (Fischhoff et al. 1979):

(a) *Technological risks which are presently being classified as acceptable by some group are regarded as upper limit in assessing the acceptability of new risk sources with similar risk profiles.* It has to be taken into account here, however, that not every risk is right away acceptable simply because there is another risk source with a risk which is just as high or higher and which is generally accepted. For in that case, we could infinitely increase the total life risk by adding small risks. Rather, the total risk accepted in a life-form may serve as a standard for including new risks, i.e., options have to be judged according to whether they significantly change the total risk regarded as acceptable in a life-form.

(b) *Risks that have been taken over by a society in the past* (e.g. risks of energy production from coal) *are regarded as a standard for the reasonableness of risks of new technological lines* (e.g. nuclear power plants). In many cases, it is not the explicit approval but only the revealed preference of a risk which is taken as a basis here (Starr 1969, 1234; a critical opinion on this: Fischhoff et al. 1985, 268f). When historically accepted risk sources are taken as a standard for acceptability, this does not depend on the situational circumstances of the historical decision, as is often presumed in literature (Fischhoff et al. 1985, 270f; Meyer-Abich 1990, 185). The question is not, for example, whether one would have allowed the introduction of the automobile, had one known about the fatal accidents we are experiencing today. For the risk-risk comparison it is only important whether those agents who accept road traffic by explicit approval or by revealed preference act in a pragmatically consistent way. Such a consistent action naturally does not rule out that one wants to systematically and lastingly reduce the risks of road traffic.

(c) *Technological risks are regarded as acceptable if they lie significantly below the risk degree of „natural risks“*. One version of this consideration is to compare a technological risk with the variation range of a natural risk. In other variants, technologically caused risks in a certain area are compared with (accepted) risks of a natural source in another area (Fischhoff et al. 1985, 273ff). The comparison with natural risks must not be based on the premise that natural risks are already acceptable because they are natural (see chapter 1.3.3). This premise would, for example, make it seem insensible to take measures to reduce the risk of infections. Incidentally, in this case the assertion that natural risks are acceptable would include a naturalistic fallacy. Rather,risk-risk comparisons are a check for reasonableness: *if* natural risks are regarded as acceptable by revealed preference, *then* technological risks with a lower degree of risk are also reasonable. The leading premise of this argumentation is a normative one, namely the postulate of pragmatic consistency. Therefore, this is not a naturalistic fallacy.

The correct application of risk-risk comparisons depends on a number of conditions and constraints:

(a) In usual risk-risk comparisons, the dimension of the utility of the risky action is ignored as a ceteris-paribus condition (Crouch & Wilson 1982, 3ff). So, for example, when coal energy is compared with nuclear energy, it is presumed that the utility of the unit electric energy is the same for the consumer in both cases. Nevertheless, it is necessary in many cases to explicitly take into account the different utility of two actions which are to be compared. In these cases, actions are not only assessed according to the dimensions of probability and damage, but also according to utility. The higher the expected utility of an action, the greater the risk involved may be (Fritzsche 1986, 425ff).

(b) If a risk-risk comparison is made in order to study the possibilities of reducing a risk, the costs of the reduction have to be taken into account for the action in question. From the point of view of costs, the comparison between coal energy and nuclear energy might have turned out differently before the introduction of nuclear technology and after its installation. So, under certain conditions, the risk-risk comparison is to be made by considering a further dimension (Merkhofer 1987, 590ff).

(c) A risk-risk comparison often only refers to the expected value of the risks that are to be compared. Such a confinement to the average damage to be expected is problematic, however, if the confidence intervals of the risk estimations vary among the risks to be compared (different degree of statistical uncertainty) or if the temporal order of the damage eventuation is relevant for the assessment (different danger potentials with the same expected value). In this case, the expected values and the distribution parameters have to be entered into the acceptability judgment as multi-dimensional evaluations (Merkhofer 1987, 595).

Due to these three constraints, risk-risk comparisons can never claim absolute validity but have to be integrated in the multi-dimensional decision matrix as one of the decisive dimensions (Wilson & Crouch 1987, 271ff; Fritzsche 1986, 478; Merkhofer 1987, 598f). Other dimensions, such as utility, distribution effects, cost-effectiveness of the reduction of risk and other relevant criteria, should always be considered in the decision process.

Apart from their basic importance in the context of normatively judging the reasonableness of risks, risk-risk comparisons have an important orientation function. Risk-risk comparisons are therefore often made use of in our every-day life if the risk of actions has to be assessed within one field of action (see chapter 1.3.3). The risks arising in one field of action will be called „risk class“ here. There are, however, many aspects according to which actions can be assigned to a certain risk class, like the purpose of the action (e.g. production of energy), objective characteristics of the risk source (e.g. equal effects), social characteristics of the action (e.g. voluntary nature), or distribution aspects (e.g. identity of persons bearing the risk and of beneficiaries) so that there is practically no limit to the definition of classes. Ultimately, each action should be comparable with another one under some classification aspect; it is a pragmatic question whether some relevant common aspects can be found for a comparison in a certain situation. In the case of class-internal comparison situations, risk comparisons may perform the following orientation functions:

(a) *Illustration of relative endangering.* In the communication process with decision makers or in public discussion, risk calculations are often hard to convey. The comparison with known risks, on the other hand, may illustrate the extent of endangering by some still unknown risk source without, at the same time, answering the normative question of acceptability (Renn 1988, 114ff).

(b) *Reduction of uncertainty about long-term effects.* If risks due to a new activity are identical with already existing natural or technological risks, experiences with the existing risks may be used to be able to make statements about the long-term effects.

(c) *Setting of priorities in risk reduction.* Since risk-risk comparisons themselves are inexpensive, they may be used for „screening" technological risks. All risks for which the expected value of some damage is relatively high compared to other, accepted risks should have priority for studies. On the other hand, risks that lie within the variation range of accepted (technological or natural) risk sources may be neglected for the time being (National Academy of Sciences 1975).

5.3.2 Cost-Effectiveness Procedure

In the cost-effectiveness procedure, measures to increase safety are judged according to how many scarce resources have to be invested to reduce a given risk. The rule is, for example, that for each dollar which is spent on reducing a risk, the reduction potential has to be exploited optimally (Merkhofer 1984, 183ff; Fritzsche 1986, 488). To carry out the corresponding optimization procedure, the marginal costs per unit utility (e.g. saved human life, prevention of a cancerous disease, saved biotope) have to be determined for all relevant risks. Then the budget which is to be spent on risk reduction is distributed among the risk sources in a way that altogether the greatest utility effect is achieved (Smith 1986, 32f). In the case of energy production, this procedure leads to the following consequence: the radiation risk due to nuclear plants is to be reduced by means of a respective budget until this budget reaches a level at which more people would be saved if it were spent on the reduction of the emissions of coal power stations. Applied to environmental standards, this principle means that the risk has to be reduced to a point at which the marginal costs needed to achieve one unit utility equal the marginal costs needed to achieve the same unit utility by means of another environmental standard in question.

The normative presumption underlying this procedure is that the risks remaining after an optimization consideration are acceptable. Acceptability, however, cannot be furnished by the cost-effectiveness procedure. The acceptability of the remaining risks has rather to be judged according to other, external criteria, for example through comparison with other risks which have already been accepted (Schrader-Frechette 1984, 275ff). It makes no sense to reduce, at high costs, the risk resulting from a certain way of acting if, with the same expenditure, the risk of another way of acting could be reduced more effectively.

This consideration obviously only applies if the damages referred to in the comparison and the probabilities of their eventuation are commensurable. Thus it is clear that the efficiency consideration is of normative force only if the principle of pragmatic consistency (or another principle of the same capability) is presupposed.

With the help of the cost-effectiveness procedure for risk reductions, a number of difficulties become manageable which arise when judging risks (Merkhofer 1984; Schrader-Frechette 1984; Morgan 1990):

- This method is able to calculate with the value of a human life without having to substitute another value for it, e.g. a monetary one.
- Values are assigned the same weighting respectively; so, a human life is equal to a human life. Thus, the procedure can easily be brought into accord with egalitarian ideas, such as the equal distribution of risks.
- Any deviation from the cost-effective solution would mean an increase in damage. Decision makers are thus under pressure of legitimation if they do not put the cost-effective solution into practice, since that would mean, for example, a willingness to sacrifice more human lives than necessary with the best possible solution.

The cost-effectiveness procedure, however, depends on assumptions and presumptions which limit its application to certain problem classes (Morgan 1990, 21). First of all, it is presumed that the budget for reducing risks is an invariable quantity. In reality, a society can decide, however, which sum of money it wants to make available for the reduction of risks. The size of the security budget cannot be determined according to the procedure of cost-effectiveness, since that would mean an endless regression. Yet, this does not altogether invalidate the importance of the procedure as a mental experiment. If we take a risk-risk comparison as a foundation for the basic acceptability of a risk, then it is often sufficient to assume just any budget to systematically select the respectively acceptable risk options according to the marginal risk reduction costs and to thus reduce the total risk. Still, three basic problems remain to be considered:

(a) *The cost-effectiveness procedure works right away only if it refers to one damage or cost dimension.* If there are several dimensions of costs or damages, which is most often the case in reality, the individual dimensions first have to be melted into one quantity, i.e., they have to be weighted among one another. Thus, the problem becomes part of the decision-theoretical questions which are to be discussed later (5.4) (Keeney & Raiffa 1976, 20).

(b) *The procedure of cost-effectiveness can produce counter-intuitive results in those cases in which risky actions are morally evaluated differently* (Schrader-Frechette 1984). It makes a considerable difference in moral evaluation whether a toxic substance is emitted into the air without the approval of the person concerned and without being of any advantage to the person bearing the risk, whether the same substance is applied in the context of a mutually signed employment contract, or as a means to commit suicide. Assuming that it were cheapest to reduce the risk of the suicide, for many it still seems to be of higher priority to protect the „innocent“ victim from health damaging emissions. Accordingly, it is furthermore accepted that the standards for health damages at the work-place are less rigid than for the population in general. Such counter-intuitive results can only be avoided if the cost-effectiveness procedure refers only to actions with the same „moral“ dimension. The definition of these moral dimensions obviously depends on culture and context.

(c) *In the collective outlook, the procedure of cost-effectiveness produces results which sometimes seem implausible, looked at from a sub-collective point of view.* Seen from a global point of view, every mark invested in Germany for security measures is invested suboptimally with regard to the Third World. Therefore, according to the principle of cost-effectiveness, risk reductions in industrial countries ought to be abstained from altogether. But analogous arguments may also be used within industrial societies themselves. Thus, investment in intensive care units is suboptimal if compared to the effect that could be achieved by handing out free airbags to all car-drivers. Graham and Vaupel have calculated that in the American society, the range of money to save a human life ranges from less than 170'000 dollars to more than 3 million dollars (Graham & Vaupel 1980). The examples show that the principle of equal treatment of all cases, irrespective of circumstances and social evaluation, which is presupposed in this procedure, is implausible in many cases. Obviously it makes a difference for many people whether money is invested for extremely needy people (such as senior citizens or ill persons) or for normal persons who can take care of their security themselves. Similar difficulties arise if it is taken into account that persons endanger themselves negligently (smokers), or that society benefits from some persons more than from others (important decision makers or artists). The cost-effectiveness procedure is thus to be confined to those areas in which there is a consensus about every potential person concerned having the same right to safety (Schrader-Frechette 1984).

5.3.3 Cost-Benefit Analysis

Both cost-benefit analysis and welfare-theoretical approaches, which are to be discussed afterwards, consist of an explicit evaluation of the benefit of a risk reduction, by comparing this reduction with the expenditure required for achieving it. In cost-benefit analysis, this is done on the basis of a comprehensive monetarization of all utility and cost categories (Fischhoff et al. 1985, 266ff). The benefits of different environmental standards are translated into monetary units with the help of various methods (shadow prices, willingness to pay); the same is done with the expenditures (costs, organizational input, conflict costs, costs for decision making etc.). The assessment of environmental standards is then possible on the basis of a simple calculation. If the risk can continuously be reduced, then that reduction level has to be chosen at which the total-cost function (sum of expenditure costs and costs caused by the remaining risk) reaches a minimum. If such a continuous reduction is not possible, then that discrete option has to be chosen for which the difference between benefit and cost is greatest (Fischer 1973, 230ff; Hansmeyer & Rürup 1975, 65ff). At the same time, cost-effectiveness of the standard as compared with other standards which are already in force can be checked with the help of the respective benefit-cost difference (Rowe 1977, 366ff).

What makes cost-benefit analysis interesting is mainly the fact that environmental standards can be judged without direct consideration of subjective preferences. If the respective market prices are used for a comparison of various options, the monetarization directly reflects the social utility of the respective evaluated consequences. For, in an ideal-type market economy, prices are intersubjective indicators for the marginal benefit connected with a product or a service.

Including external effects and evaluating collective goods, however, causes difficulties. Here, shadow prices which simulate the market value, have to be deduced indirectly. Even though scientific procedures have been developed for this, the variation range due to such procedures is so great that, in many cases, no clear results can be reached (Harvey 1985; Fischhoff et al. 1985, 268). The question about discounting such prices over time constitutes a further difficulty (Hansmeyer & Rürup 1975, 72; Smith 1986, 23; Horowitz & Carson 1990, 110). While for market prices the usual market interest rates are adopted, discounting is difficult and often ambiguous when external effects are monetarized, especially those of a non-material nature. While it makes sense to assign a negative interest rate to gains which are only expected in the distant future, it is not very plausible to rate the victim of a future ecological damage as less „valuable“ than the victim of a present pollution.

Both difficulties mentioned are particularly conspicuous with respect to the problems of environmental standards. Which monetary value corresponds with an increase by x per cent in the risk of dying of cancer? How does this monetary value change if the damage will occur only in 20 years? How many trees are worth one lost job?

So far, there are no convincing procedures to conclusively answer such questions within the framework of cost-benefit analysis (Baram 1980; Kelman 1981). The willingness-to-pay approach favored by most economists, according to which persons are asked how much money they consider an improvement or a deterioration of their risk level by x per cent to be worth, remains unsatisfactory despite the progress that has recently been made in this specific field. First, the answers diverge depending on whether one asks about the monetary equivalent of a risk improvement (willingness to pay) or of a risk deterioration (willingness to suffer) (Smith 1986, 16). Second, the question, how can different risk preferences be aggregated to a social utility function, remains unanswered (Külp 1976, 25).

Despite these problems, cost-benefit analysis fulfills an important function in economics, politics and in lawsuits, especially when the costs arising in an economy are to be compared with the benefit, independent of the subjective preferences of the members of a society. Indirect indicators, such as prices for additional insurances, costs for restoring health etc., can help to provide useful approximations of the monetary expenditures and receipts which are actually to be expected as result of an environmental standard (Fischer 1973). Thus, for example, the damage caused by acid rain can be operationalized by losses in the wood industry and in the tourism trade (Wicke 1990, 214). The aesthetic and the ecological damage, however, are not considered in such an approach (Baram 1980).

Obviously, the procedure of cost-benefit analysis remains confined to such areas where a monetarization of benefit and costs according to intersubjectively clear criteria is possible. Especially with respect to environmental standards, this precondition is often not fulfilled in politically explosive cases.[3] This problem shows that the normative force of cost-benefit analysis is also based on an implicit commensurability consideration. The actions to be compared under the aspect of their cost-benefit difference are considered to be commensurable with regard to their monetary value. As to the monetary values, moreover, a consistency principle holds: decision makers shall give preference to those actions the benefit of which surpasses the costs.

5.3.4 Welfare-Theoretical Approaches

Welfare theory is based on the utilitarian principle of granting satisfaction of needs to the highest possible number of people in a society, to a degree which cannot be further improved in relation to the attained level of economic development. Accordingly, a welfare-theoretical judgment of environmental standards includes the investigation of whether the environmental standard will lead to a situation which would not be improved by any other solution. In order to determine this optimal point, first of all, those options for which an improvement of the utility level of some individual or party is not possible without harming others are eliminated. After this selection, those possibilities remain which, as „pareto-optimal solutions", cannot be improved any further unless the potential beneficiary fully compensates the persons bearing the risk (Kaldor-Hicks criterion). As long as the payment of a compensation means a utility loss for the potential beneficiary lower than the utility gain resulting from the choice of the standards, it is worthwhile to compensate the persons bearing the risk. The ideal standard is then found when the potential beneficiaries are willing to pay exactly the same amount of compensation to the persons potentially bearing the risk as they can gain in utility from the choice of the respective standard (Külp 1976, 109ff; Just et al. 1982; Merkhofer 1984; Smith 1986, 17).

The so far unsolved problem with this approach is the aggregation of utilities as well as a satisfactory, i.e. feasible and uncontradictory solution to the distribution problem (Külp 1976, 25). Which risk is acceptable if one person considers the value of a slightly increased life-risk to be negligible, but another person regards it as particularly problematic? Average values are obviously not suitable for solving this problem. Collective decisions, such as in the form of votes, often lead to inconsistencies and suboptimal solutions (Arrow 1951). For example, 51% of the persons taking part in the vote might achieve a positive benefit worth 1 dollar, whereas 49% would suffer a damage worth 100 dollars.

Due to this difficulty, welfare-theoretical considerations are not feasible in most relevant cases. If, for a relative weighting of the target dimensions, we start from a quantitative measurement of utilities and a subjective determination of preferences, then welfare-theoretical considerations change into decision-theoretical procedures.

5.3.5 Establishing Environmental Standards According to the Principle of Precaution

The procedures discussed so far started from the assumption that the effects of the actions to be judged are known, so that both extent and relative frequency of the eventuation of damages can be specified. Roughly speaking, the consequences of an action are employed to assess its desirability. Such a method is no longer possible if one knows nothing about the effects of an action under discussion, or if there is so much uncertainty that one does not venture a respective assessment of the action. This type of ignorance is basically to be distinguished from that uncertainty which finds its expression in forecasting the eventuation of consequences with a certain probability. Uncertainty leads to reckoning with a certain variation range of the eventuation of effects; ignorance leads to disregarding at least one of the dimensions of the consequences of an action (v. Winterfeldt & Edwards 1984, 90ff).

One solution to the problem of ignorance seems to lie in demanding a general minimization of the effects of an action. Germany has granted this way special recognition within its environmental law in the form of the precaution principle (Rehbinder 1976, 365; Hartkopf & Boehme 1983). According to this principle, emissions shall basically be avoided or reduced even if, due to a lack of scientific knowledge, negative effects are not known, but cannot be ruled out either. The rule of precaution against still unknown effects of emissions is given effect according to two different principles:

(a) *The principle of the lowest possible emission „ALARA-principle“.* According to this principle, each emission has to be reduced as far as possible, with the limit of what is possible being set by the highest reduction cost that is still justifiable from an economic and social point of view. The exact definition of this reduction level is a matter of discretion. In Germany, as in many other countries, the ALARA-principle is being applied in the field of radiation protection. Every contamination by ionizing radiation shall – if by any means technologically and economically justifiable – be reduced to a minimum far below the allowed limit.

(b) *The state of technology.* According to this principle, every emission that could be avoided with a tried and tested retention technology available on the market has to be avoided. A variant of this demand is the „state of science and technology“ laid down in the German law on atomic energy, requiring that new scientific concepts for the retention of harmful substances or for the improvement of technological safety have also to be implemented, even if they have not yet been technologically tried and tested.

The application of both principles easily leads to suboptimal solutions, since expenses and benefit are not systematically compared with one another (Rowe 1979, 330ff). Theoretically, when applying the principle of the state of technology, a potentially harmful substance may be emitted in great quantities if the corresponding retention technology is not available on the market. More realistic is the reverse case, namely that valuable economic resources are used for the reduction of a completely harmless substance or for the reduction of a harmful substance far below the threshold value, simply because there are enough technologies which allow of a substantial reduction.

As to the diffusion of new technologies, risks are often greater in the initial stage of development than they are later, when the technology is fully developed. Thus, new technologies may promise a risk reduction in the long run, but they cannot be implemented according to the state of technology, because the old, fully developed technologies involve lower risks than the new ones in the initial stage of their development. Thus, new reactor types which have a higher inherent safety level can no longer be licensed according to the state of technology, because not yet enough experiences with their operation in practice are available to prove an improved safety level compared to the reactors already in use.

Similar considerations hold for the ALARA-principle: at which point a reduction is no longer reasonably justifiable (indefinite legal principle) follows either from a formal analysis of the systematic weighing of utilities and risks or as the result of a holistic judgment. In the first case, one relies on rational procedures, in the second case on the judgment of the decision maker. The problem is mainly that the minimization rule may entail costs which bear no reasonable relation to the risk reduction. If the application of the ALARA principle prevents technical options which can fulfill their risk-minimizing function only in the future, the postulate of temporal distribution justice is infringed. In certain cases, this may be reasonable and justified; yet application of the ALARA principle does not make such an infringement explicit and prevents a systematic balancing of this infringement against the utility gain on another target dimension, namely temporary risk reduction.

These difficulties illustrate that the application of the precaution principle only makes sense if one of the three following preconditions is given (see Fritzsche 1986, 72ff):

(a) The effects of the respective pollutants are not or only very little known, but it has to be suspected that harmful effects will appear in the long run.
(b) The expenditure for retention of pollutants varies considerably, depending on the situation, so that with the ALARA principle one can react flexibly, i.e.,

take the expenses for a reduction into account. But the determination of what is reasonably justifiable has to be based on a systematic weighing of facts.

(c) The demand for state of technology is made in addition to the rational setting of standards, in order to reduce pollutants even if this does not seem to be absolutely necessary with regard to their effects but is technologically and financially acceptable.

When establishing radiation protection standards, precondition (a) is certainly not given. Hardly any other sphere of activity has been studied more thoroughly than the effects of ionizing rays. Precondition (b), too, is at best given in case of catastrophic events. After the Chernobyl accident, the contamination degree of food was different in different German regions and states. According to the ALARA principle, it would therefore have been economically justified to withdraw food with an extremely low contamination degree from the market in a region with relative low general contamination, whereas in other regions with substantially higher contamination, this would have led to unreasonably high economic costs. All the same, the psychological effect of different intervention standards would have been disastrous. Hardly anybody would understand that the consumption of lowly contaminated milk is prevented in one state and permitted in the other. According to the ALARA principle, this division is absolutely rationally justifiable; but the omission of the political effectiveness as a target dimension for establishing standards highlights the limitedness of this approach. Thus, it is more appropriate to establish identical standards, for the sake of the equality principle and of political acceptance, even in cases of catastrophes where different degrees of food contamination result in different cost-benefit relations for identical standards (Renn 1990).

When the principle of precaution is applied in addition to a rational setting of limits (c), the suboptimal allocation of resources appears to be justified if there is a special political interest in the reduction of emissions, independent of their effects. Yet it has to be determined whether the installation of reduction measures itself does not cause more risks than it actually reduces. Thus, Black, Niehaus and Simpson proved that some of the safety measures which are stipulated for nuclear plants cause more risks during construction of this safeguard technology than they help to avoid during their existence (Black et al. 1979).

So, in sum, *the principles of precaution shall mainly hold in those cases in which the possible consequences of an action are largely unknown.* If this precondition is not given, they may serve as an additional incentive to reduce risks, but they necessarily lead to a suboptimal allocation of social resources. Such a deviation from the optimal model may be quite justified, if criteria not related to the effects are included in the catalog of set targets.

But such a justification can again only be made within the framework of a comprehensive decision analysis. For this reason, a justifiable way of dealing with the precaution principle is only possible if this principle is included in a comprehensive decision analysis as one of several decision dimensions.

5.4 DECISION ANALYTIC PROCEDURES OF ESTABLISHING STANDARDS

In many sciences in which problems of rationally reconstructing human decisions are relevant, the instruments of formal decision analysis are successfully made use of (*decision logic*). Especially in economics, sociology and philosophical ethics, these instruments have been extended and refined in recent years for the various demands of the respective scientific questions (Edwards 1954; Gäfgen 1963; Raiffa 1973; Behn & Vaupel 1982; Weber 1983; v. Winterfeldt & Edwards 1986). In this section, we will examine the applicability of these procedures on decision problems in the context of setting standards.

The procedures of formal decision analysis provide a number of *possibilities* to improve the precision and transparency of the decision procedure. However, due to their methodical analytic approach, they are subject to a number of *limitations*, the consideration of which keeps us from overestimating them (Hansmeyer & Rürup 1975, 93ff; Fritzsche 1986, 506; v. Winterfeldt & Edwards 1986, 18ff).

The possibilities are mainly

- the chance to make an optimal choice among different options of action, according to subjective preferences, if objectives and knowledge are given (*utility optimization*);
- the opportunity to ensure the transparency of a decision process for the public by laying open target dimensions, target evaluations and weightings and thus to make an important contribution to the *justification of environmental standards*;
- the ability to better judge given problems by decomposing the complex matter of decision as well as the decision process into small workable steps and to avoid intuitive misinterpretations (*increased level of precision*).

On the other hand, the inherent limitations are:

- the difficulty to plausibly explain and politically justify the chosen weightings of the respective dimensions due to the necessity of including subjective preferences („*decisionism*“);

- the eventuality of neglecting certain interactive influences or excluding holistic impressions, through the division of complex matters into workable individual problems („*atomism*“);
- the difficulty of satisfying minimum conditions of rational decisions, through the strict formal demands of consistency and transitivity of all judgments which are to be taken into account. Therefore, however, the reconstruction cannot claim to reflect the political reality of the decision process, in which these requirements are very often not fulfilled („*idealism*“).

In particular, it would be a misunderstanding if one expected decision analysis to be a „mechanical“ or an „algorithmic“ procedure that is fed with the necessary information as its „input“, which then processes this data and gives the „proper“ decision as its „output“. Decision analysis can only offer coherent conclusions from the information which the decision maker provides by his/her preferences and his/her state of knowledge. Thus, it merely offers a *formal framework* which becomes effective if new information, target adjustments and, especially in group decision processes, negotiations and deliberations are included in this analysis (Raiffa 1973, 265ff; Edwards 1977). The possibility of corrections, for instance of target concepts or evaluation standards, i.e. the possibility of an objection on the part of decision makers or persons affected by the decision, is facilitated precisely by the transparent depiction of the decision process and is thus the reason for the superiority of such procedures compared to holistic judgments or mere additions of individual decisions.

In the reconstruction of a decision process in the decision discourse, a necessary *idealization* of the decision process takes place, i.e., only a certain aspect of reality, which directly or indirectly refers to the decision problem, is reconstructed in a „laboratory situation“, so to speak. This idealization is necessary in order to arrive at a comprehensible field of study by reducing the level of complexity.

5.4.1 Decision Analytic Way of Proceeding for Single Decision Makers

5.4.1.1 The Decision Situation

A decision problem is given if, at a certain point in time, a person has to choose between several (at least two) options of action with different results. This person shall be called *decision maker* (also: decider), the action options shall be called *options*, the results at a certain point in time shall be called *consequences* (Phillips 1979).

The decision maker evaluates the desirability of the consequences according to the extent to which they comply with his/her objectives. Objectives are measuring staffs by means of which one can judge the desirability of options (see section 5.1). Decision situations in which consequences of the options in question affect several (often contrasting) dimensions are called *„multi-attributive"*, i.e., the consequences of the action alternatives cannot be described with merely one attribute, but only with several concurrent attributes (Edwards 1977; Watson 1982). For decision problems in connection with environmental standards, multi-attributivity can be assumed without exception.

Moreover, decision situations under *certainty, uncertainty and ignorance* have to be distinguished (v. Winterfeldt & Edwards 1986, 90ff). For classification into one of these categories, the information level of the decider is pivotal. If a decision maker believes that he or she has complete information about all components of the decision problem, i.e., if he/she can estimate, on the basis of his/her knowledge about causalities, which consequences will follow, the decision is one *under certainty*. If there is no certainty about resulting environmental conditions but if at least probabilities can be determined for the occurrence of consequences, the decision is one *under uncertainty*. If there are no such probability estimations, i.e., if the decision maker knows which consequences could possibly occur, without, however, being able to estimate the probability of their occurrence, we talk about a *decision under ignorance* (v. Winterfeldt & Edwards 1986, 94; Häfele et al. 1990, 398f).

The problem that is to be discussed in a decision situation has first to be phrased in such a way that it becomes amenable to a decision analytic procedure. That means in particular to outline, step by step, the complete possible course of the decision. The overall strategy can be described in seven steps (see Weber 1983, 7; Keeney & Raiffa 1976, 5; Phillips 1979, 3; Renn 1986, 168). These seven steps can be regarded as a further subdivision of the division into the three steps of defining objectives, collecting data and weighing, which was made at the beginning of chapter 5. The steps are:

I. *defining objectives:*
- making out objectives and judgment attributes
- determining the options

II. *collecting data:*
- prognosticating the consequences
- estimating the probabilities of the consequences

III. *weighing:*
- determining the degree of achieving the objective
- converting consequences into units utility
- evaluating the options by weighting the target dimensions
- deciding

5.4.1.2 Defining Objectives

The decision maker evaluates the desirability of the consequences according to the extent to which they comply with his or her objectives. Therefore, in a first step of the analysis, these objectives should be formulated explicitly. Relevant for the decision analytic process are those objectives which are related to the particular section of the real situation, the decision-analytically constructed „horizon" which is the basis of the decision analysis up to a certain time in the course of events. This horizon may change in the course of the decision process. It may be extended by the inclusion of new objectives or it may become smaller when objectives are discarded.

The objectives may be formulated at different levels and with different degrees of abstraction. At the beginning of a decision analysis, a system of objectives is elaborated, which, in most cases, is based on creating a hierarchy of objectives. Thus, „head objectives", as, for example, in the context of air pollution the objective „improvement of the health of the population", are split into sub-objectives, like, for example, „reduction of the emission of pollutants" and „improvement of the citizens' attitude towards the issue of air quality". Some of these may again be split, for example into single objectives related to emission reduction of individual substances. In this case, one has reached a target level which permits assigning these objectives attributes with a scale which shows the degree to which such an objective is achieved (Keeney & Raiffa 1976, 39). In some cases, definite respective scales already exist, for instance „amount of emitted pollutants per year in tons", in other cases, they have to be constructed, as in the case of the citizens' attitude towards air quality. Assigning an attribute with a scale becomes necessary from a decision analytic point of view, because the „translation" into utility units is done on the basis of these scales and scale values which indicate the achievement degree. Once the scale values of the attributes of a consequence are ascertained, the consequence can be described by means of a vector the components of which are these scale values, e.g., consequence $K_l : K_l = (x_1, ..., x_n)$, with x_i denoting the degree of achieving the i-th objective.

This provides the basis for an *evaluation* of action options (which then is done by mapping the attribute scales on a utility interval from 0 to 1, or from 0 to 100).

The consequences of each option cannot be evaluated with regard to the achievement of one single objective, be it because this objective is too general and too comprehensive itself (an objective with a high degree of abstraction), be it because there are several different, maybe contrary objectives of the decision maker.

Specification in structuring objectives and, if need be, in the creation of hierarchies is to be handled pragmatically, i.e., the number of sub-objectives still has to be manageable and the degree of specification has to be purpose-oriented, that is, geared to the next step of the decision analysis (Keeney et al. 1984, 29f). The attributes are to be chosen in a way that the decision maker, knowing the attribute level of a certain situation, has a clear understanding of the extent to which the connected objective is achieved.

Attributes with their scales have to be such as to (a) allow of probability statements about different levels of the attributes, i.e., statements about the probability of the performance being x_i on the attribute scale, and (b) enable the decider to make preference statements, i.e., to state which utility he or she assigns to a value x_i reached on the attribute scale. In order to keep the system of objectives from getting out of hand, the decider is asked with respect to each objective whether a fictitious optimal course of action would change if this objective were left out. If the answer is „yes", the objective is included, otherwise not (v. Winterfeldt 1980; Weber et al. 1988; Keeney 1988).

Dynamic processes and corrections should also gain entry into the elaboration of a system of objectives. If, for example, it should prove in the course of a decision analysis that an objective thought to be significant does not have this importance after all, it can be omitted; in the same way, objectives can be split further if this seems necessary for a consideration of all important aspects of a decision situation.

5.4.1.3 Determining the Options

For a representation of a decision problem in a model, at first, all options of action arising in the course of the decision process should be listed. Here it will often be the case that options initially thought to be actual possibilities of action will prove impracticable, or that, due to a simplified selection strategy, suboptimal options can be identified at an early stage and excluded from the set of options to be analyzed.

A *systematic reduction* of the conceivable options may take place according to two procedures:

(a) „Satisfycing strategy“. According to this procedure, the individuals check, with respect to every consequence, to what extent a minimum threshold value considered necessary is exceeded (Simon 1976). Once minimum criteria for each target dimension are determined, those options are selected which exceed the minimum requirements for all relevant objectives.
(b) „Elimination by aspect“. All options are checked for the extent to which they successively meet a hierarchically constructed list of target requirements (Tversky 1972). In this way, subdominant solutions can be identified with relative reliability and can be deleted from the list of options.

Many decision analysts have pointed to the special relevance of the search for options (v. Winterfeldt & Edwards 1986, 26ff; Keeney et al. 1984, 21). Often options that may serve to reach an objective are searched in too narrow a field. With respect to the question about the choice of a means of transportation, for example, the options „travelling on public transportation“ or „taking a taxi“ are often not considered at all, because the problem is structured only in terms of private transportation.

Especially in the field of ecology, it makes sense to include functionally equivalent alternatives to environmental standards in the analysis, because often (if not always, as discussed in detail in chapter 6.2) these alternatives offer utility advantages compared to the rigid standards.

5.4.1.4 Determination of the Consequences

Prognoses on consequences are statements about cause and effect; they denote the *results of actions and courses of events*. This may just take place in an undifferentiated way. For example, in the form of: „The construction of this power plant will result in a deterioration of air quality.“ Decision analysis specifies such cause-effect relations. This is done on the basis of splitting head objectives into sub-objectives and of assigning attributes with scales (Keeney & Raiffa 1976, 39ff). For example, the above-mentioned general statement could be specified with regard to the attribute „SO_2-emission“ and the connected scale „SO_2-emission in tons per year“ as follows: „The construction of this power plant will result in an increase in the SO_2-emission by so and so many tons per year.“

To say it in a more formal way: a set of consequences $K=(K_1, ..., K_j)$ is to be determined further with regard to its achieving the formulated objectives of the set $Z=(Z_1, ..., Z_n)$. To each objective Z_i, an adequate attribute x_i is assigned, for which a measuring scale already exists or for which a scale can be constructed. The values y_i denote the degree of achievement of Z_i shown on the scale with respect to a certain consequence.

With respect to the choice of attributes, it has to be decided whether to construct a scale or whether to use an existing one. That depends mainly on the formulated objectives. If for some decision maker, e.g., SO_2-emission is the head objective, then a scale has to be constructed to evaluate the consequences with regard to this head objective. If improvement of air quality is the head objective, and hence SO_2-emission is a sub-objective, one can use an existing scale for this sub-objective, e.g. tons of SO_2-emission per year (Keeney 1981, 29).

It has also to be considered whether it makes sense at all to generate a scale for the head objective. For, it may be so general that a rational evaluation is almost impossible to reach; all aspects inherent in this objective remain implicit. Once adequate attributes have been assigned to the objectives, the consequences in a multi-attributive case can be regarded as vectors $y=(y_1, ..., y_n)$, with y_i denoting the achievement degree of the attribute, measured with respect to the objective Z_i in this consequence.

Now these consequence vectors have components which indicate their values in different units of measurement. A further step of decision analysis is then evaluation (see section 5.4.1.7), in order to make these different quantities commensurable and thus to be able to indicate *one* value for each consequence.

5.4.1.5 Calculating the Probabilities of Consequences

If decision processes are depicted in the form of decision trees, any alternative can be traced on its respective paths up to its consequences. If new decision situations arise in the course of the path, they are depicted by branchings and pursued further. At many points in a decision process under risk, situations arise in which there are several possibilities of a further course. If the occurrence of these possibilities does not depend on the decision maker, these situations are depicted in the decision trees as random branchings. So, consequences do not definitely occur, but only with a certain probability (Raiffa 1973, 27; Huber 1974, 432). Figure 5.1 shows diagrammatically a decision situation under risk.

The square node in figure 5.1 is a branching representing a decision situation; the circled node is a random branching. In the diagram, action A leads to a random branching where consequence K_1 will occur with probability p_1 and consequence K_2 will occur with probability $1-p_1$. The action option B definitely leads to consequence K_3.

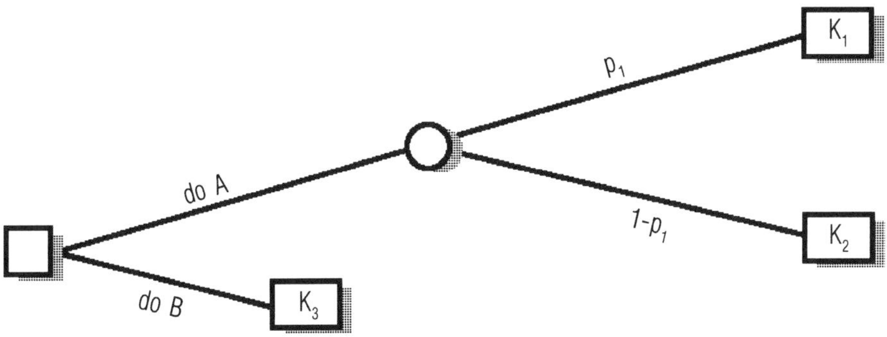

Fig. 5.1

In decision processes in which further decision situations with several options arise after the starting decision, the possible action sequences which might be chosen are called strategies and are pursued on their respective paths in the decision tree. A strategy is hence a sequence of actions in a certain order, with a certain consequence. Thus, an evaluation is not only based on the starting option but on the whole strategy.

Calculating the probabilities for different consequences is then easy if there is enough data from the past, and if, moreover, the external conditions under which the cause-effect relations inherent in the data are valid remain largely stable. This year's number of deaths on the road, for example, can be relatively reliably derived from last year's number of deaths on the road. In complex technological systems, known malfunction probabilities of parts of the system or of components are often synthesized to a system probability (Häfele et al. 1990, 378). In all of these cases, the calculated probability is based on the relative frequency of an event per unit time.

In connection with the application of new technologies, there is often a lack of historical data. Frequently, data are of an unacceptable quality or are not representative for future events. In these cases, one has to fall back upon *subjective probabilities* (v. Winterfeldt & Edwards 1986, 93). Basically, the same holds if there are not enough empirical values for the action options that are to be dealt with, or if these empirical values scatter so widely that statistical confidence intervals do not allow a sensible statement (Häfele et al. 1990, 402ff). Subjective probabilities are not to be regarded as pure speculation but as the formalization of empirical values. For this, those persons should be questioned who, due to their experience with similar systems or their familiarity with the system components, are able to make such estimations better than others (Winkler 1968). In the present context, these judgments are to be made. in particular, on the basis of *expert consultations* (Raiffa 1973, 274ff).

With the assignment of probabilities to the consequences of the action alternatives, the consequences evaluated in units utility can be weighted with their probabilities (see section 5.4.1.7). To say it in a formalized way: the *expected utility* (EU) is the sum of products of values (utilities) and probabilities of the consequences (Raiffa 1973, 24)

$$EU(A) = \sum_{i=1}^{n} p_i \times u_i$$

with i denoting the i-th consequence.

In figure 5.1 above, the expected utility of the alternative „do A“ is:

$$EU(\text{do A}) = p_1 \times u(K_1) + (1 - p_1) \times u(K_2).$$

In this equation, $u(K_1)$ and $u(K_2)$ are the values (utilities) of the consequences. Important here is the evaluation of the consequences, on the basis of which eventually the assessment of the action option occurs. Before this evaluation can be made, however, the consequences have first to be evaluated according to the degree of their achievement and then be converted into units utility. In a multi-attributive decision, this will be done along the lines of the following analytical process:

a) determining the extent of the consequences, i.e. establishing the degrees of achievement on the attribute scales;
b) generating a utility function for each attribute;
c) determining weighting factors;
d) aggregating the individual utilities of each consequence to the compound utility of the option;
e) determining the expected utility of the options.

5.4.1.6 Assessing the Performance of Each Option

In order to select from a set of options the one whose consequence, with a high probability, would mean the maximum performance of target concepts, an evaluation of the consequences is necessary. Here, two cases are first to be distinguished: (a) decisions under certainty and (b) decisions under uncertainty.

(a) In the easier case, that of *decisions under certainty*, each action alternative can be pursued with certainty on its respective paths to its consequence/s. An option has several consequences in the certainty model if in the decision process there are further decision situations (in decision trees: branchings). In order to evaluate the options, only the consequences have to be evaluated according to their utility gain (see section 5.4.1.7). The utility of an option is then equal to the utility of the consequence which will certainly ensue. If an alternative has several consequences (this is usually the case), combinations of consequences are evaluated. The utility of a combination is then equal to the sum of the utility values of each consequence, weighted with its relative importance.

(b) In the case of *decisions under uncertainty*, consequences occur with a certain probability. In order to evaluate the options, in a first step the consequences have to be evaluated like in case (a). In a second step, we calculate the expected utility of the last point in the sequence of actions from where on the further course could only be determined by probabilistic data.

In figure 5.1, the last point in the sequence of actions is identical with the random branching. The expected utility at this point is (see last section):

$$EU(\text{do A}) = p_1 \times u(K_1) + (1 - p_1) \times u(K_2).$$

In figure 5.1, this value can be be adopted right away as expected utility of the alternative, since in the course of events on the path of this alternative no further random branching occurs. The more complicated case shall be depicted by an extension of figure 5.1.

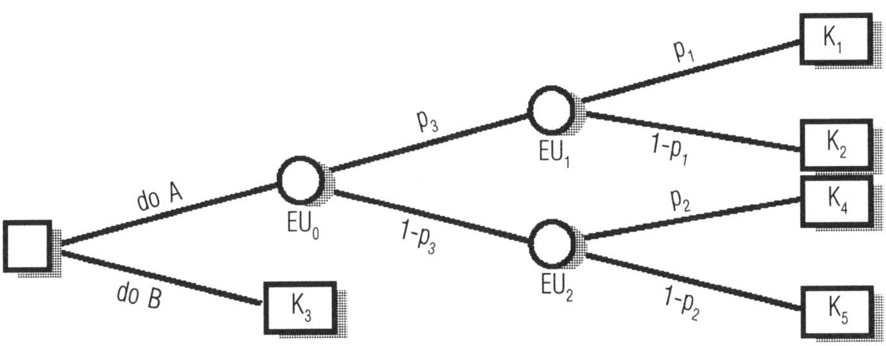

Fig. 5.2

In the case depicted in figure 5.2, the so-called „roll-back analysis“ (Raiffa 1973, 39) was carried out. For this, the decision course, including the probabilistic data and the utility values of the consequences, has to be given. In order to evaluate the options, we successively determine from right to left the expected utilities of the random branchings in the decision tree, on the basis of the given data about probabilities and consequences. In the course of events depicted in figure 5.2, we will first determine the expected utilities of the two last random branchings (i.e. the ones further on the right side).

$$EU_1 = p_1 \times K_1 + (1 - p_1) \times K_2$$

$$EU_2 = p_2 \times K_4 + (1 - p_2) \times K_5$$

From these values and from the probabilities p_3 and $1\text{-}p_3$, the expected utility of the preceding random branching can be calculated:

$$EU_0 = p_3 \times EU_1 + (1 - p_3) \times EU_2$$

Since there is no other point preceding this one, the expected utility of the option „do A“ is equal to EU_0. This is now compared to the expected utility of the option „do B“, like in the model above (Raiffa 1973, 39ff).

After having distinguished these two models, we will now have to look at the further course of the evaluation of options. This is based on the *evaluation of consequences*. For this, the individual consequences have to be converted into units utility.

5.4.1.7 Generating a Utility Function

With the help of utility theory, utility values are assigned to the consequences in their decomposed form, i.e., to each individual attribute; thus, the various measuring scales of the attributes are mapped on a single utility scale, the values of which lie between 0 and 1 (or 1 to 10, or 1 to 100). This mapping shall reflect the preference order of the decision maker. The mapping rule is therefore generated by questioning the decider, e.g., by assigning minimum or maximum values to the worst or the best consequence, and intermediate values to all other consequences. Once enough points are ascertained, the mapping rule is deduced (by a curve in the coordinate system) and in a second step, the correctness of this mapping rule is tested by presenting the calculated values to the decision maker for a check.

The utility function is then corrected according to his or her answers (Farquhar 1984; v. Winterfeldt & Edwards 1986, 241ff). This procedure is carried out for every attribute. With these utility functions, each consequence can then be represented in its components through units utility (e.g., every value on the scale „SO_2-emission in tons per year" can be expressed in units utility).

The enquiry can be carried out directly or indirectly. With the *direct method*, the decision maker is asked how many units utility he or she assigns to the scaled performances, e.g., which utility he/she attaches to a pollutant emission of so and so many tons per year. If the *indirect procedure* is applied, the utility function is generated by working with lotteries. The decision maker is asked with which lotteries (details about the lotteries with payments and their respective probabilities) he or she becomes indifferent between the lottery and a specific value of the attribute, i.e. which value of an attribute he/she equates with a specific lottery (Behn & Vaupel 1982, 27ff).

This enquiry is repeated until an indifference curve is attained which has to meet the conditions of consistency and transitivity. In case of inconsistencies, the enquiry has to be continued until the curve is corrected in this respect. This indifference curve can be interpreted as a utility function by identifying these lotteries with utility values (Raiffa 1973, 174ff). Working with lotteries has the advantage that the individual attributes, and thus the objectives, are placed in the background so that they lose a bit of their potential emotional charge.

5.4.1.8 Determining Weightings and Compound Utility

Up to this point in the analysis, the consequences are decomposed into attributes. In order to assign a compound utility to each consequence, now the relations among the decision maker's objectives according to his or her preferences have to be ascertained. Are the objectives equally important or are some of them more significant than others? If the latter is true, which ones, and how can the weighting be expressed? The point is therefore to consider the *preference structure* of the decider, so as then to be able to generate an aggregation function which combines the individual utility values to an aggregate utility value. Then, according to this value, the total evaluation of the action options is made, hence the decision in favor of the rational action option. This decision follows the rationality postulate: „Choose *that* option with which the greatest utility can be expected with the given preferences and the given knowledge structure" (see Keeney & Raiffa 1976, 6).

Since the seventies, various methods have been developed to determine the compound-utility functions, only one of which shall be outlined here as an example

(Keeney & Raiffa 1976; Schoemaker 1982; Weber 1983; v. Winterfeldt & Edwards 1986). The starting point of many methods to determine the utility function are representations of consequences which, in decomposed form, already indicate the degrees of achievement of the attributes in units utility. We will now attempt to establish the correlation between the vector y = $u(y_1)$, ..., $u(y_n)$ and u(K), the compound utility of a consequence.

Some methods consist in generating a weighting system (K_1, ..., K_n) which reflects the preferences of the decision maker. The relative weighting of objectives is an explicit demand of all decision procedures according to the MAUT principle (Multiattribute Utility Theory). However, weightings can only then be elicited sensibly by means of enquiries if the potential variation range of the performances is precisely defined. For this reason, the utility transformation for each attribute has to take place before relative weights are distributed for each attribute.

As an example, we will now present a class of procedures which start from linear utility functions and are confined to finding out the weighting system of the decision maker (Weber 1983, 102ff). For other utility functions, the so-called trade-off procedures have proved successful (Keeney & Raiffa 1976; Raiffa 1973). There, the decision maker is asked which amount of a certain attribute he would do without, in order to reach a level one unit higher for another attribute. From the answers of the decision maker, as many equations are set up as are necessary to determine the weighting factors.

The class of procedures which is to be presented basically consists in the attempt, based on the given achievement degrees of the consequences, to generate a weighting system of the objectives (k_1, ..., k_n) that corresponds with the preference order of the decider. For each objective, a weighting factor is to be determined that represents the importance of the objective for the decision maker. Determining the weighting factors is done via the attributes, because these are scaled and the performances of the consequences are already known. So, we receive the evaluation of a consequence by summing up the weighted values of the individual attributes already mapped on a utility-value scale.

$$u(K) = \sum_{i=1}^{n} k_i \times u(y_i)$$

with $u(y_i)$ indicating the individual utility of the attribute in its performance y_i and k_i indicating the weight of the corresponding objective Z_i; u(K) is the compound utility of a consequence which results from the sum of the weighted individual utilities. Preconditions for this procedure are additive independence and preference independence of the objectives (Fishburn & Keeney 1974). The advocates of this method are very well aware of the simplification, but at the same time they can commend the easy application of the procedure.

The formulated objectives are ordered according to their importance; the least important objective is given the number 10. Alternatively, one can also mark the best option (on an attribute) with the utility number 100 and the worst with the utility number 0. The decision maker has to judge how much more important the second-least important objective is to him compared with the least important one (or how much less important the second-most important objective is to him compared with the most important objective). He then assigns a number to the second objective, which reflects this relation. The number which is assigned to the third-least important objective shall result from the comparison of this objective with the already considered objectives. The decision maker might have to carry out several revisions by the time he reaches the most important objective. The weighting system results from a standardization of the relative weights to 1. Here is an example to illustrate this:[4]

> In a department, a professor is to be appointed. The nomination committee has to choose from a number of applicants (in this case, these are the options). The committee has set three objectives as a basis for its evaluation:
> - z_1: high publishing activity
> - z_2: pleasant personality
> - z_3: high commitment to teaching
>
> The following attributes are assigned to these objectives:
> - y_1: number of publications
> - y_2: personal impression
> - y_3: number of lectures
>
> In a first step, the order of the attributes (and thus of the objectives) is determined: $y_2 \geq y_1 \geq y_3$. According to the procedure, the number 10 is assigned to attribute y_3. The decision maker decides that y_1 is twice as important as y_3; so $y_1 = 20$. Again, y_2 is twice as important as y_1; so $y_2 = 40$. The decision maker thereby accepts transitivity, namely that y_2 is four times as important as y_3. If this were not the case, the weightings would have to be revised until the condition of transitivity is met.
>
> The thus ascertained values are standardized to 1:
> k_1=20:70; k_2=40:70; k_3=10:70.

From multiplication of these weighting factors with the individual utilities and summation of the products, the compound evaluation of a consequence is obtained. In case of decisions under uncertainty, these compound evaluations of the consequences are linked to the corresponding probabilities before summation.

By means of sensitivity analyses, the influence of the respective utility transformations and of the weightings of objectives can be checked. In a linear system, only proportional changes of output are to be expected after such changes of input. Nevertheless, a sensitivity analysis has proved valuable in practice, because it forces the decision maker to check his or her input and to test the robustness of the results (v. Winterfeldt & Edwards 1986, 399f).

5.4.1.9 An Example: Buying a Car

The abstract procedure of the decision analysis can be illustrated with a simple example. Let us assume that the decider wants to buy a new car; his or her decision sequence would be as follows:

(a) *Defining the objectives:* The decider should make up his or her mind about the objectives before choosing the options, because otherwise the intuitive preferences for certain options (like the red sports car in the window) would indirectly influence the selection of objectives. In our case, the buyer chooses the following objectives: inexpensiveness (subdivided into purchasing price, resale value and fuel consumption), reliability (subdivided into servicing frequency and freedom from surprises), durability, engine performance (subdivided into horsepower and acceleration), comfort, and aesthetics (subdivided into shape and color).

(b) *Determining the options:* Since the purchase of a private car is concerned, other means of transportation, such as public transportation or a bike, are out of the question. Therefore, the number of options is confined to private cars. Since there are hundreds of possibilities, the decider first makes a preliminary selection according to the satisfycing procedure. He/she determines minimum requirements which every candidate has to meet in order to be considered at all. These requirements could be: minimum number of seats, a price span from $15'000 to $25'000, an at least acceptable result in an automobile test published by a technical journal, etc. All cars which meet these requirements are put on the short list.

(c) *Determining the consequences:* Each car has now to be evaluated on each attribute. For this, information, as far as available, has to be gathered from the respective sources. Some attributes, such as the purchasing price, are definitely given and can be included in the matrix without modification; with other attributes, conversions are necessary. The resale value, for example, is not a fixed quantity (especially for a new model) and has therefore to be subjectively estimated or worked out by expert consultation. In both cases, subjective probabilities, i.e. the intensity of one's own or the expert's conviction, play an important role. Other attributes, such as the evaluation of the shape, require new scales, since this attribute does not entail any obvious measuring unit. In this case, the decider could set up a subjective aesthetic scale from 0 to 10 and evaluate each option relative to the other (0 for the ugliest car and 10 for the best-looking one, the other candidates between these extreme values).

(d) *Calculating the probabilities:* Some consequences of the purchasing decision will come to light only in the future, so that there is uncertainty on the true performance of each option on the respective target dimension. In this case, one has either to fall back upon relative frequencies (if existing) or upon subjective judgments. Let us assume that the servicing frequency of each car has been tested in the past ten years. Then, a distribution function for defects can be generated from the test data of 20 or 30 tested cars of the same type. This function indicates how often each car had to be serviced during the period under consideration. From this, an objective expected value can be calculated, which can be taken as a basis for the decision matrix. Depending on the risk aversion of the decider, one could also insert the lowest value of the confidence interval or the two-thirds value instead of the expected value, in order to reduce unpleasant surprises, or to be on the safe side, because one thinks that whatever can go wrong, will go wrong. But it is absolutely necessary to apply this rule to all options equally, in order to ensure equal chances.

(e) *Evaluating on the target dimension:* Often prognoses of consequences cannot directly be measured on the scale assigned to each objective. In this case, the consequences have to be estimated on the scale. This is particularly difficult when the consequences have been weighted with subjective probabilities. This can be illustrated with the example of the target „comfort". The decider has test-driven all cars and made notes on the comfort of driving. From experience, however, he or she knows that a good judgment on comfort is only possible after a long drive. For this reason, estimations have to be given for this value. For some cars, additional information is available, e.g. from friends who own the same car, or from car magazines which have tested the comfort. How does he/she arrive at judgments on the basis of this cursory information? In the MAU-analysis, one proceeds as follows: first, the various levels of comfort are defined and utility values are assigned to them (like: very uncomfortable = 0 points, a little uncomfortable = 10 points, relatively comfortable = 40 points, very comfortable = 60 points, and extremely comfortable = 100 points). On the basis of all this previous information, each car and each category are assigned subjective probabilities (car A [with much information]: 0 points 5%; 10 points 10%; 40 points 70%; 60 points 10%; 100 points 0%; car B [with little information]: 0 points 10%; 10 points 20%; 40 points 30%; 60 points 20% and 100 points 10%). The scale value for both cars then results from the sum of the utility values multiplied with the probability of their occurrence. In our example, the calculation is as follows:

car A: (0 x 5) + (10 x 10) + (40 x 70) + (60 x 10) + (100 x 0) = 3500
(35% achievement of the objective)

car B: (0 x 10) + (10 x 20) + (40 x 30) + (60 x 20) + (100 x 10) = 3600
(36% achievement of the objective)

The result shows that the utility value of the risky alternative turns out a little better than that of the less risky alternative. The slight deviation between the two options makes clear, however, that further criteria are needed to make a reasonable choice. Yet, the advantage of the risky option is remarkable, because it could not have been expected intuitively. Of course, also risk aversion (i.e., a particularly negative weighting of uncertain consequences) can be taken into account in utility calculation. Partly, this has already been done by choosing an asymmetrical point system. Moreover, the percentage of achievement of the objective indicates how far-off the evaluated option is from the ideal state (extremely comfortable). This information helps to look out for new options in case of low achievement of an objective (Keeney et al. 1984, 21).

(f) *Transforming into units of utility:* While many of the subjective scales, such as the evaluation of shape and color, are already given in subjective utility values, the objective scales have to be transformed into utility values. Such a transformation is necessary because different intervals of the same size may have different utility values in different places. Thinking of the target „durability in years“, for example, it makes no difference for a driver who never drives his car longer than for four years whether the car will last for seven or for eight years on average, but it does make a big difference whether it will last for three or for four years. For this reason, the obvious measuring units cannot be equated with the units utility. The decider has either to generate a continuous utility curve for the target „durability“ and then read the respective utility values off this function, or estimate the utility of the various options relative to each other on the basis of a scale from 0 to 100. At the end of this step, all options on all attributes will have to be transformed into utility values.

(g) *Weighting the target dimensions:* For the decision, all dimensions have to be made commensurable. This is best done by a relative weighting of each attribute. It is useful to weight first the attributes of each individual objective and then assign weights (from 0 to 1) to the objectives themselves (in doing so, the weights are successively multiplied, according to the roll-back procedure, from attributes to sub-objectives and from there to objectives). Although measuring of utilities and weighting are independent operations, it is sensible to consider the scattering range of the options on each attribute during weighting. If the options differ only slightly from one another on one attribute, it makes sense to keep the weight of this attribute low, even if the connected objective is classified as important. At the end of this step, the decider has an options matrix in which each attribute has a numerical weight and each option has a utility value on each attribute.

(h) *Aggregating the utility values:* The aggregation of the utility values takes place according to the simple formula (5.2):

$$u(\text{car A}) = \sum_{i=1}^{n} k_i \times u(y_i)$$

with $u(y_i)$ denoting the utility value of the i-th consequence, weighted with its probability, k_i denoting the relative weight of the i-th attribute, and u(car A) denoting the compound utility value of car A. The application of the sum formula, however, depends on the condition that all attributes are independent of each other and interactive effects between the attributes can be neglected. This postulate can often only be met by introducing auxiliary attributes which are designed especially for combination effects. Thus, e.g., the interaction of color and shape may constitute its own quantity, which, apart from the two individual components „shape" and „color", plays a role in the evaluation of the aesthetic quality. By skillfully selecting auxiliary attributes and reformulating attributes (in order to avoid redundancies), the postulate of independence can usually be met. At the end of this step, for each option a compound utility value can be specified which reflects the accumulated utility values on each attribute. By a sensitivity analysis, the result can then be checked for its robustness towards slight changes of the input.

5.4.2 Decision Analytic Way of Proceeding in Group Decisions

In the preceding sections, a formal framework for the evaluation of action options was introduced under the condition that the consequences of these possibilities of action are not only to be evaluated with regard to one, but with regard to several objectives of the decision maker. While so far we assumed *a single decision maker,* now *plural group decisions* shall be discussed. Group decisions are especially relevant in the context of political decisions, because they are characterized by diverging interests of decision makers when it comes to realizing certain objectives. In the following, a formal framework for decisions will be constructed, in which the decision maker consists of several parties which want to reach a joint decision (for formal analysis, see Raiffa 1982, 44ff).

So the purpose of the following strategy considerations is to make a group consensus possible by finding all chances of consensus, even if they are only latent, and by making suggestions for their optimal exploitation in terms of group consensus. The social interactions which are expedient in the individual „decision situations" are called *„deliberations"*. Here it has to be emphasized that the formal instruments of decision analysis may serve to reach transparency of the decision discourse for all parties by means of explication and reconstruction, but they cannot replace the actual deliberations by quasi-algorithmic mechanisms. In particular, there is no formal procedure by means of which a consensus can be conjured up if the conditions for reaching a consensus are not given. Therefore, first of all, attention has to be paid to the explication of these conditions, so that, based on them, strategies for reaching further consensus might be attained.

Finding a consensus may refer to two aspects: consensus on procedures and consensus on contents. A *procedural consensus* is given if all parties agree on a regulation about the method of reaching a decision referring to some issue, e.g., if they (consensually) accept a decision due to the procedure of the majority rule.

The procedures for collecting or interpreting data, too, can first be clarified consensually. Thus, the various parties may agree on the method of ascertaining immissions, or on which group of scientists shall be put in charge of which task. The legitimation of objections can also be limited from the start by means of procedural consensus.

A *consensus on contents*, on the other hand, requires that all parties unanimously agree on a certain option. Although a consensus on contents comes closest to the ideal of an aggregated pareto-optimal utility increase, such a consensus is difficult to reach in practice, because each party can claim a right to veto. Since such a right to veto often leads to an inability to decide, it is often in the interest of the parties to voluntarily constrain the principle of consensus on contents by means of procedural consensus. This is to be expected in particular if group members assign a higher preference to maintaining the discourse or to the necessity of a decision than to carrying their individual point (Bacow & Wheeler 1984, 21ff).

One of the methods of handling the consensus problem is the appropriate choice of a confinement of the issues agreed upon. For example, prohibitions are more easily agreed upon than stipulations. A prohibition may be indispensable for some members and acceptable for others, because they can quietly put up with it and it requires no active adherence or corresponding action. This is not the case with stipulations, because setting up a stipulation requires a positive position of every member of the group. Moreover, a world of stipulations requires a perfect level of knowledge about their consequences, which is not given in most cases. Therefore it is important to recognize and use this asymmetry of the readiness to reach an agreement in group decisions.

5.4.2.1 Characteristics and Objectives of Group Decisions in Case of Multiple Objectives

A group decision with multiple objectives is characterized by a *set of alternatives with at least two elements*, the characterization of the options and their consequences *by several attributes* relative to a given set of objectives, and a *decision maker consisting of several parties*. The reconstruction of group decisions is supposed to improve the „quality" of the decision, i.e., to increase transparency and acceptance and hence enforceability of the decision in the political process.

It shall, moreover, allow for taking into consideration as many interests as possible (Weber 1983, 5). In the theory of group decisions, there are two main approaches, first the *theory of collective decision*, and second the *theory of interpersonal cardinal utility comparison*. The theory of collective decision denies the possibility of an interpersonal cardinal utility comparison and tries to order the individual preferences ordinally and to aggregate them to a group preference (Arrow 1951; Külp 1976, 13ff). The theory of interpersonal utility comparison works on the basis of the individual utility functions and aggregates these by means of a weighted additive operation (Schoemaker 1982; v. Winterfeldt & Edwards 1986). This theory is presented below.

5.4.2.2 Initial Situations of a Group Decision

Relevant components for describing the initial situation of the group decision are the *options*, the *group of deciders*, the *group members' individual systems of objectives*, and the *group's aggregated system of objectives*, which, without negotiations on objectives, is equal to the union of the individual sets of objectives.

The options are combined to a set of alternatives $A = (A_1,...A_m)$, by the group itself or by a connected body. The group of deciders is represented by the set M =

$(m_1,...m_k)$. The set $Z = (Z_1, ..., Z_n)$ denotes the system of objectives on which the evaluation is based. This is established by the group prior to taking up deliberations on the evaluation of the options. It is composed of the parties' individual systems of objectives. Let us assume the following:

(a) A, M and Z are finite sets with at least two elements each.
(b) All members of the group know A and M.
(c) Each member m_j of M knows his or her system of objectives Z_j.
(d) Each alternative can be followed on its respective paths to its consequences, including determination of the probabilities of an occurrence and extent of the consequences.
(e) Each consequence of an alternative can be described by means of the system of objectives Z.
(f) Each party is able to evaluate each consequence with respect to the system of objectives.
(g) Each party has the right to weight elements of the system of objectives with zero.

A group decision process is over when the group has succeeded in finding a common evaluation, i.e. a „group"-evaluation of the alternatives.

5.4.2.3 Decision Situations

Because of the degree of complexity of a group decision with multiple objectives, a procedure will now be introduced which distinguishes different *decision situations* in the decision process. As a criterion for the reaching of a step, we use the *aggregation degree* in evaluating the alternatives within the group (Weber 1983, 32ff).

(a) A decision process is in *situation 1* if each party has an opinion about the evaluation of a consequence on each individual objective. So each party has a number of evaluations per consequence, which corresponds with the number of objectives. Formally speaking, there is an evaluation function: $\varphi\ (m_j,z_i)$:K $\rightarrow$ R, with m_j of M and z_i of Z. This evaluation function assigns a real number to each performance of an objective z_i of Z, according to the preferences of the member m_j of M. In the context of utility functions, the value range is standardized to the interval [0,1]; in the following, we will start from this presupposition. Hence it results for the evaluation function in situation 1 that: $u(m_j, z_i) : K \rightarrow [0,1]$. If we represent the consequences as vectors, the components of which indicate the achievement degree of the attributes, in situation 1 there are a maximum number of k vectors (k=number of parties) of the form $y = [u(y_1), ..., u(y_n)]$. So there is neither a consensus among the parties, nor is there a unanimous evaluation of each consequence, i.e. an evaluation that is not split up in individual performances.

(b) A decision process is in *situation 2* if there is consensus among the parties on the individual objectives, i.e., a group opinion has formed about the individual utilities. So there is one function for each objective: $u(z_i) : K \rightarrow [0,1]$, with z_i of Z; there is one vector $y^* = [u(y_1), ..., u(y_n)]$ per consequence.

(c) If each party has evaluated the options with respect to the system of objectives and has formed an opinion about the consequences, the decision process is in *situation 3*. In this situation, there are, for each option, as many evaluations of the consequences as there are parties. A higher degree of aggregation as compared to situation 1 manifests itself in the combination of the individual evaluations, that is, the evaluations concerning each objective, to one compound evaluation of each option. Formally: $u(m_j) : K \rightarrow [0,1]$, with m_j of M; in other words, there are k compound evaluations, i.e., the consequences are no longer described in a decomposed way.

(d) Finally, *situation 4* is characterized by the fact that the opinions of the parties are identical, i.e., for each option a group opinion has been found. The parties do not only agree on the individual evaluations, but also on the entire system of objectives. Thus, there is a *collective preference order*: $u : K \rightarrow [0,1]$. The expected utility values of the options can be calculated from the utility values of the consequences and the probabilities.

These decision steps are not meant to represent a mandatory course of a group decision process; they rather serve to identify and characterize a certain level of a decision in order to remove potential difficulties in the formation of a consensus. This may happen by pointing at the possibilities existing in a certain situation, so as to reach a situation with a higher degree of aggregation (with regard to group opinion). On the basis of the four mentioned situations, the following sequences of situations are *possible*:

(a) 1 - 4 (b) 1 - 2 - 4 (c) 1 - 3 - 4
(d) 1 - 2 - 3 - 4 (e) 1 - 3 - 2 - 4

Up to now, a formal framework for the decision analysis has been provided; in a second step, we will now attempt to specify strategies according to which a specific decision discourse can be carried out. Within the formal framework, we will describe how existing scopes may be used to reach decisions that are supported by the parties involved. For this, procedures will be suggested according to which the members of a group may reach consensual decisions. In doing so, we will basically adopt the steps of the individual decision course described in 5.4.1 and only enrich them by the different decision sequences.

5.4.2.4 Determining a Common System of Objectives

Since the evaluation of the action options with their consequences is based on the performance of certain objectives, a group system of objectives of the participating parties has to be elaborated first. One may first think of simply combining all systems of objectives of the parties. Yet two objections may be put forward here: first, it is possible that objectives of different parties directly contradict each other so that they cannot be realized at the same time. Second, there are objectives that do not directly contradict each other, but are not pursued by all parties. Therefore, one should not combine all objectives into a comprehensive system of objectives without taking a look at them first, since the group system of objectives is supposed to represent a consensus on which the parties will base their evaluations. For this reason, in a first step, an attempt should be made to elaborate a common system of objectives. This does *not mean an additional effort* in the group decision process; it is rather a *step to reach consensus*.

For this, all parties shall first denote their systems of objectives. Then, the intersection set of these systems is taken, so as to have, first of all, a „*core consensus*“, starting from which, perhaps, an understanding about dissentient concepts of objectives may be reached. Then, the other, non-consensual objectives are discussed. It is good to make a *test of importance* for each objective here (Keeney & Raiffa 1976, 43), i.e., the parties are asked whether the options differ on the respective dimensions to an extent that different estimates could result if this objective was excluded. If an affirmative answer is given, this objective has to be considered further, with a negative answer it may be excluded. In case of an affirmative answer, one can again try to reach group consensus on its inclusion in the objective system; if this is not successful, it has to be discussed further.

With continuing dissension, potential extreme objectives may possibly be eliminated, while parties giving up extreme objectives may denote new objectives which are closer to the cluster. Since there will also be time limits for the discourse about establishing a system of objectives, the remaining dissentient objectives should be renegotiated, and in a way that one party will possibly be ready to reformulate an objective or to give it up altogether if the other offers some return. Finally, in a voting procedure supported consensually by the entire group, the group will have to decide on the remaining dissentient objectives.

The value-tree analysis developed by Keeney, Edwards and Winterfeldt is an alternative to this procedure (Edwards 1977; v. Winterfeldt & Edwards 1986, 36ff; Keeney & Raiffa 1976). According to this procedure, all groups are first interviewed separately about their objectives and attributes, and all objectives, sub-objectives and attributes are then combined to hierarchically formed value-trees (Keeney et al. 1984, 20ff). All group-specific value trees have to be validated by the respective parties and confirmed as a representation of their structure of objectives. Once all value trees are generated, the analysts try to combine the various trees to one common super tree. In doing so, all objectives are included; only redundancies and intersections are excluded by restructuring as far as possible. The common super-tree has again to be checked by all parties for its ability to represent the respective group objectives. If this ability is given, the common super tree is taken as a basis for all further negotiations.

As a rule, there are objectives included in the super tree which are not shared by all parties. In this case, all parties have the possibility to weight with zero in the following weighting procedure the objectives or attributes which they consider irrelevant (Keeney et al. 1984, 22). This possibility of a zero weighting allows of initially including all objectives and attributes that are suggested by the parties. Thus, long-winded debates on general principles concerning the legitimacy of individual objectives are avoided. The procedure of the value-tree analysis was, for example, successfully applied in Germany in the *Sozialverträglichkeitsstudie des Forschungszentrums Jülich* (a study on social tolerability, by the nuclear research center Jülich, Germany) (Keeney et al. 1984, 39ff; Keeney et al. 1987).

Another help for structuring objectives is the *„top down" approach* in which one starts from abstract objectives and then successively develops the respective attributes. Experience shows that it is easier to reach a consensus on objectives with a high degree of abstraction than on those with a low degree. If there are, for example, several low-level objectives of different parties which may be represented as sub-objectives of an objective with a high degree of abstraction (e.g. „improvement of the health of the population"), one may try to form new and consensual sub-objectives, starting from this head objective. Conversely, it will often be possible to reach a consensus on concrete attributes (*bottom up*), like pollutant emission per unit time, and to determine the general objective or sub-objective via this attribute (Eisenführ & Weber 1986).

Moreover, one may try to formulate these objectives not only as „signposts", such as „reduction of the emission of pollutants", but right away as a *quantitative attribute*, such as „reduction of the emission of pollutants by 10% compared to the present situation" (Keeney & Raiffa 1976, 34). Such a formulation of objectives will more likely allow of a coherent evaluation of the consequences, because it is of an orientation nature and optimizes dealing with numbers. If merely a direction is given, one party can already regard a reduction of the emission of pollutants by 5% as a significant success and rate it very highly whereas the other party sees this as a minor success and rates it correspondingly low. Such differences of opinion can be prevented by formulating the objectives in the most quantitative and group-consensual way possible.

After the system of objectives has been established, the objectives should once again be examined for their degree of abstraction, and if they are defined in too general a way, they should be split up further, if need be, and possibly be phrased as quantitative objectives. The restrictions already mentioned for decisions of individuals should be respected in differentiating the objectives. When this procedure is completed, attributes with scales have to be assigned to the objectives, i.e., a set $X=(x_1, ..., x_n)$ is assigned to the set Z. Here one should, if possible, make use of already existing scales. If this is not possible, new scales have to be constructed.

Within every single system of objectives (objectives, sub-objectives, attributes), a hierarchical order is a sensible and necessary thing allowing of later calculation of probabilities and weightings from back to front by means of the roll-back procedure. A hierarchization of the objectives themselves, however, is not necessary in this step, because each objective will later be assigned a numerical weight between 0 and 1, from which the relative ranking of the objectives among each other may then be derived (Edwards 1954; Humphreys 1977).

Attributes and scales for each dimension of objectives are tied to measuring instructions. In most cases, for most attributes there are already scientifically acknowledged indicators or indices by means of which the performance of each option can be measured on the dimension of an objective. Selection of the indicators and measurement of the options should be done by experts. This measurement of options, however, does not yet render superfluous an evaluation of each measured value in units utility (v. Winterfeldt & Edwards 1986, 298ff). If, for example, the law of decreasing marginal benefit holds, then the respective utility value drops with each additional unit of a desirable good (see the car-purchase example in 5.4.1.9). The conversion into utility values is tied to subjective preferences, but it may be generalized by group consensus.

5.4.2.5 Determining the Consequences and Calculating the Probabilities

Before the deciders make an evaluation, experts should first study the course of events. This may take place in the form of event trees, for example. Each alternative should be studied down to its consequences at a certain point in time. For this, all available information is made use of; if necessary, further information is gathered. The prognosis is made according to this level of information; the consequences are specified both in their extents as well as in their probabilities. The assessment of the extent of a consequence is made by experts only for attributes with objectively measurable scales. In case of subjective scales, the assessments are made by the parties.

If there is much uncertainty about the nature and the extent of possible consequences and a high degree of interdependence among the consequences, it is often not possible to measure the consequences by intersubjectively established measurement procedures. In these cases, one has often to fall back upon the collective experience of experts. For this, a number of procedures have been suggested in literature (Winkler 1968; Raiffa 1973, 274ff; Keeney & Raiffa 1976, 599ff; Bacow & Wheeler 1984, 76ff).

Among these procedures, the *DELPHI survey* has proved particularly valuable. This procedure was developed by the RAND Co. in the mid-sixties and was initially applied for the evaluation of defense technologies. Later it was used mainly as an instrument of prognosis in the context of technology assessment (Mintroff & Turoff 1975; Benarie 1988).

The Delphi procedure consists of the following steps:

a) A team of researchers sets up a *catalog of questions* which ask about the expected consequences of some stimulus (such as environmental standards). At the same time, subjective confidence intervals or percentage values of the self-assessed judgment certainty (from 1 to 100%) are requested for each answer category.

(b) This *questionnaire is sent* to a group of acknowledged experts in the respective special fields. The experts are requested to answer all questions to the best of their knowledge and to indicate for each answer their subjective certainty (expressed in confidence intervals at a confidence level of 95%, or by indicating a percentage value of judgment certainty).

(c) The group of researchers analyzes all questionnaires and works out the *average values, the extreme values and the variance for each question*. The subjective judgment certainties can here be included as weighting factors. Extreme values are listed together with the respective judgment certainties.

(d) The original questionnaire is sent back to the experts together with the evaluation of the first survey. None of the names of the experts (even in case of extreme values) are mentioned so as to avoid influencing by status or seniority. The experts are asked to *fill in the questionnaire once again*, this time, however, with the request to regard the results of the first survey as a corrective for their own judgments. The purpose of this second survey is to reduce the variance of the possible answers and to increase the collective judgment certainty.

(e) Steps (b), (c) and (d) are repeated until the experts no longer modify their judgments. As a result of this iterative survey, the frequency distributions of the answers either shift to a steep normal distribution (if group consensus prevails) or they split up into distributions with two or several distinct maxima (representations of different scientific camps).

In the ideal case, the Delphi procedure sorts out those evaluations which the group of experts cannot agree on. Due to the iterative process of the survey and the fact that the participants are kept anonymous, the respective knowledge level can be represented without considering the prestige value of each participant in the Delphi process.

One of the grave *disadvantages* of the Delphi procedure is the lack of explanations for judgments that deviate from the median of all participants (Hill & Fowles 1975). Extreme opinions or a particularly high judgment certainty may either be due to explainable insights of a participant, which are not accessible to the other participants, or they may be a product of personal overestimation or the strategic attempt to influence the result.

For this reason, Renn and others have suggested a modification of the procedure and have tested it in various fields (Renn & Kotte 1984, 190ff; Webler et al., 1991). According to the modified procedure, the experts are gathered for a joint discussion after the first survey. In this discussion, all participants whose evaluations vary significantly from the mean of all other participants have to give reasons for their point of view and defend it in the debate. These debates are recorded on video and, if necessary, made available to the experts in the next survey.

This way, not only extreme positions which lack substantiation can be eliminated; the result of such a Delphi also provides the argumentative basis for majority or minority positions. The disadvantage of this modification is the abandonment of the participants' anonymity. But since normally all experts are acknowledged specialists in their field, the influence of the prestige of individual participants as a disruptive factor is not very relevant, especially because experts with high prestige often dominate the middle range of the opinions and therefore only act as critics of the extreme opinions in the discussion. The procedure is, however, based on the presumption that extreme opinions more strongly require an explanation than the respective prevailing theory does (Webler et al., 1991).

The evaluations reached by the Delphi can be regarded as best approximation of the existing expert knowledge at a certain time (provided that the relevant experts were asked and the rules of the Delphi were applied correctly). Although this knowledge may also be objectively wrong (and this, in fact, was often found in retrospect), as a rule it is still to be preferred to all intuitive procedures or estimations based on subjective preferences.

After the described steps of analysis, the consequences (with the reservation just mentioned) are available in a scaled and decomposed form; they thus provide the basis for further evaluation.

5.4.2.6 Agreeing on Decision Steps

When the consequences have been estimated and measured on the target dimensions according to objective or subjective scales, the next step, in an individual decision, will be the transformation of the respective measured values into units utility. In group decision, this step is much more complicated, because the respective parties make different utility evaluations for each scale value. Since often even objectives and attributes are controversial, a homogeneous utility transformation is an illusion in most cases. Therefore it is advisable to reach an agreement on the decision steps and, if possible, consensus on the agreement procedure before the utility transformation and the weighting take place. Based on section 5.4.2.3, the following decision sequences for an evaluation of the options may be distinguished:

(a) *Decision step 1–2* aims at getting from each party's evaluations of the individual objectives to a group evaluation. That means that the parties have to agree on the evaluation n times (n represents the number of attributes). This can, e.g., be done by n times forming the arithmetic mean. Other procedures will be discussed below. The use of this step lies in reducing k evaluations per performance to *one* (k denoting the number of parties). So we have as many values per alternative as there are attributes. The formula for this is:

$$f_{1/2}: [u(m_1,y_1), ..., u(m_k,y_i)] = u(y_i), \text{ with } y_i \text{ of } Y;$$

this operation has to be performed n times, i.e. for each attribute. Speaking in vector terminology, the use of this decision step lies in reducing k vectors per consequence to one vector.

(b) In *decision step 1–3*, each party forms a compound evaluation of the consequence by aggregating its evaluations of the individual performances and its weightings of the objectives. Hence, we have k evaluations of a consequence. Formally speaking: in this case there is no homogeneous, consensual function $f_{1/3}$, but k different ones, each of which represents the individual decision step of one party. In this decision step, procedures for „decisions with multiple objectives“ are applied.

(c) In *decision step 1–4*, the group *directly* agrees on the evaluation of the options. In such cases, the explicit decision analytic procedures introduced here are not applied. The evaluation of the options is, for example, formed by deriving means from the parties' evaluations of the individual performances. Formally:

$f_{1/4}$: $[u(m_j, y_i)] = u$, with m_j of M, y_i of Y

(d) In *decision step 2–3*, each party forms an opinion about the option, based on group evaluations of the performances in situation 2. That means, each party aggregates the evaluations, on the basis of its own weighting system, to a compound evaluation of the alternative. Formally:

$f_{2/3}$: $[u(y_1), \ldots, u(y_n)] = u(m_j)$, with y_i of Y

This decision step differs from decision step 2–4 by the individual weighting system of the objectives. Decision step 2–3 can thus be understood as each party making a suggestion with respect to a weighting system, which has then to be agreed upon. So 2–3 could be a strategy of how to reach a common weighting system of the objectives.

(e) In *decision step 2–4*, the parties agree on a system of objectives and then reach a group evaluation of the alternatives by means of procedures for decisions with multiple objectives. Formally:

$f_{2/4}$: $[u(y_1), \ldots, u(y_n)] = u$, with y_i of Y

Starting from situation 2, it is useful to find a common weighting system of the objectives in order to then reach situation 4, if necessary via situation 3, with decision analytic procedures.

(f) In order to carry out *decision step 3–4*, the k opinions on the alternatives have to be developed into one group opinion; this can be formally expressed as follows:

$f_{3/4}$: $[u(m_1), \ldots, u(m_k)] = u$, with m_j of M

Here again, the arithmetic mean would be possible; but further procedures will be discussed in the section on strategies.

5.4.2.7 Decision Steps in Group Decisions

(a) *Sequence 1–2:* According to the described procedure of making a prognosis, there are a maximum number of k different vectors per consequence (k denoting the number of parties). Thus, the decision process is in *situation 1*. The aim is now to map these different values on a utility-value scale, i.e. on values between 0 and 1 (or 1 to 100), in a group-consensual way. A number of difficulties arise here, which shall be described in the following. Let us first take a look at the easier case, when there are objective scales, hence performance degrees are objectively measurable. In that case, the values indicating the performance degrees are the

same for all parties (they were established by experts when the course of events was determined, for decisions both under uncertainty and under certainty). These values have now to be mapped on utility values between 0 and 1. There are two possibilities for this:
(a) each party carries this projection out on its own or
(b) the group tries to jointly generate a mapping rule.

If the first possibility is chosen, the utility values of all groups have to be aggregated in a second step, according to a procedure which is to be determined jointly (for example, by forming means), whereas the second possibility yields consensual utility values. In case of objective scales, the second possibility is advisable, since otherwise one would go back again from common values to individual evaluations.

If, however, the assignments of utility values differ heavily, separate utility evaluation by each party is recommendable. Here, by means of enquiries (either directly or with lotteries) each party assigns utility values to the performances. Attention has to be paid to the fact that this assigning must be complete and transitive, i.e., if an option B is preferred to an option A and option A is preferred to an option C, then option B has to be preferred to option C. In case of a negative marginal utility (in other words, a utility reduction through an increasing performance degree), the respective attribute should be reformulated in a way that it is either monotonously ascending or descending. Thus, the objective „cleanness of water", which, from a certain degree of cleanness on, would entail utility losses, could, for example, be replaced by the attribute „water quality", in which case water that was too clean would have to be rated as of minor quality.

Moreover, for different consequences of the same objective, we must not follow different assigning rules. Hence, these assignments are done for each objective by considering the differing performances in the various consequences and mapping them on the utility-value scale.

The above observations on the utility transformation referred to objective scales. Still, the case of *constructed, not measurable scales* remains to be discussed. In this case, no homogeneous values for the performances are given. There are two possible ways of proceeding: first, one could try to achieve common values for the performances, so as to then establish the mapping on utility values; or each party could first carry out the mapping on utility values and then find common utility values in a further step. In both cases, the measurement refers directly to utility and thus needs no transformation. If, however, the units utility are established for each party separately, then aggregation becomes an even greater problem.

The utility values thus obtained will now have to be compared and discussed among the parties. Different opinions concerning the utility values should be removed as far as possible by means of deliberations and negotiations. These two strategies of reaching a consensus will be the issue of 5.4.2.8.

(b) *Sequence 1–3:*
This decision step is an individual decision, as already described in 5.4.1.

(c) *Sequence 1–4:*
This decision step is best carried out with a vote. The decision sequence 1–4 seems to make little sense, however, since it is not clear why each party should first evaluate each consequence with regard to each performance and then, in a next step, vote on the evaluation of the compound consequence. Here, many components, such as weighting factors, evaluation criteria etc., remain implicit. Thus, this procedure provides *little transparency* and should therefore not be pursued further.

(d) *Sequence 2–3:*
This way of proceeding is also rather inexpedient, since here the laboriously achieved consensus as a result of decision step 1–2 is not developed any further. Although the ascertained group utility values are applied, each party establishes, based on them, its own weighting system for the objectives and aggregates this data to a total evaluation of the consequence. Therefore, there are again k opinions per consequence, which have to be negotiated in a further step 3–4 among the parties. So it seems more practicable to try, starting from situation 2 of the decision, to establish a common weighting system for the objectives (decision step 2–4). Sequence 2–3 may, however, be of special interest in case of non-linear individual utility functions.

(e) *Sequence 2–4:*
This decision step aims at getting from consensual evaluations of the individual performances to a consensual total evaluation of the consequences. For this, a weighting system for the objectives has to be established. To do so, all objectives may be ordered if this makes cardinal weighting easier. The aim of the weighting should be, however, to assign a cardinal weight to each objective (on a scale from 0 to 1). Therefore, each party should make suggestions about weightings, which the group then again is supposed to discuss, correct, and bring closer to a consensus, by means of the deliberation strategies mentioned. The process of establishing weights of objectives entails also the opportunity to draw upon patterns of practiced social behavior, which allow of conclusions about the hierarchy of the objectives.

The result of this process is an ordering of the objectives indicating the weights of attributes and objectives, as already described in the context of individual decisions. Here, of course, transitivity has also to be checked and, if necessary, corrected. With this, the weighting factors of the objectives can be determined according to the procedure described. The compound utility of a consequence results from multiplication of the weighting factors with individual utility values and summation of the products. In the case of decisions under uncertainty, the expected utility values of the options are calculated from the utility values and the probabilities of the consequences according to the roll-back procedure.

(f) *Sequence 3–4:*
The starting point of this decision step are the k evaluations of each consequence, corresponding with the k parties participating in the decision. The aim is a homogeneous evaluation of all consequences on the part of the parties. In case of major evaluation differences, the compound-utility values should first be checked for consistency. This requires information about the data from which the parties started and about the procedure according to which they formed their opinion. If irregularities are found, they are to be corrected in accordance with the postulates of consistency and transitivity.

If the opinions of the parties in situation 3 were not reached by an aggregation of individual utilities with the help of a weighting system but by direct formulation, hence by a rather intuitive formation of opinion, and if dissension cannot easily be reconciled, it may be necessary to start the decision analysis again, this time in a more detailed way. That would mean that first a system of objectives would have to be established, individual objectives would have to be determined and attributes and corresponding scales would have to be defined. Such a case may occur if a decision analysis like the one described here was not planned initially, but is then used as an analytical method to reach some result in a decision process that got stuck. Often, however, a correction of the evaluations towards a group consensus is already brought about by asking the parties about their weighting systems and evaluation criteria and then checking and discussing these with the help of the deliberation strategies.

5.4.2.8 Strategies of Reaching a Consensus

For the transformation of scale values into units utility and for the weighting of attributes and objectives, processes of reaching a consensus are necessary in group decisions. These processes can be facilitated or accelerated by deliberation and negotiation strategies.

(a) *Static strategies:* Strategies that make a party change its opinion without necessarily affecting the opinions of the other parties are called static strategies. This may happen according to the following procedures:

- *Determining equal assignments:* Before the various utility assignments are discussed, the identical ones should be stressed as fixed points, so to say, on the basis of which the still controversial evaluations are negotiated.
- *Recourse to common head objectives:* Differences in the assignment of utility values can often be overcome by drawing on a consensual head objective and formulating new, compatible sub-objectives (moral principle by Lorenzen/ Schwemmer; (Lorenzen & Schwemmer 1972, 118ff)).
- *Cluster formation and elimination of extremes:* On the basis of cluster formation, it shall first be checked if and where a broad consensus on the evaluation is given. So all parties first give their estimation; then it is ascertained for which evaluations clusters develop. Then, on the basis of this core-consensus, a discussion with the dissentient parties about their evaluation may take place. In case of extreme values in the evaluation, one should ask about the assignment rule and its justification, and in addition, the assignment should be checked for consistency, transitivity and completeness. Moreover, it should be made sure that all parties involved have approximately the same information level on the matter that is to be evaluated. If extreme evaluations still persist within the group after these corrections, one should try to agree on the elimination of extremes for the sake of the decision formation.

(b) *Formal strategies:* Formal strategies are actions that facilitate the settlement of a conflict on the basis of rules that were established beforehand. These formal strategies are necessary in almost every case and are often needed as a supplement to static and interactive strategies. The following procedures are to be mentioned as formal strategies:

- *Averaging:* If no further change of opinion is to be expected from the parties, the arithmetic mean of the utility values may be formed.
- *Majority decision:* If no consensus can be reached along the lines of the strategies mentioned, i.e., if no party is ready to change its mind, one should draw on the method of majority decision, in consideration of the limited time of such decision processes. There are many variants here; for example, there may be qualified majorities, graded authorities etc. But majority decisions may lead to different results if there are more than two options, depending on the order in which the options are presented for a vote (Arrow 1951; Raiffa 1982, 44ff).
- *Common utility transformation:* Conjointly, the parties assign utility values to the performances of each consequence. In case of dissension, again the above-mentioned deliberation strategies are to be applied. The advantages of this procedure are that there is one less procedural step, and thus common assignment rules are available right away. The parties can be requested to justify

their assignment suggestions directly during the procedure and not only afterwards, so that they can correct their evaluation criteria immediately. A disadvantage of this procedure is that it can only be carried out in small groups. It should be considered whether one should not try to form coalitions if a large number of parties are participating in the decision, so that the k opinions could be reduced to a smaller number. For the formation of a consensus, it is indispensable that the parties are aware of the type of conflict.

- *Supra decision maker:* In this case, the parties delegate the aggregation of utilities to an arbitrator (benevolent dictator) who tries to aggregate the various utility functions of the individual parties in a way that either a pareto-optimal solution is reached or the Kaldor-Hicks criterion is met (utility losses can only be accepted if the utility loss of one party is compensated for by a utility gain of the other party) (Keeney & Raiffa 1976, 539f)
- *Participation:* In this case, the aggregation of the utility values is carried out by delegating the decision to the persons affected by the measure. Here it is difficult, especially if risks and utilities are distributed unequally, to identify the number of persons concerned or to select a representative sample of the affected persons. In some cases, citizens chosen by a random selection procedure may make such evaluations (Renn 1986). The evaluations themselves, however, have then again to be carried out by means of one of the procedures described in this section.

(c) *Interactive strategies:* Strategies in which the opinions may change each other mutually, are called interactive strategies. Among the interactive strategies, above all, negotiations have to be mentioned. Negotiations refer to several objectives at the same time. They correspond with market-conformable strategies. For example, one has to find out to which extent a party is willing to change its evaluation (utility assignment) in one objective, if, in turn, another party changes its evaluation of another objective (Raiffa 1982, 310ff; Bacow & Wheeler 1984, 21ff).
So we cannot always start from the assumption that each party insists on its evaluations under any circumstances, and that situation 1 can only be changed by persuasion, outvoting etc. Even if one sees moral and/or legal limits for this kind of negotiations, there is still a large field in which negotiations are successful (e.g., apartments near noisy streets are cheaper; jobs with higher risks are granted danger pay). If a group first sees no possibility of consensus on the evaluation of certain objectives, it may delegate the decision to another group/institution or to a kind of arbitrator, as is usual in wage bargainings. Here, the following *models of delegating decisions* can be distinguished (Raiffa 1982, 218ff; Bacow & Wheeler 1984, 156ff; Neale & Bazaerman 1985):

- *Negotiations without an arbitrator* (self-organization of the participants);
- *Negotiations with an arbitrator without a right to vote.* Here, the arbitrator tries to induce the participants to reach a compromise, but he or she himself/herself has no power to enforce this compromise (role of facilitator);

- *Negotiations with an arbitrator who has the right to vote.* Here, the arbitrator has the casting vote in a stalemate situation; in parity bodies, this may mean a strong power position. Such a position of the arbitrator can only be introduced if the participating groups agree (role of decisive negotiator);
- *Negotiations with an arbitrator who determines the result of the negotiations on his own authority.* The arbitrator or the court of arbitration hears the participating groups and determines the outcome of the negotiations according to rules established beforehand. This model again can only work if the participating groups agree to this competence of the court of arbitration (role of mediator).

No matter which procedure one tries in order to reach an agreement, it is advisable to first diagnose the type of conflict, since different forms of conflict settlement are required, depending on the type of conflict. According to von Winterfeldt and Edwards (1984), the following classification of conflicts can be made:
- *Conflicts about facts* (finding a consensus by defining a procedure for ascertaining the facts; appointing a scientific expert committee to clarify the issue; hearing experts etc.).
- *Conflicts about the interpretation of facts* (finding a consensus by hearing experts; referring to the practice in other countries; referring to accepted interpretations in similar situations; perhaps also risk-risk comparisons).
- *Conflicts about dimensions of objectives* (finding a consensus by acknowledging all objectives with the simultaneous possibility of zero-weighting; selecting on the basis of legal or political plans; determining a threshold-value for relevance of dimensions; combining dimensions in clusters, etc.).
- *Conflicts about weightings of objectives* (finding a consensus by eliminating subdominant options, compensating parties whose weightings of objectives are not taken into account; forming means of weights; calling a court of arbitration, etc.).
- *Conflicts about decision procedures* (finding a consensus by an open discourse; applying to the present issue a procedure that has stood the test in other situations; integrating critics of the procedure into the procedure, etc.).

These five conflict levels are not ordered according to their sequence in the decision process, but according to their degree of integrating problem-independent criteria into the process of finding a consensus. Accordingly, fundamental debates are to be expected more often in conflicts about the decision procedure than, for example, in conflicts about facts.

5.4.3 Decision Analytic Procedures in the Context of Risk-Risk Comparisons

Risk-risk comparisons may be integral parts of a decision analytic procedure. Decision analysis serves to make overly complex decision problems comprehensible and commensurable by means of information on preference orders and probability distributions, so as to allow for comparisons with correspondingly adapted, already accepted risks. The action alternatives with their respective consequences, adapted according to decision logic, as well as the likewise adapted already accepted risks, are then used in setting standards as an essential rule about the acceptability of a risk.

On the basis of decision analysis, which results in group-consensual evaluations of consequences, such comparisons can be carried out. These comparisons are characterized by drawing on critically tested convictions as a standard for the reasonableness of risks. Here, risks are not abstractly compared with other risks, but all relevant options are compared with one another on different dimensions, like in decision analysis, with the risk – in other words, the probability of a damage – presenting an essential evaluation category.

Since in group decisions these risk values are assigned different utility values, one can rely on the formal decision rule that risks may then be regarded as reasonable when the respective parties explicitly accept risks of the same or of a smaller order or if their acceptance can be concluded. The conditions necessary for risk comparisons (equal utility, exhausted cost-effectiveness of risk-reducing measures, remaining uncertainty in risk assessment, risk distribution, etc.) do not have to be taken into account here, because they are separately included and evaluated in other objectives or attributes. By this decomposition, the risks, both the already accepted ones and those underlying the considerations about standards, can be compared, since they are *mutually commensurable* through inclusion of the conditions into other dimensions. Together with the demand for pragmatic consistency (see chapter 1), a basis for rationally judging the reasonableness of risks is thus given. Here, the consistency postulate is applied to both collectives and individuals.

5.4.4 Application of Decision Analysis in Political Practice

The procedures of decision analysis are logical and rational. They ensure that, with given preference structures and given knowledge level, the option is chosen which promises the highest net utility. Moreover, they improve the transparency of decisions and make the decision procedure comprehensible for persons not involved. Still, the question remains whether they are also practicable in the political decision process.

Political decisions are often determined by a strategic way of proceeding. The parties involved are often not willing to disclose their objectives or to lay their preferences open to the public (Linnerooth 1983). They often try to give effect to their particular interests by linking generally acknowledged objectives (e.g., improvement of environmental quality) with group-specific attributes (e.g., more money for their own interest group). Likewise, they are not interested in integration of the respectively best expertise but try to filter input data according to their own interests (v. Winterfeldt & Edwards 1986, 517ff). Is a rational decision procedure enforceable at all under these circumstances?

As was explained in detail in chapter 4, the social reality of reaching a decision is by no means congruent with the normatively sensible decision procedure. This does not mean, however, that normative suggestions to improve the situation should therefore be dispensed with. Such a conclusion could only be justified if the application of decision analysis were impossible under the prevailing political conditions.

Indeed, decision analytic procedures are rarely to be found in German politics. If, however, we take the U.S. and other countries as a standard of comparison, it becomes clear that crucial political measures were prepared and often even taken with the help of decision analysis.[5] A few cases, like the choice of the location of the airport in Mexico City, have acquired a model character for many other political decisions since.

In the field of environmental policy, the Environmental Protection Agency (EPA) and most of the U.S. State Agencies have consulted decision-theoretical models ever since their institutional foundation. This can be observed mainly in the policy of preservation of clean water, and not only in the U.S. but also in many internationally active organizations (Major 1978; Loucks & Somlyody 1986). Here, the decision analytic approach (in its form of a multi-criteria analysis) has become standard by now.

A particularly interesting application is the evaluation of different sites for the disposal of nuclear waste, which was conducted by Merkhofer and Keeney on behalf of the National Academy of Sciences in the U.S. (Merkhofer & Keeney 1987, 173ff). Here, the analysts chose to proceed according to the MAU-technique described above. Various sites were evaluated according to the criteria „geologic appropriateness“, „costs“, „short-term and long-term risk for health and environment", „sociopolitical effects on the surroundings" etc. Utility aggregation and weighting took place through negotiations among different interest groups. In the case of irreconcilable conflicts, the team of analysts became active as an unbiased arbitrator. The input data was obtained by consulting experts. On the basis of this analysis, a sequence of suitable sites was determined. Number one on the list was Yucca Mountain in Nevada; this site was, in fact, eventually chosen, although the U.S. Department of Energy (DOE) initially wanted to give preference to Hanford in Washington State.

The examples show that decision analysis is practicable in democratic countries. Not all applications were successful and not all problems can be solved with this method (v. Winterfeldt & Edwards 1986, 517ff). But it seems sensible to grant the decision analytic procedures more room for application in German politics, too, in view of the possible devastating effects of wrong political decisions.

5.4.5 Summary

From the above considerations, we may derive elements for a procedural suggestion in case of problems of multi-attributive group decisions in the field of environmental politics. These elements are not to be regarded as a normative guideline, but as a mental aid or decision support in structuring and preparing the setting of standards.

In a first step, a common system of objectives should be established; attributes with scales on which the performance degrees of the objectives can be indicated have to be assigned to the objectives. The objectives with their corresponding attributes have to be chosen in a way that, on their basis, the consequences of the action alternatives can sufficiently be described.

In a second step, all action options that seem feasible at this time should be listed and considered with respect to their possible course of events. That means that each action alternative is followed on its respective paths; it is taken into account where further decision problems occur in a string of actions and where there are random branchings. At these random branchings, probabilities have to be assigned to the possible consequences. If all action alternatives have thus been pursued to their consequences, values have to be assigned to these consequences.

In the field of environmental politics, it should be expert groups who carry out the assignment of values and probabilities. Here, the Delphi procedure or other group surveys may be used. Conflicts among experts should, if possible, be settled by expert discourses and not by political alliances (advocate model).

According to the above considerations it is advisable, in case of group decisions, that the parties proceed according to the situation sequence 1-2-4. If the decision discourse is already in situation 2, the decision step 2-4 should be carried out; if the discourse starts in situation 3, situation 4, which is aimed at, may perhaps only be reached by a partial step back to situation 2, namely by splitting up compound evaluations and trying to first reach a consensus on individual evaluations of objectives.

If the decision discourse has led up to situation 4, i.e., if consensual compound evaluations of the consequences with their probabilities are available, the expected utility values of the action options can be computed. Such expected values can be established on each dimension. Even though decision analysis only evaluates options relative to one another, it may often be necessary in the political process of finding a consensus to set up acceptance rules for each dimension.

As regards the dimension „risk“, it is thus sensible to check whether the risks of the respective options lie below or above risks that are rated as acceptable by revealed preference. As regards the dimension „utility“, only such options that result in a positive net utility can be accepted. Further decision rules can be established, which facilitate a judgment of the respective option. Eventually, however, the decision has to be made by making trade-offs, i.e., assigning relative weightings to the respective dimensions. In doing so, not only the various expected utility values can be considered but also, in more detail, the individual consequences in their split-up evaluations. Thus, for example, it can be determined how high a certain utility has to be with respect to one objective so that a potential damage with respect to another objective is accepted.

Based on the evaluations of the individual objectives, the individual options can then be compared with one another and the option that promises to provide the greatest utility can be chosen. The assignment of weightings is not scientifically determined; it has to be elaborated in the political process by means of consensus of the groups involved. Here, negotiation strategies or arbitration strategies may be pursued.

This decision sequence may be illustrated in a more formalized way with the following process scheme:

I Determining objectives

1. Establishing a system of objectives

a) risk reduction (subdivided in individual and collective risk),
b) utility gain (subdivided in monetary utility and intangible utility),
c) distributive justice (subdivided in risk distribution and risk-utility distribution),
d) sociopolitical acceptance (subdivided in conflict potential and institutional controllability) and
e) adaptability (subdivided in flexibility and resilience).

2. Determining options

a) no action or regulation,
b) emission limitation standards (with different limits),
c) immission limitation standards (perhaps bubble concept),
d) technical procedural regulations,
e) subsidies or environmental levies,
f) tax reduction or tax increase,
g) selling of pollution rights
h) reduction of ecological consequences (e.g. liming forests, resettlement of population etc.),
i) production bans.

II Collecting data

3. Determining the consequences of each option

a) expert survey,
b) Delphi,
c) expertises (commissions).

4. Assigning probabilities to each consequence

a) relative frequencies for statistical data,
b) synthesized probabilities for complex systems,
c) subjective probabilities for „educated guesses".

5. Transforming into utility values

a) situation sequence 1–2,
b) utility transformation by expert body,
c) hearing of interest groups and persons affected.

III Weighing

6. Acceptability criteria within each dimension

a) for dimension risk: risk-risk comparisons as standard,
b) for dimension utility: positive cost-benefit balance,
c) for risk distribution: applicability of a justice criterion or compensation of the persons at a disadvantage,

d) for sociopolitical acceptance: conflict costs lower than utility; institutional controllability,
e) for adaptability: reversibility of decisions in case of altered knowledge.

7. *Weighting the respective dimensions*
   a) situation sequence 2–3 and 2–4,
   b) negotiations in the decision body,
   c) hearing of interest groups,
   d) possibility of appeal or revision.

8. *Aggregating the expected values on each dimension*
   a) testing the independence of dimensions,
   b) utility calculation by summation formula,
   c) sensitivity analysis.

ANNOTATION:

1 Further examples illustrating the importance of the reference quantity can be found in Wilson & Crouch 1990.

2 See the discussion referring to Rawls 1971, documented in Höffe 1977; also Höffe 1987.

3 Baram 1980, Kelman 1981; see also the reply to Kelman in Butters et al. 1981.

4 Taken from Weber 1983, 103.

5 A whole series of case examples for the application of decision theory can be found in the books of Keeney & Raiffa (1976, 354–479), Behn & Vaupel (1982), v. Winterfeld & Edwards (1986, 448–528), Susskind et al. (1979, 40–75), and Talbot (1987)

LITERATURE:

Arrow, K.J. (1951): Social Choice and Individual Values, New York, Basic.

Bacow, L.S., Wheeler, M. (1984): Environmental Dispute Resolution, New York, Plenum.

Baram, M. (1980): Cost-Benefit Analysis: An Inadequate Basis for Health, Safety, and Environmental Regulatory Decisionmaking, Ecology Law Quarterly 8, 473–531.

Behn, R.D., Vaupel, J.V. (1982): Quick Analysis for Busy Decision Makers, New York, Basic.

Benane, M. (1988): Delphi and Delphilike Approaches with Special Regard to Environmental Standard Setting, Technological Forecasting and Social Change 33, 149–158.

Black, S., Niehaus F., Simpson, D. (1979): How Safe Is ‚Too' Safe? Report WP-79–68, Laxenburg, International Institute for Applied Systems Analysis.

Butters, G., Califee, J., Ippolito, P. (1981): Reply to Steven Kelman, Regulation 5, Nr. 2, 41–42.

Crouch, E.A.C., Wilson, R. (1982): Risk-Benefit Analysis Cambridge, Ballinger. Dalkey, N., Helmer, O. (1963): An Experimental Application of the Delphi Method to the Use of Experts, Management Science 9, Nr. 3.

DeFina, R. (1977): Public and Private Expenditures for Federal Regulation of Business, St. Louis, Center for the Study of American Business.

Derby, S.L., Keeney, R.L. (1981): Risk Analysis: Understanding How Safe Is Safe Enough, Risk Analysis 1, 217–224.

Diederichsen, U. (1990): Risikobewältigung durch Jurisprudenz in: Risiko und die Herausforderung der industriellen Welt, M. Schüz (ed.), Vol. 1, Pfullingen, Neske, 150–171.

Edwards, W. (1954): The Theory of Decision Making, Psychological Bulletin 51, 380–417.

Edwards, W. (1977): How to Use Multiattribute Utility Measurement for Social Decision Making, IEEE Transactions on Systems, Man, and Cybernetics, SMC-7, 326–340.

Eisenführ, F., Weber, M. (1986): Zielstrukturierung: ein kritischer Schritt im Entscheidungsprozeß, Zeitschrift für betriebswirtschaftliche Forschung 38, 907–929.

Evans, L., Frick, M.C., Schwing, R.C. (1989): Is it Safer to Fly or Drive? General Motors Research Laboratories, GUR–6722, Warren, Mich., General Motors.

Farquhar, P.H. (1984): Utility Assessment Methods, Management Science 30, 1283–1300.

Fiorino, D.J. (1989): Technical and Democratic Values in Risk Analysis, Risk Analysis 9, 293–299.

Fischer, D. (1973): Kosten-Nutzen-Analyse–Hilfe für rationale Entscheidungen, Zeitschrift für die gesamte Staatswissenschaft 129, 230–245.

Fischoff, B., Watson, S.R., Hope, C. (1984): Defining Risk, Policy Sciences 17, 123–139.

Fischhoff, B., Slovic, P., Lichtenstein, S. (1979): Weighing the Risks, Which Risks are Acceptable?, Environment, 21, 17–20, 32–38.

Fischhoff, B., Slovic, P., Lichtenstein, 5. (1985): Weighing the Risks, in: Peribus Progress: Managing the Hazards of Technology, R.W. Kates, C. Hohenemser & J.X. Kasperson (eds.), Boulder, Westview, 265–283.

Fishburn, P.C., Keeney, R.L. (1974): Seven Independent Concepts and Continuous Multiattribute Utility Functions, Mathematical Psychology 11, 294–327.

Frankena, W.K. (1963): Ethics, Englewood Cliffs, Prentice Hall.

Fritzsche, A.F. (1986): Wie sicher leben wir? Risikobeurteilung und -bewältigung in unserer Gesellschaft, Köln, Verlag TÜV Rheinland und Gäfgen, G. (1963): Theorie der wirtschaftlichen Entscheidung, Tübingen, Mohr.

Graham, J., Vaupel, J. (1980): The Value of Life: What Difference Does It Make?, Risk Analysis 1, 89–95.

Häfele, W., Renn, O., Erdmann, G. (1990): Risiko, Unsicherheit und Undeutlichkeit, in: Energiesysteme im Übergang–Unter den Bedingungen der Zukunft, W. Hafele (ed.), Landsberg, Poller, 373–423.

Hansmeyer, K.-H., Rürup, B. (1975): Staatswirtschaftliche Planungsinstrumente, Tübingen, Mohr.

Hartkopf, G., Boehme, E. (1983): Die Ausgangssituation unserer Umweltpolitik. Aus Politik und Zeitgeschichte 42, 3–12.

Harvey, C.M. (1985): Decision Analysis Models for Risk in Distant Futures, In: Risk Assessment and Management, L.B. Lave (ed.), New York, Plenum, 509–558.

Hattis, D., Kennedey, D. (1990): Assessing Risks from Health Hazards: An Imperfect Science, in: Readings in Risk, T.S. Glickman & M. Gough (eds.), Washington, D.C., Resources for the Future, 156–163.

Hill, K.Q., Fowles, J. (1975): The Methodological Worth of the Delphi Forecasting Technique, Technological Forecasting and Social Change 7, 179–192.

Höffe, O. (1977): Über John Rawls Theorie der Gerechtigkeit, Frankfurt a.M. Suhrkamp.

Höffe, O. (1987): Politische Gerechtigkeit. Grundlegung einer kritischen Philosophie von Recht und Staat, Frankfurt a.M., Suhrkamp.

Hoerger, F. (1987): On the Evaluation of Risk Assessment and Risk Management and Future Needs, in: Risk Assessment and Management, L.B. Lave (ed.) New York, Plenum, 649–556.

Horowitz, J.K., Carson, R.F. (1990): Discounting Statistical Lives, Journal of Risk and Uncertainty, 3, 403–413.

Huber, G.P. (1974): Methods for Quantifying Subjective Probabilities and Multiattribute Utilities, Decision Sciences 5, 430–458.

Humphreys, P. (1977): Application of Multi-Attribute Utility Theory, in: Decision Making and Change in Human Affairs, H. Jungermann & D. de Zeeuw (eds.), Dordrecht, Reidel, 265–205.

Hyman, E.L., Stiftel, B. (1988): Combining Facts and Values in Environmental Impact Assessment, Boulder, Westview.

Just, R.E., Health, D.L., Schmitz, A. (1982): Applied Welfare Economics and Public Policy, Englewood Cliffs, Prentice Hall.

Keeney, R.L. (1981): Measurement Scales for Quantifying Attributes, Behavioral Science 26, 29–36.

Keeney, R.L. (1988): Structuring Objectives for Problems of Public Interest, Operations Research 36, 396–405.

Keeney, R.L., Raiffa, H. (1976): Decision with Multiple Objectives. Preferences and Value Tradeoffs, New York, Wiley.

Keeney, R.L., Renn, O., Winterfeldt, D. von, Kotte, U. (1984): Die Wertbaumanalyse. Entscheidungshilfe für die Politik, München, HTV Edition „Technik und Sozialer Wandel".

Keeney, R.L., Renn, O., Winterfeldt, D. von (1987): Structuring West Germany's Energy Objectives, Energy Policy 15, Nr. 4, 352–362.

Kelman, S. (1981): Cost-Benefit Analysis: An Ethical Critique, Regulation 5, Nr. 1, 33–40.

Kip Viscusi, W. (1984): The Influence of Legislative Mandates on the Oversight of Risk Regulatory Agencies, in: Risk Analysis, Institutions, and Public Policy, S. Hadden (ed.), Port Washington, N.Y., Associated Faculty Press, 117–131.

Krewski, D., Birkwood, P.L. (1987): Risk Assessment and Risk Management: A Survey of Recent Models, in: Risk Assessment and Management, L.B. Lave (ed.), New York, Plenum, 339–406.

Külp, B. (1976): Wohlfahrtsökonomie 1: Wohlfahrtskriterien, Tübingen 1976, Mohr.

Lagadec, P. (1987): From Seveso to Mexico and Bhopal: Learning to Cope with Crisis, in: Insuring and Managing Hazardous Risks: From Seveso to Bhopal and Beyond, P. Kleindorfer & H. Kunreuther (eds.), Berlin, Springer, 13–46.

Linnerooth, J. (1983): Die Risikoanalyse im politischen Prozeß, in: Risikoanalyse und politische Entscheidungsprozesse, H. Kunreuther & J. Linnerooth (eds.), Berlin, Springer, 219–258.

Lorenzen, P., Schwemmer, O. (1972): Konstruktive Logik, Ethik und Wissenschaftstheorie, Mannheim, Bibliographisches Institut–Wissenschaftsverlag.

Loucks, D.F., Somlyodi, L. (1986): Multiobjective Assessment of Multipurpose Water Resources Projects for Developing Countries, National Resources. Forum 10, 61–77.

MacLean, D. (1986): Social Values and the Distribution of Risk, in: Values at Risk, D. MacLean (ed.), Totowa, Rowman and Allanheld, 75–93.

Major, D.C. (1978): Multiobjective Water Resource Planning, Washington, D.C., American Geophysical Union.

Mayntz, R. (1990): Entscheidungsprozesse bei der Entwicklung von Umweltstandards, Die Verwaltung 23, Heft 2, 137–151.

Merkhofer, L.W. (1984): Comparative Analysis of Formal Decision-Making Approaches, in: Risk Evaluation and Management, V.T. Covello, J. Menkes & J. Mumpower (eds.), New York, Plenum, 183–220.

Merkhofer, L.W. (1987): The Use of Risk Comparisons to Aid the Communication and Interpretation of the Risk Analyses for Regulatory Decision Making, in: Risk Assessment and Management, L.B. Lave (ed.), New York, Plenum, 581–607.

Merkhofer, L.W., Keeney, R.L. (1987): A Multiattribute Utility Analysis of Alternative Sites for the Disposal of Nuclear Waste, Risk Analysis 7, 173–194.

Meyer-Abich, K. (1990): Zulassung von Risiken für die Allgemeinheit, in: Risiko und Wagnis: Die Herausforderung der industriellen Welt, W. Schüz (ed.), Pfullingen, Neske, 172-191.

Mintroff, 1.1., Turoff, M. (1975): Philosophical and Methodological Foundations of Delphi, in: The Delphi Method, H.A. Linstone & M. Turoff (eds.), Reading, MA, Addison-Wesley, 17–36.

Morgan, G. (1990): Choosing and Managing Technology-Induced Risk, in: Readings in Risk, T.S. Glickman & M. Gough (eds), Washington, D.C., Resources for the Future, 17-28.

National Academy of Sciences (1975): Decisionmaking for Regulating Chemicals in the Environment, Washington, D.C., National Research Council. National Academy of Sciences, Committee on the Institutional Means for Assessment of Risks to Public Health (1983): Risk Assessment in the Federal Government: Managing the Process, Washington, D.C., National Research. Council.

Neale, M.A., Bazaerman, M.H. (1985): Perspectives for Understanding Negotiation: Viewing Negotiation as a Judgmental Process, Journal of Conflict Resolution 29, 33–35.

Phillips, L.D. (1979): Introduction to Decision Analysis, Tutorial Paper 79- London, London School of Economics and Political Science.

Raiffa, H. (1973): Einführung in die Entscheidungstheorie, München.

Raiffa, H. (1982): The Art and Science of Negotiation, Cambridge, Harvard University Press.

Rawls, J. (1971): A Theory of Justice, Cambridge, Harvard University Press.

Rehbinder, E. (1976): Umweltrecht, Rabels Zeitschrift für ausländisches internationales Privatrecht 3, 365–372.

Renn, O. (1986): Decision Analytic Tools for Resolving Uncertainty in the Energy Debate, Nuclear Engineering and Design 93, Nr. 2&3, 167–180.

Renn, O. (1988): Evaluation of Risk Communication: Concepts, Strategies, and Guidelines, in: Managing Environmental Risks, Air Pollution Control Association (ed.), Washington, DC., APCA, 99–117.

Renn, O. (1990): Risk Management and Communication in the Aftermath of Chernobyl: A Comparative Perspective, Environmental Psychology 10, 135–147.

Renn, O., Kotte, U. (1984): Umfassende Bewertung der vier Pfade der EnquêteKommission auf der Basis eines Indikatorkatalogs, in: Energie im Brennpunkt, G. Albrecht & H. U. Stegelmann (ed.), München, HTV Edition „Technik und Sozialer Wandel“, 190–232.

Renn, O., Kals, J. (1990): Technische Risikoanalyse und unternehmerisches Handeln, in: Risiko und Wagnis: Die Herausforderung der industriellen Welt, M. Schüz (ed.), vol. 1, Pfullingen, Neske, 60–80.

Rowe, W.D. (1977): An Anatomy of Risk, New York, Wiley.

Rowe, W.D. (1979): What is an Acceptable Risk and How Can It Be Determined? in: Energy Risk Management, G.T. Goodman & W.D. Rowe (eds.), London, Academic Press, 327–344.

Rushefsky, M. (1984): Institutional Mechanisms for Resolving Risk Controversies, in: Risk Analysis, Institutions, and Public Policy, S.G. Hadden (ed.), Port Washington, Associated Foreign Press, 133–148.

Schoemaker, P.J. (1982): The Expected Utility Model: Its Variants, Purposes, Evidence, and Limitations. Journal of Economic Literature 30, 529–563.

Schrader-Frechette, K.S. (1984): Risk-Cost-Benefit Methodology and Equal Protection, in: Risk Evaluation and Management, V.T. Covello, J. Menkes & J. Mumpower (eds.), New York, Plenum, 275–296.

Schrader-Frechette, K.S. (1985): Risk Analysis and Scientific Method: Methodological and Ethical Issues with Evaluating Societal Risks, Dordrecht, Reidel.

Schürmann, H.J. (1978): Ökonomische Ansätze zu einer rationalen Umweltpolitik und wirtschaftspolitische Konsequenzen, 2. Edition, München, R. Oldenbourg.

Simon, H. (1976): Administrative Behavior: A Study of Decision-Making Processes in Administrative Organizations, 3. ed., New York, Basic.

Slovic, p. (1987): Perceptions of Risk, Science 23, 280–285.

Smith, V.K. (1986): A Conceptual Overview of the Foundations of Benefit-Cost Analysis, in: Benefits Assessment: The State of the Art, J.D. Bentkover, V.T. Covello & j. Mumpower (eds.), Dordrecht, Reidel, 13–34.

Starr, C. (1969): Social Benefit Versus Technological Risk, Science 165, 1232–1238.

Susskind, L., Richardson, J.R., Hildebrand, K.J. (1979): Resolving Environmental Disputes, Cambridge, Massachusetts Institute of Technology.

Talbot, A.R. (1987): Settling Things: Six Case Studies in Environmental Mediation, Washington, D.C., The Conservation Foundation und the Ford Foundation.

Tversky, A. (1972): Elimination by Aspects: A Theory of Choice, Psychological Review 79, 281–299.

Watson, S.R. (1982): Multi-Attribute Utility Theory for Measuring Safety, European Journal of Operational Research 10, 77–81.

Weber, M. (1983): Entscheidungen bei Mehrfachzielen, Wiesbaden, Gabler.

Weber, M., Eisenführ, F., Winterfeldt, D. von (1988): The Effects of Splitting Attributes on Weights in Multi-Attribute Utility Measurement, Management Science 34, 431–445.

Webler, T., Levine, D., Rakel, H., Renn, O. (1991): The Group Delphi: A Novel Approach to Reducing Uncertainty, Technological Forecasting and Social Change 39, Nr. 3, 253–263.

Wicke, L. (1990): Der Ökonomische Wert der Ökologie, in: Risiko und Wagnis: Die Herausforderung der industriellen Welt, M. Schüz (ed.), vol. 1, Pfullingen, Neske, 210-227.

Wildavsky, A. (1979): No Risk Is the Highest Risk of All, American Scientist

67, Nr.1, 32–37.

Wilson, R., Crouch, E.A.C. (1987): Risk Assessment and Comparisons: An Introduction, Science 236, 267–270.

Wilson, R., Crouch, E.A.C. (1990): Risiko-Abschätzung und -Vergleiche, in:

Risiko und Wagnis: Die Herausforderung der industriellen Welt, M. Schüz (ed.), Pfullingen, Neske, vol. 1, 42–59

Winkler, R.L. (1968): The Consensus of Subjective Probability Distributions, Management Science 15, B61-B75.

Winterfeldt, D. von (1980): Structuring Decision Problems for Decision Analysis, Acta Psychologica 45, 71–93.

Winterfeldt, D. von, Edwards, W. (1984): Patterns of Conflict about Risk Debates, Risk Analysis 4, 55–68.

Winterfeldt, D. von, Edwards, W. (1986): Decision Analysis and Behavioral Research, Cambridge, Cambridge University Press.

6

INSTITUTIONAL AND ORGANIZATIONAL PERSPECTIVES

The preceding analyses dealt with problems concerning substance, instruments and institutions relevant for the establishment of environmental standards, and, in doing so, also pointed at related deficiencies. Overcoming these deficiencies means finding a procedure of social risk precaution which has to meet the following requirements:

- The procedure is supposed to adequately integrate scientific contributions into the process of achieving a decision. This includes a distinction of that scientific expert knowledge which comes up to internal standards of scientific disciplines. Scientific competence has to be made controllable with regard to those standards and with regard to the respective state of science.
- The procedure is supposed to unfold within the framework of social structures. Yet social formation of opinion is not to be adopted uncritically, it has rather to be investigated for its rational core, according to the procedures described in chapter 5.
- The procedure has to be an integral part of the democratic political institutions and their functioning.

Elements of such a procedure will be presented below. But first, we will make some introductory remarks about public influence on the establishment of environmental standards.

Chapter 4.1 described the cognitive heuristics which individual persons use in forming their intuitive judgments. Included were examples of deviation of a subjective risk assessment from the objective size of the risk. So, for instance, easily accessible memories, vivid ideas, personal experiences and easily comprehensible opinions of others play a particularly accentuating role in personal judgment formation. Furthermore, the formation of judgments is influenced by the dominance of current reference systems, with the formulation of the decision problem playing an important role. Finally, the wish to achieve assurance quickly favors the latent tendency towards self-confirming wishful thinking.

In chapter 4.1 these aspects were related to the individual intuitive assessment of the risk of nuclear energy, in chapter 4.2 to the social and institutional level. The selective social assimilation of public topics and the formation of political arenas, however, prevent us from employing the concept of risk evaluation in society. What dominates in everyday life are notions with a medium range, which interlink subjectively comparable situations. Characteristic of the social handling of the risk problem is the selective choice of risks with regard to which the uneasiness becomes manifest. Social conflicts have general causes or issues. Environmental problems offer a rich spectrum of potential risks and problems which may spark off social conflicts. All factors mentioned in section 4.2.2 hold for the whole field of environmental problems:

- With increasing knowledge, society has an increasing number of instruments and institutions at its avail to recognize risks of civilizatory action and their consequences. With increasing sensitivity of measurement equipment, ever smaller amounts of problematic substances can be measured. Both these facts lead to an increasing necessity of making decisions in situations which, in former times, were understood as fate or were not perceived at all.
- The relative portion of natural dangers has strongly diminished in industrialized societies whereas the portion of man-made risks has increased. Damages to the environment are seen as dominant risk factors today.
- Modern technology reduces the risk in terms of probability of damage multiplied with extent of damage, but this risk reduction is not reflected in social perception. On the one hand, the strong reduction of the probability of an eventuation results in many people perceiving the extent of damage in an overproportional size. On the other hand, the potential extent of damage involved in the application of some new technologies has actually increased so much that the probability of an eventuation must be reduced considerably.
- The individual marginal utility of material prosperity has decreased against the marginal utility of general health and clean environment.

Therefore, it is obvious that the environmental issue provides material for social conflicts. It is furthermore understandable that the public influences the establishment of environmental standards to a considerable degree, directly (e.g., via citizens' initiatives) and indirectly (e.g., via influencing political decisions through polling data). As individual and public assessments are subject to the described constraints and do not yield optimal results in terms of rational decision making, the danger arises that social resources are used in a wrong way if public opinion is taken into account in an unrestricted – or better: unreflected – way, and that the community does harm to itself by its own decision mechanisms.

An improvement of this situation of decision making is required both for decisions of medium importance in the frequent handling of pollution issues and for major problems due to large-scale industrial applications, e.g., in the sectors of chemistry, pharmacy and energy production. This process of decision making has to include elements informing public opinion, which is constitutive for our democratic system. Correspondingly, the process of decision making has to be designed in a way that it may change public opinion according to the rules by which the latter may be influenced.

6.1 PRINCIPLES AND PROBLEMS OF THE PRESENT PRACTICE IN GERMANY

6.1.1 The Diversity of Environmental Standards

Environmental standards, or the principles on which they are based, are laid down in laws, decrees, technical directions and administrative ordinances.

(a) Examples from the field of *laws on dangerous substances:* (aa) *Law on chemicals* with decree on dangerous substances: protection against noxious effects of dangerous substances – basis for MAK-values and TRK-values, obligation to have new substances tested and registered, possibility to ban substances (even scraps). (ab) *Law on fertilizers*: environment-friendly fostering of plant growth – obligatory type approval on the basis of expert opinions given by the advisory board on issues of fertilization. (ac) *Law on plant protection* with decree on pesticides: protection of cultivated plants and averting of environmental hazards – among other things obligation of licensing; prohibition of applications. (ad) *Decree on maximal levels of pesticides:* establishment of maximal admissible concentrations in food, e.g. concentrations of DDT. (ae) *Law on DDT:* interdiction of handling DDT; permission of several exceptions. (af) *Law on food and consumer goods:* protection against health damage – authorization to establish maximal admissible levels of xenobiotic substances in food and others. (ag) *Decree on drinking water:* precaution against health damages – limits and minimization rule for biological and chemical pollution. (ah) *Law on pharmaceutical products:*

special regulation supplementing the law on chemicals – denial of license in case of unjustifiable (side) effects; absolute liability.

(b) Examples from the field of *laws on immission protection:* (ba) *Federal law on immission protection* (so far with the technical directions on air and noise): principles of protection against noxious environmental impacts – concretization of precaution and averting of dangers by emission and immission standards in technical directions. (bb) *Federal decrees on immission protection* (examples): emission limitation for small-scale firing plants, by, e.g., standardization of burners and fuels, emission limitation of highly-volatile halocarbons, limitation of the sulfur content of certain fuels, limitation of wood dust emission, reduction of the noise of lawn mowers by restricting type approval and admissible time of daily use, emission limits for large-scale firing plants. (bc) *State decrees on smog:* protection against noxious environmental impacts during weather conditions with little exchange between air layers – setting of limits for smog alarm; restriction of motor vehicle traffic in certain regions and further measures reducing emission. (bd) *Law on lead in gasoline:* reduction of air pollution by lead compounds – establishment of maximal admissible contents in gasoline; tax reductions for unleaded gasoline. (be) *Law on air traffic noise:* protection against dangers and material discomfort: establishment of noise abatement zones, building bans and restrictions as well as rules for passive measures of noise abatement.

(c) Examples from the field of *laws on water protection:* (ca) *Law on water supply* and state laws on water: protection against impairments of water bodies – special decrees on handling substances which endanger water; absolute liability. (cb) *Law on waste water levies:* minimization of discharge of waste water by a „taxation“ according to units damage, based on different reference criteria. (cc) *Law on detergents* with legal decrees: prevention of impairment of the quality of water bodies – maximum levels of tensides and phosphates; possibility of prohibition of phosphate if a substitute is available.

(d) Example from the field of *laws on waste: Law on waste* with technical directions on waste and hazardous waste: avoidance or reduction of waste – preference for recycling against deposit; principles of orderly and controlled waste disposal; concretization in the technical directions.

(e) Examples from the field of *laws on atomic energy and on radiation protection:* (ea) *Law on atomic energy* and decree on radiation protection – promotion of nuclear energy and protection against its hazards – among other things definition of principles of protection. Establishment of limits in the decree on radiation protection (StrlSchV). (eb) *Law on preventive radiation protection:* provisions for precautionary protection of the population in extraordinary situations – e.g., authorization to set limits and to initiate measures that will minimize exposure to radiation.

Apart from these German provisions, there are numerous directives of the European Union: EU directives are binding for the member states insofar as they have to be transformed into national laws within a certain period of time. One example is the EU directive on drinking water of 1980, on which the limits of the above-mentioned decree on drinking water are based.

The diversity of these regulations demonstrates the various methods of handling the task of establishing environmental standards. Standards may be established, for instance, by the German federal parliament, or, in the form of decrees, by the national government, but also, at an international level, by the European Commission; they may have the character of limits or consist in procedures; they may refer to substances or to environmental media (air, water, soil) and they may concern emissions, immissions or the putting into circulation of substances. Finally, they may be based on impact criteria, on basic natural exposure, or on criteria of realizability, and they may serve as a precaution against or a method of averting dangers. The principles according to which environmental standards are formed vary in Germany from state to state, and in Europe from nation to nation.

The respective category may be chosen on the basis of rational considerations and be adequate to the problem; yet a recognizable systematics underlying such a choice is lacking. Whether a code of environmental law, which is – as rumor has it – already under preparation, may help to make the systematics of environmental standards more transparent and thus also contribute to tighten them, remains to be seen.

Radiation protection standards, as they are preventively established in the German law on atomic energy and the radiation protection decree, form a kind of red thread through this study. They are characterized – to outline it once again – by three principles:

- There exists both, exposure to natural radiation and exposure to man-made radiation which are not restricted by standards, the levels of which largely depend on the circumstances (e.g., medical applications), which are well-known in their variation ranges, and which are significantly higher than exposure to radiation additionally admitted by set standards. Moreover, it was found that exposure to natural and to anthropogenic radiation can be compared as to their levels and impacts. So, exposure to natural radiation may serve as a reference system for radiation protection standards.

- There is a scientific basis for radiation dosimetry and radiation effects, which, in the course of decades of research, has attained a degree of completeness and certainty which allows of a risk assessment of sufficient reliability for the evaluation of limits.
- There is a rule, but also a far-reaching consensus among science, users and authorities, that any exposure to radiation which is referred to in the radiation protection decree, notwithstanding its observing the given limits, should be justified by its benefit and be minimized in its level within reasonable borders, hence virtually be optimized.

These three principles cannot simply be transferred to other environmental standards, or have not been transferred so far, which is reflected in the documented diversity of existing environmental standards:

- As a rule, there is no – or at least no appreciable – natural exposure to other environmental noxae that are to be regulated. Nor is it therefore possible to directly refer to such an exposure, which is exactly what makes radiation standards so understandable.
- A less fundamental but equally important difference is the more limited knowledge about, on the one hand, effective doses at the affected place, or, on the other hand, dose response relations. With regard to these two aspects, knowledge in the wide field of chemicals differs largely but is mostly poor as compared with knowledge about radiation.
- Rules of minimization exist for exposure to chemicals as well, yet usually they do not supplement limits but replace them, when limits cannot be established, for reasons of insufficient knowledge about dose response relations, or for principal reasons. The idea of combining limit and optimization rule is not yet rooted in the practice of defining toxicological and ecotoxicological regulations.

6.1.2 Priorities Regarding the Establishment of New Standards

With all the diversity of existing environmental standards, many conceivable standards are still lacking, in particular limits concerning chemicals. This should not be automatically regarded as a serious deficit. On the contrary, environmental standards should exist only for those cases in which there is a rationally justified need for averting dangers, or else, perhaps, a need for precaution. The need for averting dangers and need for precaution must be oriented towards the pragmatic objective, defined in chapter 1, of protecting the environment from further damages and of abating already existing damages as far as possible. In cases where individual regulations are needed although no standard is set, though, the corresponding scope creates legal uncertainty. In particular, there is the danger that figures which are pragmatically used in regulatory practice become somehow definitive as standards without a legitimized process of establishing standards having taken place.

In view of the many potential impairments of the environment, there is a problem of priorities in establishing new environmental standards; and this all the more so since a first phase of establishing standards has been completed in Germany. With respect to air, the decree on large-scale firing plants, the three-way catalyst and the use of low-sulfur fuel oil and of natural gas have either already led to a reduced emission of the large-scale pollutants SO_2 and NO_x by one order of magnitude, or they will do so. Of dominant significance will then be the import of these pollutants. Another example of a pollutant, the emission of which will be reduced in the foreseeable future, are the CFCs. With regard to water, waste-water purification plants, waste-water levy, registration of indirect discharge and more stringent discharge conditions have led to a visible and measurable recovery of flowing waters, so that the objective of optimum water quality has been reached or has become reachable.

Of course, much remains to be done. Since, however, no major problems of establishing standards seem to be left on the national level after this phase, it becomes ever more difficult to set priorities in environment protection and hence in providing legal prerequisites. At the same time, it becomes ever more urgent to act, following the extent to which public problem awareness and growing environmental administrations increase the general pressure. This should not make us forget that setting priorities necessarily includes setting posteriorities, and this means nothing else but renouncing action in some cases. Incidentally, doing this deliberately and reasonably may often better contribute to an efficient environment protection than some undistinguished actionism.

6.1.3 Establishing Non-Effect Exposure Standards

As a rule, exposure limits should be based on dose response relations. If a threshold for the impact can be ascertained, the exposure standard should be set according to this value. If no threshold value can be proven to exist, any exposure standard means taking a certain risk into account. In such cases, the objective must be to quantitatively estimate this risk, in order to enable a discussion about its acceptability.

But the present-day pressure to take environment-political action has also led to the establishment of such limits for which knowledge of the dose response relation either is not sufficient for establishing an impact-related standard or has not been taken as its basis. Such settings, which must essentially be understood as a precaution, usually proceed from the state of technology, namely either from the state of retention technology or from the state of analytical detection technology.

Due to the vast progress of chemical analytics, the latter mostly means establishing zero emission or zero immission as environmental standard.

One example of an establishment of standards based on the state of retention technology are the emission standards of the decree on large-scale firing plants. Here, values have been established, above all for SO_2 and NO_x, which can be reached with available technology. This was, at least with respect to SO_2, a rather effective, if expensive contribution to an improved quality of air, which certainly is generally welcomed, but whose measurable impact on the environment cannot sufficiently be appraised so far.

An example of a rather doubtful establishment of standards, which is justified on the grounds of established detection technology, is the limit for pesticides in the decree on drinking water. This value, which is prescribed by a European regulation, amounts to 10^{-7}g/liter for any single substance and to 5 _ 10^{-7} g/liter for any mixture. When, in 1984, adoption of EU limits by German legislation began, industry was forced to start an expensive research program to ensure that routine methods of analysis with the necessary sensitivity were available at least for the most important substances.

Apart from the fact that cost-benefit considerations were left aside in this case in establishing standards, such extremely restrictive standards may also put a heavy strain on public confidence, since they often cannot be met. In the present case, an estimated 30% of the waterworks had to get exceptional permissions in order to be able to ensure supply of drinking water at all. Exceeding the limits as it is permitted here is far from being relevant for human health, according to toxicological judgment; but this can hardly be explained to the average citizen, once much lower limits have been established.

6.1.4 Optimization Deficiencies

As has already been mentioned in chapter 5, the multidimensionality of decisions in environment protection requires optimization or at least balancing considerations. Many deficiencies which can be found today in the practice of environmental standards arose since no scientific standards of assessment were available and since only one dimension of the problems was considered. The case of the limit for pesticides in drinking water is a typical example of a non-optimized establishment of standards, which results in an undue feeling of insecurity on the part of the population when limits are exceeded. Only one single aspect was considered here, namely that of avoiding pesticides in drinking water as completely as possible. Establishing the limit for pesticides in the decree on drinking water served this precaution principle exclusively, it does not directly refer to the strain on health caused by individual pesticides. So, the established limit is, as to known substances, far below concentrations which could mean an impairment of health. All other aspects, like costs, technical feasibility, psychological consequences of noncompliance, comparison of the risk with other risks, were not taken into account.

Environment-political decisions take place at two levels, the level of establishing standards, which is the proper subject of this study, and the level of enforcement. Decisions are not only made at the first level but also at the latter, where they do not lead to standards of general validity but regulate individual situations. Since they are made for the individual case only, they may be mere administrative acts, which underlie other procedural criteria than the actual establishment of standards. Yet, exactly for the reason that here the individual case with its very specific combination of parameters is dealt with, it will often be easier to consider it in a balancing way, whereas standards have always to be established according to average considerations. Such practice, applied to radiation protection, resulted in the fact that actual values remain regularly – and mostly to a considerable extent – below established limits. Therefore, the principle of combining standards and optimization rules, which is valid in radiation protection, should generally become the principle of choice in environment protection. We will leave open to debate the question whether, with such a combination, less restrictive standards than those in radiation protection could be established, since standards then would virtually be relevant in exceptional situations only.

Essential aspects of such an optimization are:

- the multidimensionality of decisions, which requires consideration of every dimension, if possible,
- the connective structure of a system, which may lead to the situation that a benefit in one respect causes an unforeseen or unjustifiable damage in another respect,
- the marginal utility, which should not become too small. Here, this means in concrete terms that expenditures are appropriate only to the extent that an appreciable additional reduction of exposure is reached,
- the different efficiency of different ways of reaching a desired exposure reduction.

These aspects have, of course, to be already taken into account when standards are established, at least in the sense that there is enough scope for them in the enforcement of standards.

The rule of optimization or balancing consideration must not become a mere rule of minimization; that is, the letter „R“ in the American ALARA-principle (**A**s **L**ow **A**s ***Reasonably*** **A**chievable), which holds in radiation protection, must not be omitted. Nevertheless, a rule of minimization would very much comply with the state of consciousness of many German citizens. The readiness to take this state of consciousness into account is paradigmatically expressed in the slogan, originating from the political debate on nuclear energy, „safety comes before economic efficiency“. This, however, fails to notice the fact that there are always possibilities to further reduce exposures, if only enough expenditures are made. If here the point is attained where the respective product or technology reaches the limits of economic efficiency, it should at least be considered which consequences of reduced safety in other respects a renunciation would entail. This is often overlooked in the endeavor to submit to anticipated conditions of public acceptance.

Basically, balancing consideration or optimization should be achievable in the enforcement of standards rather than in their establishment, but optimization is by no means a matter of course in ordinary regulatory practice. Attempts to optimize are often labeled as search for a bad compromise. This holds, first of all, for taking multiattributivity of decisions into consideration (cf. chapter 5.4). Taking connective structures into account is often limited by lack of insight. Another obstacle, finally, consists in the fact that the most efficient way of reaching an objective is often not the way of least resistance. But in view of the increasing density of the net of regulations, balancing considerations and optimizations will probably become ever more necessary in order to maintain a social capacity to act.

Some examples from the field of enforcement of environment protection may illustrate optimization deficiencies and their consequences.

(a) An example of consequences of connective structures: politically desired ever more stringent limits for polluting substances originating from waste recycling lead to increasing, politically undesired, amounts of waste:

- The recommendation on the part of UBA and BMU to renounce, in the medium term, deposit of sewage sludge on pastures because it contains traces of dioxin, leads to an increase in the amount of sewage sludge to be processed as waste.
- The recommendation of the SSK threatens the use of ammonia from uranium processing as fertilizer, although the concentration of radioactivity here is possibly no greater than radioactivity of natural origin in conventional mineral fertilizers; this could result in additional „radioactive“ waste.
- The SSK-targets to constrain utilization of material from demolished nuclear plants, even below given limits, could also lead to a considerable increase in the amounts of „radioactive“ waste.

(b) The marginal utility becomes too low if extremely increased demands on results of purification processes lead to corresponding resultant burdens of waste:

- In rehabilitation of former production sites (dangerous waste from the past), extreme demands on restoration of the original state result in expensive renunciations to use opened-up sites, e.g. for industrial installations.
- Ever higher demands on purification of flue gas lead to increasing amounts of waste concentrate, which will possibly have to be disposed as special waste.

(c) The most efficient, i.e. optimal ways with respect to cost and effect, often presuppose that psychological or political obstacles are impossible to overcome; so less efficient ways are preferred:

- reduction of pollutants from garbage incineration by means of purification of flue gas, instead of having garbage presorted by the user,
- protection of the ground water from nitrates and phosphates by a tertiary treatment in waste-water purification plants, instead of regulating the use of fertilizers in agriculture,
- preservation of clean air by exploiting the entire progress achieved by retention technology in Germany, instead of taking international measures to reduce transboundary transportation of pollutants.

As to establishing standards, the question about optimization arises in an exemplary way, as has been stated several times (cf. chapter 1.3.4 and 2.1.6), when definite exposure limits are established. If a threshold value is derived from a dose response relation, the limit will, as a rule, remain below this threshold. This is reasonable, since – at least in principle – the threshold marks the border between no damage and certain damage. Of course, in principle, a case is also conceivable where compliance with such a threshold value would cause such high costs that it would be recommendable to allow exposures above this threshold value. Examples hereof exist only for very specific situations, above all for deliberate therapeutical exposures in the field of medicine or for accepted high exposures when on duty in an emergency case, but these are optimizations as well. After careful balancing considerations, one evil is accepted in order to avoid or diminish another one. Those cases have the advantage that balancing considerations refer to risks and chances of the same dimension, namely health.

We are faced with the necessity to make multiattributive balancing considerations, particularly when no threshold value exists, that is, in cases of stochastic effects (cancer, mutations). Since innocuousness of finite exposures cannot be reached here, as a matter of principle (at least we proceed from that hypothesis), we can only choose between the alternatives of complete avoidance of an exposure (dependent on civilization) – and hence, usually, renunciation of some application – and acceptance of a certain risk. The order of magnitude of an acceptable risk for the population can then only be the result of a process of rational balancing considerations.

Such an establishment has not occurred so far in Germany. Radiation protection limits for the population have been derived from the variation range of exposure to natural radiation, according to the principle of pragmatic consistency. As to exposure of the population to carcinogenic and mutagenic chemicals in the environment, there are no standards so far, but only rules of minimization; up to now, establishing an acceptable risk has been avoided. Yet it begins to show – above all through developments in the U.S. – that risk limitations will have to be introduced in this sector as well by establishing limits. Indications regarding this issue were given in chapter 3.3.

6.2 ENVIRONMENTAL INSTRUMENTS OF MARKET AND LAW

6.2.1 Environmental Instruments of the Market Economy

In democratic societies with market economies, development and employment of environmental instruments can basically be realized by two different forms of control: the market system and the system of law. While market instruments determine the availability of the environment within a legally given frame, to be changed continuously if need be, according to readiness to pay for a certain benefit or readiness to compensate for a lost benefit, governmental prescriptions in form of prohibitions and rules constrain the action scope of individuals, institutions and organizations, and administratively determine by standards the extent of permitted utilization of the environment (Züricher 1989, 1; Mohr 1990, 8; Steger et al. 1990, 200). All free democracies adhere to the principle of abstinence of the state from all regulations which could be replaced by voluntary agreements or market-compatible instruments. If we confront market-compatible and regulatory instruments of environmental policy here, this is, however, not to be misunderstood as a mutual exclusion of these two approaches. It is only by a framework of governmental regulations that markets can develop at all. Some markets were only created by governmental boundary conditions, e.g. by the legal form of a joint-stock company and by the law on stock exchange. If we talk about market-compatible versus regulatory instruments here, this is not to say that conformity with the concept of market means abstinence from any governmental regulations, but rather the self-organizing balancing of supply and demand according to marginal benefit and marginal cost. Regulatory instruments according to this concept are governmentally set amounts or prices.

To regulate environmental quality by prescriptions understood in this sense requires a specific justification, according to the principle of priority of market-compatible procedures. The necessity of governmentally established standards is proved only if it can be demonstrated that, compared to regulatory procedures, the market cannot yield equivalent achievements. In the following, we will discuss the efficiency of market instruments and compare them with advantages and disadvantages of a regulatory solution.

6.2.1.1 Principles of Market-Compatible Instruments

Market-compatible instruments are founded on the principle of optimum allocation of social resources according to the degree of their scarcity as evaluated by the economic subjects. Four principles of economic action are of relevance here (Baumol & Oates 1975; Siebert 1987a; Smith 1986; Steger et al. 1990; Renn & Kals 1990):

- the principle of *cost effectiveness* (production factors have to be combined in a way that a given objective can be reached with the least possible expenditure of natural resources, labor and capital),
- the principle of *marginal utility evaluation* (the value of a good or a service is measured by the marginal utility which the availability of the good or service means to the demander),
- the principle of *market equilibrium* (goods and services are exchanged on the market until the marginal utility of the demanders matches the marginal costs of the suppliers),
- the principle of *raising welfare* (goods and services are supplied or exchanged only if all parties at least maintain their existing utility level and if at least one party achieves a utility gain).

If these four principles were taken as measure for environmental regulation, the instruments of governmental regulation policy would perform badly. In the first place, regulations are not cost-effective (Züricher 1989, 2ff): standards force companies to remain below a certain threshold value, without taking the costs of a reduction of pollutants into account, which may be different for different companies. For one company A, pollution reduction below the threshold value may be less cost-intensive than meeting the threshold value is for another firm B. If company B undertook to pay the costs of company A for pollution reduction below the threshold value, the same environmental quality would be achieved, and at lower costs. Yet also transaction costs have to be taken into consideration. If a market-compatible organization of environmental regulation entails more indirect costs than it saves in direct costs by replacing regulatory prescriptions, it is sensible to renounce a market solution, even from an economic point of view.

The level of environmental standards is not determined by balancing marginal utility and marginal costs, but by an act of political will. The latter may be based on a cost-utility analysis, but only in exceptional cases would we assume the marginal utility of any unit money spent on environment protection to match the marginal costs for financing the observation of standards. Often, huge amounts of money are spent for only marginal improvements of safety, while big utility gains are not sufficiently exploited (Smith 1986, 13f).

How much public welfare is raised by environmental standards is doubtful as well. On the one hand, handling limits flexibly according to the specific cost-efficectiveness of a situation could result in a reduction of total costs with pollution remaining constant, and hence in an improved general welfare. On the other hand, fixed standards do not provide incentives for company-specific innovations, since there is no financial advantage in improving environmental quality below given standards (Züricher 1989, 3; Siebert 1987b; Mohr 1990, 8).

6.2.1.2 Preconditions for Market Efficiency

The use of market-compatible instruments is dependent on certain preconditions, among them the precondition of measurability. In the first place, participants in the market must have sufficient information about the kind and quality of products, in order to be able to subjectively evaluate them (Koslowski 1983, 169). If, for example, there are noxious substances in certain products, which are harmful to health but are not perceived by the consumer, this ignorance might falsify the true, i.e. benefit-optimized, demand. Moreover, fixed prices on the market also presuppose that both supplier and consumer can exclusively claim the benefit arising for both from the act of buying (Tietenberg 1984; Mohr 1990, 6). If third parties are affected positively or negatively (external effects), or if the benefit cannot be privatized (public good), it comes to suboptimal solutions, when negotiations are inadmissible or too expensive.

Finally, goods which are understood as utility gain by everybody are often not supplied at all, because they are of interest for the individual only if they are demanded by all others or at least by the majority of the others (Sen 1967; Tullock 1976). If people are not sure about this general demand, the individual demand is usually omitted. The reverse is also problematic: if the demand for some good (e.g. a building site in the country) leads to a utility loss when all people satisfy their demand (spoliation of the countryside), losses in welfare occur, although each individual optimizes his or her utility.

It is obvious that the environment is a good which does not meet these conditions (Steger et al. 1990, 197). The environment is a scarce good, and therefore amenable to the economic process, but it is a public good, for which rights of exclusive utilization can often be claimed only under difficult circumstances (Kloepfer 1989, 196; Steger et al. 1990; Tietenberg 1988). This is due to the following factors:

(a) Pollution is usually an *external effect*, i.e., in most cases it does harm to third parties not involved in the transactions concerning goods and services (Baumol & Oates 1975; Renn & Kals 1990, 61). But the price of a good can only govern the behavior of buyers and sellers, not the utility loss of third parties. Without state measures ensuring compensation to third parties (for example by levies on the price), this problem cannot be solved by the market alone.

(b) Environment is a suitable example of a good which can*not be used exclusively*, since demanders of a clean environment cannot exclude anybody who is not willing to pay, although he or she is enjoying the clean environment as well (Tullock 1976; Siebert 1987a). If, however, such free riders avail themselves gratuitously of the good 'environment', the demanders who are basically ready to pay will not want to pay any more either. In that case, the product is not supplied at all, although everybody would benefit.

(c) For both these reasons, external effects and non-exclusiveness of use, the price of a clean environment can*not* be ascertained *by the market alone*, i.e., by balancing marginal utility and marginal costs (Schürmann 1978). The costs for environment restoration or for redressing a damage to the environment are suitable indicators for such a price (Wicke 1990), but they do not reflect the utility gained or lost (subjective assessment) through the environmental damage. Therefore, the price (or the amount) has to be fixed externally, for instance by the government.

(d) The *utility loss* of environmental damages often appears only in the future, whereas the *utility gain* by consumption takes place in the present (Koslowski 1983, 177). If the present utility gain is irreversibly linked to a future utility loss,and if this utility loss is not transparent to the economic subjects, the price established by the market may give false signals and entail long-term losses in welfare. So, in the interest of the general public, the state has to intervene here in the market process and limit the use of the respective good by prohibitions or directives.

(e) Environmental pollution may lead to irreversible damages, which the economic agents either are not aware of or which they are ready to accept because they urgently need certain vital goods (Koslowski 1983, 178). The acceptance of dangerous substances, which may heavily damage health or even lead to death, are not to be exchanged against other goods for ethical reasons. The market relies on the process of learning through trial and error. If, however, an error entails unacceptable consequences, the state has the duty to take provisions to protect the individual against an irreversible error (this is not to say that the state is always free from errors in its intervening action). The German environmental law emphasizes the necessity to constrain individual liberty or property rights if major issues of public welfare or the prevention of dangers are at stake (Kloepfer 1989, 196).

Since the economic good environment is largely unamenable to the market, market mechanisms cannot exclusively operate here, governmental intervention is required. There are certain environmental products for which the above-mentioned criteria do not hold, first of all the principle of non-exclusiveness (e.g., real estate property, holidays in a climatic health resort, clean water in bottles). But even these kinds of utilizing the environment are often regulated by environmental standards for special cases. This is all the more so with regard to improvement of air and water quality, because here the market does not admit exclusive utilization rights.

For this reason, market instruments discussed in the field of environmental policy are always mixed forms, between state regulation and market incentive. Here, the point in a state regulation is to create or simulate market conditions and thus achieve market-compatible results, thereby overcoming the inflexibility and inefficiency of governmental standards.

6.2.1.3 Market Compatible Instruments of Environmental Policy

The simplest way to make use of the market as a directing instrument in enviro-mental policy is to privatize environmental goods. Since, however, the environment can be made exclusively utilizable only in exceptional cases, a complete privatization through creating environmentally relevant property and utilization rights cannot be carried out. This concept may be used for services of environmental conservation or restoration, for instance in privatization of waste disposal. Environment-hostile receptacles, for example, may be understood as property of the producers, a property relation which continues to exist after the sale of the content of the receptacle and includes the duty to dispose of the waste. As to environmental standards, however, which are of interest here, only a few cases can be imagined where it would be possible or desirable to create individual utilization rights (Steger et al. 1990, 198).

Yet the instruments of levies and emission-dependent taxes, but primarily the issuing of certificates, are proposed solutions, which theoretically would be able to replace or supplement regulatory standards (Züricher 1989, 9ff; Mohr 1990, 13ff). They are based on targets set by the government, though. Since market prices for environmental goods do not spontaneously develop on the market, the state has to define either price or amount in advance (Steger et al. 1990, 200). In the first case, polluters are presented the bill by way of a levy on emitted amount of pollutant, in the second case, the admissible amount of pollutants is established in advance by the state, which may later reduce or increase the amount of certificates by buying them or issuing additional certificates. This amount is sold, piece by piece, in a kind of auction, to the highest bidder.

Apart from that, amounts may also be limited by negotiations among the economic subjects, in which the admissible quote of pollution for each participant is negotiated under the condition that the total amount remains below a certain immission level (the so-called immission cap) (Kloepfer 1989, 194). This form of flexible compensation is already practiced in the American environmental law in four ways (Hahn & Hester 1989, 109ff; Mohr 1990, 13ff):

(a) by the *netting procedure*, according to which emissions from one production site can be set off against those from another production site in the same region, within the same corporation. In that way, old facilities, for which new equipment by environment-friendly technologies would not be cost-effective, may continue to operate without emission limitation if the new installation in the neighborhood remains that much below the legally prescribed standard that the surplus emission from the old installation is at least set off. This possibility of netting has been applied in the U.S. in about 10'000 cases so far.

(b) by the *bubble procedure*, according to which several firms in the same region can set off their emissions against each other. The procedure is the same as netting, apart from the fact that the participating plants do not have to belong to one corporation. This procedure has been applied very rarely in the U.S. so far, above all because compensation payments between autonomous firms are hard to organize on a voluntary basis. Besides, the regional bubbles were often defined too small to offer enough incentives for business cooperation.

(c) by the *offset procedure*, according to which emissions of plants of one corporation at different sites can be set off against each other. This procedure as well has been rarely applied so far, and is, moreover, controversial even among supporters of market-compatible solutions, since a compensation independent of regions means a systematic disadvantage with respect to environmental quality to regions with a high concentration of old facilities.

(d) by the *banking procedure*, according to which non-exploited amounts of admissible emission may be „saved“ and used in excess of admissible emissions in the future. This procedure has scarcely been employed in the U.S. so far. For one thing, a firm will hardly meet with public benevolence if it designs higher emission of pollutants for future plants than for existing ones. Furthermore, the American regulatory law does not warrant this permission for higher emissions, since any pollutant might be assessed differently in the future, on the basis of new insights, and consequently, emission standards might change.

The possibility of compensation according to the bubble and netting procedures has been adopted meanwhile in the German „TA Luft“ (technical directions for the preservation of clean air) of 27 February, 1986 (Kloepfer 1989, 194). Accordingly, older facilities may differ from regulatory targets with respect to precaution (but not with respect to danger), provided that technical measures in other installations of the same firm or of other firms in this region ensure an altogether further-reaching emission reduction of the respective pollutants or of pollutants which cause the same effect than would have been achieved with observing the regulatory requirements. So far, this possibility has only rarely been made use of in Germany.

Another possibility of market-internal regulation is the extension of liability law (Steger et al. 1990, 200f). Persons whose health is impaired by pollution may claim compensation from the originator. As compensation means costs, and to a growing extent, loss in reputation for the firms, they will try to avoid damage and relate costs for taking environment-friendly measures to costs arising from potential claims for compensation. As will be shown (in chapter 6.2.3), this solution does not provide the desired directing function.

Finally, economic incentives to reduce pollution can also be established by the deposit system. A deposit is levied on recyclable products, which is repaid when the products are returned to the producer or seller (Züricher 1989, 14ff). This procedure will not be considered further here, since it is not applicable to the field of environmental standards. Nevertheless, there have been positive experiences with the deposit system in several countries, which could probably be extended beyond the traditional range concerning bottles to other potentially polluting products (e.g. batteries).

Fiscal taxes, subsidies or levies independent of emissions, on the other hand, are not market-compatible instruments, since they do not exert a directing function on the allocation of environmental goods. Incentives and levies have an effect on allocation only if environment-friendly behavior is rewarded on the market and environment-hostile behavior is punished, namely according to the costs avoided or arising for the whole economy. Any market-compatible instrument can only meet its directing function if it provides an incentive to set off costs due to environmental damages against costs necessary to reduce these damages.

6.2.1.4 Advantages and Disadvantages of Market Compatible Instruments

As compared to fixed regulatory standards, the above-mentioned market compatible instruments, especially emission-dependent taxes and certificates, have the advantage of meeting, on principle, the four criteria of an economic way of proceeding: They are cost-effective, since it is up to each firm to either reduce emission or pay levies to the state or buy certificates in the desired quantity, according to its own cost-benefit calculations (Siebert 1987b). Certificates and levies may be divided into such pieces that any firm may set off marginal costs of levies or certificates against marginal costs of environment reduction. The marginal unit of the good 'environment' is used exactly in that application where it entails the highest utility (Mohr 1990, 5). An equilibrium on the market is reached by the process that those economic agents with the highest readiness to additionally renounce environment organize an exchange of utility with those economic subjects with the highest readiness to additionally buy environmental utilization rights (Mohr 1990, 6). Competition for marginal units utility by purchase and sale of pollution rights also increases welfare, because the trade with these certificates is profitable to buyers and sellers. Both achieve a utility gain, due to their different assessment of the marginal unit environment. It has to be kept in mind, though, that this offset presupposes that the collective utility of the economic good environment was adequately determined when prices or amounts were externally established by the state.

In spite of the impressive advantages, a set of serious problems is connected with market-compatible instruments, which partly depends on the nature of environment protection itself and partly on the necessity to simulate a market for the application of these instruments. Among these problems, the following have to be mentioned first of all:

(a) *Measurement of environmental damages:* The efficiency of the directing function of market-compatible systems depends on the condition that the costs of environmental damages are roughly known and that levies and certificates are linked to an emitted substance which causes these damages. Both preconditions are often not satisfied (Steger et al. 1990, 199ff). In many cases, there is considerable uncertainty about the negative impacts of environmental noxae, which complicates an evaluation. Frequently, one has to refer to indicators like the costs of restoration of the former state or economic losses due to pollution (Wicke 1990, 214ff). Still harder is a decision on the cause (Mohr 1990, 11). Often, it takes the simultaneous existence of several pollutants to cause a certain damage (e.g., in the formation of ozone). These problems equally occur in regulatory procedures, but there, no optimal allocation of resources is claimed from the beginning.

(b) *Establishment of amounts or prices by the state:* As to levies, the state has to establish the price per emitted unit pollutant, as to certificates the admissible total amount of the respective pollutant. Both targets may lead to serious distortions of allocation (Züricher 1989, 13). In the levy alternative, the amount of emitted pollutants is determined by the price. If the price is too high, losses in welfare are the consequence, since the production impediment causes higher costs than the costs saved by a clean environment. If the price is too low, there are losses in welfare as well, since the costs of pollution in that case exceed the costs saved through less stringent environmental conditions. As the price for improvements of the environment or damages to the environment cannot find its optimal equilibrium in the course of market processes, and as, moreover, no clear indicators for an evaluation can be given in retrospect, governmental setting of prices for the economic good environment does (with a few exceptions) not lead to an offset of marginal costs and marginal utility. In the certificate procedure, the amount is restricted, and thereby the objective of a limitation of the total amount of pollutants is fixed, similarly as with regulatory standards. With low elasticity of demand, i.e., with little possibilities of consumption alternatives, an effective limitation of amounts may lead to considerable price increases (Züricher 1989, 13). This is all the more problematic as interest groups may artificially tighten the certificate market, for reasons of speculation or of environment protection (Kloepfer 1989, 193f).

(c) *Control of damages to the environment:* It is difficult to control the observance of general limits, even in the case of regulatory prescriptions. In the case of levies or certificates, each individual firm receives differentiated limits for each emitted pollutant. The expenditure necessary to penalize each individual exceeding of limits is probably much higher for many pollutants (e.g., organic furans and dioxins) than the utility gained by the introduction of an amount of pollutants per firm, differentiated with respect to marginal costs.

(d) *Distribution of damages to the environment:* As was mentioned already in the description of the offset procedure, market-compatible systems may help to reduce a given total amount of pollutants at lowest costs, but the remaining amount of pollutants is not necessarily equally distributed (Kloepfer 1989, 194). On the contrary: regions with a high share of old installations are in most cases placed at a disadvantage, since an expenditure for environment protection is least profitable there. Regions with new facilities, on the other hand, will be particularly environment-friendly, since here, firms will reduce emissions beyond the standard.

(e) *International and global environmental impairments:* Damages to the environment are, to an increasing extent, no longer regionally limited, instead they spread far across national borders (Mohr 1990). Examples of this are acid rain, or the damage done to the ozone layer by CFCs. As it is much more difficult to harmonize market-compatible systems at the international level than to set internationally accepted standards, international agreements about environmental damages will, in future as well, be accomplished on the basis of limits or other standards. It may, however, be left to each individual country to comply internally with its international duties (e.g., to reduce CFCs) by means of regulatory or market-compatible instruments.

(f) *Indivisibility of the economic goods life and health:* The theory of marginal utility starts from the assumption that goods can be subdivided in single units and be exchanged against units of other goods (Steger et al. 1990, 201ff). This view, however, loses its validity when goods are indivisible or unexchangeable. If, for instance, the emission of a certain amount of a pollutant causes acute and severe damages, the government will have to prohibit this emission according to the German constitution, independent of the level of economic damage (Mohr 1990, 16). It is for this reason that in Germany the spectrum for applying market-compatible instruments is restricted to the field of precaution; from the field of danger those instruments are excluded (Kloepfer 1989, 194). The postulate that life must be unharmed holds even if the person concerned should agree to this harm.

6.2.1.5 Conclusions for Environmental Policy

No doubt, the introduction of market-compatible instruments in environmental policy contributes to better cost-effectiveness. Yet the efficiency of these instruments does not result from the free market as a matter of course but can only be the consequence of regulatory targets and state control. The relation between state and market in this field of politics is therefore not so much a substitutive one but rather a complementary one. The crucial question is to what extent market-compatible instruments are supposed to operate within the regulatory frame.

One essential limitation of market-compatible instruments becomes manifest in the field of danger. The loss of life and health cannot be exchanged against other goods. If there are acute dangers, the German state has, according to its legal system, the duty to protect its citizens. Such a limitation is not immanently given with levies or emission-dependent taxes; with certificates, the claim for reduction below a certain limit can be achieved by limitation of the totally admissible amount of pollutants. So, market-compatible instruments should be restricted to the field of precaution.

Another limitation is given by the distribution effects of market-compatible instruments. Optimal allocation does not necessarily lead to a desirable distribution of effects. Large-scale application of certificates or compensations will probably mean a systematic disadvantage regarding environmental quality to regions with a high share of old installations. Technologies whose environmentally damaging effects are to be expected only in future will probably be favored against other technologies the harmful effects of which occur now. It is for these reasons that regulations are better suited to achieve a fair distribution of environmental burdens with respect to regions, social strata and future generations.

A third limitation of market-compatible instruments follows from the necessity to redefine limits if new insights about the effectiveness of pollutants are gained. If pollutants turn out to be less innocuous than was assumed, levies may be lowered or the amount of certificates may be increased. A reduction of the amount of pollution rights is harder to achieve. If more stringent environmental standards are necessary, the state has to repurchase certificates or to increase levies. In the first case, the state may be extorted financially, as it is forced to reduce the amount of certificates for legal reasons (e.g. to avert danger); hence its elasticity of demand equals almost zero, and suppliers who are strong in the market may demand more or less any price for their unused certificates. In the case of levies, the state can never be sure that increased levies will bring about or even exceed the desired reduction of amounts.

Finally, the discussion of problems in section 6.2.1.4 shows that, for a large spectrum of situations, market-compatible solutions might not be able to fulfill their directing function either. It should be mentioned that plant-specific rules of minimization are incompatible with market systems. The problems of environmental valuation, the difficulties in measuring environmental damages and in attributing them to an originator, the possibility that interest groups or speculators might dominate the market in certificates, and the impossibility to register transboundary pollution, necessarily require an accompanying regulatory control, similar to regulations by the antitrust law and other regulatory directions referring to goods and labor markets. In many cases, the less efficient but simpler procedure of establishing regulatory standards may be more practicable and even more cost-effective than an application of market-compatible instruments, if costs of control are included.

The limitations of market-compatible systems is clearly one reason for the fact that no industrialized nation has applied these procedures on a larger scale so far. The U.S., and, in a restricted way, Germany as well, permit compensatory solutions, in which environmental burdens are distributed in a more cost-effective way, where regulatory law sets the frame for the range of compensation. To sum up, it may be stated that market-compatible instruments may well claim a position within environmental policy, since they are better able to steer production and consumption of environmental goods according to cost-effectiveness and welfare increase than regulatory law. Yet their application is restricted to the field of precaution and requires further elaboration and limitation by regulatory measures. The priority of regulatory policy has not only proved to be useful in practice, but is to be preferred to a comprehensive market solution also for theoretical reasons.

6.2.2 Environment Protection in Civil Law and in Public Law

The preceding section 6.2.1 shows that market-compatible environmental instruments are well suited to solve some problems of environment protection, but assigning them the leading role, either alone or to a predominant degree, is to overrate them. Two reasons, above all, speak against that:

- Governmental structuring and intervening is indispensable for the environmental market, as it is for any market.
- The distribution of the economic good environment according to market principles alone can only be accepted below the thresholds of hazards or harms. Market mechanisms alone do not guarantee observance of these thresholds.

For both reasons, the state has thus to intervene with the instrument of legal regulation, in order to create an environmental market and to ensure thresholds of hazards and harms. In principle, regulations both on the basis of civil law and on the basis of public law come into consideration here.

In Germany, regulations concerning environmental protection are mainly found in public law. Almost all environmentally relevant fields of life are covered by decrees of public law, from the law on air traffic noise, the law on conservation of nature, the federal law on immission protection to the law on atomic energy. But regulations concerning or allowing for environment protection are found in civil law as well.

6.2.2.1 *German Private Environmental Law*

The German civil law does not include an independent, complete environmental law; it rather consists of norms which also provide regulations for environmentally relevant problems.[1] The core norms of private environmental law are found in particular in private neighbor law (§§ 906ff, § 1004 BGB) and in tort law (§§ 823ff BGB). Attention has to be paid to the fact that private neighbor law so far relates basically to immovables.[2]

6.2.2.1.1 Claims for Prevention and Compensation in Detail

a) The direction on immission of § 906 BGB, which regulates immission of gases, vapors, odors, smoke, soot, heat, noise, vibration and similar immissions originating from another property, concedes a claim for prevention of these impairments, in connection with § 1004 Abs. 1 BGB.[3] According to the prevalent opinion[4], however, the concept of immission of §§ 906ff BGB does not include negative effects like deprivation of light, air, sun and wind.

The claim for removal according to § 906, in connection with § 1004 Abs. 1 BGB, exists independently of fault, but is restricted by tolerance duties.[5] On the one hand, the rightful claimant is bound to tolerate an impairment which is insubstantial[6] (§ 906 Abs. 1 BGB). On the other hand, even a substantial impairment has to be tolerated if it is due to a use of the property which is customary at that place and cannot be prevented by economically reasonable protection measures. In the latter case, § 906 Abs. 2 S. 2 grants monetary compensation.

Problems arise with regard to public law environment protection about criteria of what is „substantial“ and what is „customary at a place“.
How far public law regulations and the environmental standards laid down there may determine what is substantial and what is customary at a place remains to be discussed.[7] So, in particular, the limits laid down in the TA Luft are interpreted as indicators of the acceptability of an impairment, yet only to the extent to which they are assigned an indication effect as to what is customary at a place and what is not.[8] The concept of 'being customary at a place' refers to the character of a certain delimitable area, which results from its actual state and the respective utilization of the properties.[9] Basically relevant in this respect is the property from which the emissions originate.[10] Here, the question arises again how far the judgment about what is customary at a place is influenced by public law norms

and public law planning.[11] An argument in favor of such an influence is the fear that private neighbor law could annul public law planning so that planning objectives could no longer be realized. According to the prevalent opinion regarding civil law[12], however, only the actual character of an area is relevant and not targets set by planning and utilization determination according to public law.

Since basic decisions of public law must not be corrigible by means of civil law, some norms of public law referring to the licensing of installations superimpose private law claims for prevention (so, for instance, § 14 BImSchG, § 7 AtG, § 11 LuftVG) and thus constrain private environmental law from the beginning. A person suffering from impairments may, however, claim protective provisions which aim at avoidance of the impairment. If such provisions are economically unjustifiable according to the respective state of technology, the person concerned is granted a right to compensation (cf. § 14 Abs. 1 S. 2 BImSchG).

Whether a person may raise a claim depends absolutely on property, according to the directions of § 906 in connection with § 1004 Abs. 1 BGB. That is, only the owner of a property or persons of the same status (owners of a hereditary right to build or persons with a right to permanent residence) are entitled to claim, which is another limitation of private environmental law.[13] The present protection function of the claim for prevention therefore refers, primarily, to situations where the impairing operation or installation needs no license or does not comply with a respective license.

(b) *Environment protection by tort law.* The central norm of civil liability law, § 823 Abs. 1 BGB, secures goods worthy of protection, like life, body, health and comparable rights (e.g., personal rights). But it knows neither the good 'environment' worth being protected nor individual environmental goods.[14] Therefore, § 823 Abs. 1 BGB has to be discarded as a norm for claims referring to environment protection. Attention has to be paid, though, to the attempt to strengthen delictual environment protection by extending duties of securing traffic[15] insofar as the state of technology, or local circumstances, or reference to observing the official licensing conditions alone do no longer justify a damage.

Traffic duties, which, in this context also always orient by environment protection duties of public law, are supposed to ensure a further-reaching protection against impairments in the individual case than do directions of public law, but only in cases where it is understandable to the emitter that mere observation of admissible emission values cannot suffice to avoid inadmissible immissions.[16]

As another delictual liability standard, § 823 Abs. 2 BGB is applied, which grants a right to compensation in case of an infringement of a protection law. Public law norms of environment protection may be consulted as protection law here, as far as they have the character of protecting third parties[17] (for instance, § 5 Nr. 1, 2 BImSchG, § 8 WHG in connection with laws of the German states).

6.2.2.1.2 Problem Fields of Private Environmental Law

The fact that private environmental law is inadequate to ensure effective environment protection is evident in particular with respect to questions of causality and evidence.[18]

The above-described claims for prevention and compensation often fail for reasons of the civil law burden of proof, according to which the person suffering from the damage has to prove causality, i.e. the causal link between emissions and damage. The sufferer has to take the risk that this might not be clarified. Moreover, he or she has to be able to definitely name the adversary of the claim, the polluter. This is hardly possible with complex long-term damages. It is almost impossible in the case of so-called summed-up immissions, where environmental damages result from a multitude of polluting causes, which, moreover, accumulate due to long-distance transportation of pollutants from a large region and become effective only for that reason.[19] The present German civil law has no solution to offer to this problem field;[20] attempts for a solution proposed in literature[21] cannot be realized without a basic change of the present civil law.

It is jurisdiction[22] which tries to improve the situation of a thus inefficient private environmental law by facilitating proof and reversing the burden of proof. Facilitated proof up to a reversed burden of proof shall be possible if it can be shown that the polluter has exceeded the limits laid down in the administrative directions (of public law).[23] Moreover, it is the polluter who has to prove that he or she has taken the necessary measures to constrain immission.[24]

But it is still up to the sufferer to prove causality with respect to the damage.[25] This, however, will probably be next to impossible in most cases (summed-up emissions). So, no distinct improvement of private environmental law by facilitated proof alone can be noticed, at least so far.

6.2.2.2 German Public Environmental Law

6.2.2.2.1 General Principles

In Germany, ensuring environmental protection is almost exclusively the task of public law. The core of public environmental law includes, in particular, the law on dangerous substances, the law on immission protection, the law on water protection, the law on waste, the law on atomic energy and radiation protection, and the law on conservation of nature and preservation of the landscape. Certain leading principles[26] of environmental protection have crystallized, which are generally relevant in any environmental law and determine its structures. These are:

- the principle of precaution[27], which, apart from preventing danger, is supposed to ensure a foresighted environment protection and a careful use of resources;
- the principle of causation[28], according to which the originator of pollution and environmental damages has to pay the costs. Attention has to be paid to the fact that the notion of originator in environmental law does not always denote the person, at least not the person alone, who causes an environmental effect whatsoever, e.g., an emission. The environmental damage, for instance noise or odor pollution, may possibly arise only due to the person who moves into the vicinity of the emission source. Hence, the principle of causation in environmental law may well be a reciprocal one;
- the principle of cooperation, according to which an effective environment protection has to be the task of all social forces concerned, not that of the government alone.[29]

6.2.2.2.2 Instruments of Public Environmental Law

The instruments of public environmental law[30] may be subdivided into planning instruments, instruments of direct behavior control and instruments of indirect behavior control.

(a) *Planning instruments.*[31] Environmental planning serves to manage environment protection and is one of the core aspects of the precaution principle, where not only environmental planning norms in certain sectors are of relevance (like, e.g., landscape planning [§§ 5ff BNatSchG], plans concerning clean air [§ 47 BImSchG], official approval of plans for repositories for nuclear waste [§ 96 AtG], plans for waste disposal [§ 6 AbfG]), but also planning in sectors which do not merely relate to environment protection (like, e.g. planning of truck roads [§§ 17ff FernStrG], planning of airports [§ 96 LuftVG]), since environment protection has to be taken into consideration here. Aspects of environment protection are furthermore considered in area planning, for instance in urban development planning, where environment protection has to be taken into consideration both as a general planning objective and in individual planning guidelines[32] (§ 1 Abs. 5 S. 1, 2 BauGB).

(b) *Direct Control Mechanisms.* The major part of norms securing the environment are direct governmental ordinances, i.e., regulatory instruments. Characteristic of these instruments is the obligation to behave according to certain rules, prohibitions and environment-protecting duties.

The most important instrument is the *prohibition with the reservation of granting permission*[33], which serves danger prevention. An action which potentially damages the environment may be permitted with the reservation of particular preconditions being fulfilled (e.g. §§ 4, 6 BImSchG, § 12 AbfG, §7 AtG). Besides, there is the repressive prohibition with the reservation of granting exemption, which, in exceptional cases, allows actions which are legally generally disapproved of. There is no claim to permission here, the authority giving the respective permission can only be claimed to correctly use its scope (so, e.g., §§ 7, 8 WHG; § 31 BNatSchG, § 9 BWaldG-prohibition of clearing forests).

The action scope of the administration is considerably increased by the instruments of additional regulations[34], which allow of the adaptation of already given permissions to new states of danger or knowledge. Of particular importance here are the instruments of supplementary ordinances, besides directions of withdrawal and cancellation.[35]

Supplementary ordinances (e.g., § 17 BImSchG, § 17 AtG; § 5 WHG) enable the administration to more rapidly and more flexibly adapt duties related to operating a firm to changing environmental conditions and advanced scientific-technological insights.[36] Moreover, supplementary ordinances are a milder kind of intervention than complete cancellation or withdrawal of permission. They do not need an official reservation, and may therefore be established immediately on the basis of the respective law.[37]

Among the regulatory competences is the competence of environmental monitoring (e.g. §§ 3–27 StrlSchVO, § 46 BImSchG – measurements, § 21 WHG, § 19 AtG), which gives control rights[38] to the administration (rights to enter and to inspect, rights to claim information, rights to examine documents etc.).

Notification obligations, on the one hand, serve an opening control of an environmentally harmful activity[39], and on the other hand, the monitoring control of moderations of an environmentally harmful activity, which is not yet subject to permission (§ 16 BImSchG, § 11 AbfG, §§ 4ff ChemG).

Further instruments are the legally regulated but relative *duties to omit environmental damages altogether* or to avoid them (§ 1a Abs. 2 WHG, §§ 5, 22 BImSchG, § 1 AtG in connection with StrlSchVO). Among them are duties of preserving, of conserving, and of cultivating, which refer, above all to possession or property of environmental goods, so, e.g. § 11 BNatSchG, §§ 9, 11 BWaldG. Furthermore, duties of precaution, monitoring and securing have to be observed in case of environmentally harmful utilizations, so, e.g., the duties of private monitoring (§ 7 Abs. 1 Nr. 3 BImSchG), the malfunction provision duties (§§ 3ff StörfallVO), and duties of warning.

In addition, there is the obligation that environmentally dangerous firms have to appoint firm commissioners for environment protection[40] (institutionalization of responsibility of environment-protective matters), which are supposed to ensure and control observance of environment-protective directions within a firm.

Part of direct behavior control are also the *penal regulations*[41] or *regulations concerning administrative offense*[42], which are regulated in most laws and decrees on environment protection and in the criminal code (cf. §§ 324–330 d StGB). These regulations are also supposed to ensure observance and enforcement of environment-protective standards.

Instruments of indirect control[43] are of growing importance in recent times, since environmental concern has generally increased, and since it has become obvious that an effective environment protection cannot be reached by governmental rules and prohibitions alone.

Environmental levies[44] are important instruments of indirect control, as they contribute to avoidance of environmental damages by the burden of levies on environmentally damaging behavior, but moreover they serve the financing of environment protection[45] (waste water levy[46] as the most important environmental levy).

Environmental subsidies[47] are payments of the state for environment-friendly behavior. Apart from direct financial payments, in particular, tax reductions[48] and fee reductions[49] are important instruments.

Informal instruments may have environment-protective effects as well, and it is exactly the advantages of informal action, namely to save costs and time, to be flexible and to avoid litigations, which may contribute to their effectiveness. Among these instruments are, particularly, government information about environmental data and environment protection, publicity for environment-friendly products (environment sign), recommendations[50] and, especially, so-called environmental agreements[51] (agreements on self-constraint[52] or promises of self-constraint on the part of industry, e.g. regarding avoidance to use CFCs).

6.2.2.3 *The Relationship between Private Environmental Law and Public Environmental Law*

The fact that environment-protective standards are double-tracked often leads to great uncertainties and problems, so, for example, with respect to the question how far public environmental standards influence the interpretation of civil law directions[53], whether public law permissions have basically a shaping effect on private laws, and how far public planning decisions and planning targets superimpose private law.[54] The structural differences and intentions of both fields of law, especially, have to be taken into consideration to determine the relationship[55] between legal environment-protective norms in both fields of law.

Norms of private law are basically meant to regulate smaller spheres of life.[56] In accordance, they also have only a limited protection function as to the actually concerned subjective rights of individuals, and in most cases intervene only when a damage has occurred. Hence their character is merely restitutive, they refer only to individually caused and individually suffered damages. This also explains the major weakness of private environmental law. Protection against and compensation for damages to general goods (like forests, water, air) are not provided for by civil law regulations, since damages cannot be attributed to individuals.[57]

Private law is an indirect control instrument to avoid environmental damages by means of a compensation obligation in concrete individual cases, but is not at all suited for an ecologically sensible action and for damage prevention.[58] Precisely for ensuring an effective environment protection it is crucial to have damage-preventing instruments at one's avail. Once damages have occurred, they can often hardly be repaired at all, or only at high financial and scientific-technological expenditure (e.g. accidents of oil tankers, forest damages).

Moreover, difficulties of proof have to be taken into consideration, which are an obstacle to enforcing civil law claims.[59] In spite of the facilitations of proof, which, in part, are conceded by jurisdiction, insoluble problems remain, e.g. with regard to summed-up immissions and long-term damages, which hardly seem to be soluble within the system of civil law.[60]

Connected to the problems of proof is the procedural disadvantage in a civil suit. On principle, the plaintiff has to put forward every fact which substantiates his or her claim, and to prove it (cf. §§ 138, 282, 288 ZPO). Apart from higher costs and incalculable value in dispute, this leads to a high risk of litigation in a civil suit.[61] Contrary to that, public law is based on the principle of official investigation[62] and refers to lower values in dispute (through norm values in dispute).

Public environmental law provides protection and precaution mechanisms for an effective environment protection exactly through the possibility of preventive intervention. It is an exaggeration to criticize public environmental law as being shaped throughout by all too vague legal concepts, which relativize protection. On the one hand, environmental regulations have to be able to always adapt to new scientific-technological states of knowledge, for only in that way can they adjust to new potential dangers. It is just vague legal concepts like „state of technology" which allow of a quick and flexible adaptation to new insights without requiring a preceding long process of legislation. On the other hand, one very function of environmental standards is to make the vague legal concepts of environmental law more feasible and more predictable without simultaneously entailing the same, not easily changeable solidification which is the consequence of detailed legal regulations.

Moreover, public law has efficient control mechanisms at its avail in the form of possibilities of legal monitoring, which directly ensure compliance with environment-protective directions. This control is called inefficient by different sources, which mention cases of cooperation between authorities and industry, and insufficient enforcement of directions on the part of public administration. Yet it must not be ignored that these deficiencies are not structural weaknesses of public environmental law but refer to enforcement of norms in general; they occur with respect to any legal regulation. As compared to that, we have to consider the fact that a private person can renounce claims and sell them.

Today, public law does not only regulate the relationship between the state and its citizens, but it also entitles private persons to claim intervention of the authorities against third parties. So, norms of public environmental law have not only an objectively but also a subjectively legal character, that is, a third-party protecting character. They also grant an individual protection, which can be claimed by recourse to the administrative courts.

Norms[63] which also protect third parties[64] entitle persons concerned to claim prevention, protection and compensation. This public law protection of the individual enables and ensures observance and control of directions also by third parties, in addition to objectivized administrative control. Insofar, public law has adopted the preventive function of private environmental law as well.

All in all, it can be said that public environmental law allows of a more flexible, quicker and therefore more effective environment protection. Civil law with its directions designed only for narrow neighborhood ranges cannot provide sufficient protection against environmental damages, which take place at an ever larger scale, due to technological development.

6.3 INSTITUTIONALIZATION OF THE PROCEDURE OF ESTABLISHING ENVIRONMENTAL STANDARDS

6.3.1 Environmental Discourse

Establishing environmental standards is particularly problematic where scientific knowledge is very incomplete or uncertain or where there are (other than in radiation protection) no natural standards of comparison. Environmental standards are supposed to serve social risk precaution and account for the unequally distributed preferences in society, taking into consideration many evaluative points of view (multiattributivity; cf. chapter 5.4) in an existing conflict situation, and thus overcome the deficiencies of an intuitive risk assessment. In cases of cognitive uncertainty and evaluative dissension, decisions are judged to be adequate and substantiated on the basis of the procedure by which they come about rather than on the basis of their content, as was discussed in chapter 1.3.4.

To live up to these expectations, a procedure has to have the form of a publicly understandable justification discourse (cf. chapter 1.2.1) which makes the expediency of environmental standards plausible. This means, on the one hand, to point out the „state of science“ (example: chapter 2) and to integrate it into the discourse; on the other hand, the plurality of preference attitudes must have their say and be represented in the normative decision process. The latter has to take place in a rational way, e.g., considering criteria of decision analysis (chapter 5.4), and it has to be mindful of pragmatic consistency (chapter 1.3.3). A discourse attempting to meet these targets will be called *environmental discourse* here.

It is the objective of the environmental discourse *in a narrower sense* to expound a scientifically substantiated correlation between a hazardous agent and the resulting effect, and to establish environmental standards in a conclusive procedure. These standards are supposed to be included in a legally practicable procedure which meets the objectives of environment protection and which can be accepted by all persons concerned. The preceding parts of this study explained why such a result cannot be reached by knowledge and reason of the natural sciences alone, and why the reason of social sciences, jurisprudence, and ethics and philosophy is needed in this context as well.

This leads to the task of combining expert competence and procedures in a way that the objective of a reasonable environmental policy becomes realizable. It includes an optimal utilization of the respective state of scientific knowledge in the process of establishing standards. Because of the complexity of the responsibilities involved, and since the determination of objectives is dependent on interests, this can only happen in an *institutionalized form*. Therefore, institutionalized procedures have to be developed which gain social acceptance and have sufficient operational flexibility to be adjusted by means of information feedback and processing of gathered experience, without much administrative expenditure. What is required is an independent institution, the tasks of which include to systematically and with scientific competence

- ascertain the necessity of establishing a standard,
- find out what are the actual and operational presuppositions for proceeding rationally,
- formulate principles of assessment,
- express the environmental standard quantitatively and
- introduce the quantitatively expressed environmental standard as a recommendation into governmental decision procedures.

The following aspects have to be acknowledged in this process:

6.3.1.1 Formulation of the Problem and Mandate

The basis of any institutionalized procedure is a precise definition of the problem and, based thereupon, a concrete mandate to work. The responsibility for this mandate has to be taken by the political institutions in charge (parliaments). Environmental institutions with vague global responsibilities do not fit into this system and should therefore be avoided.

The mandate can be determined more precisely, the more information is available about the type of the noxa and its provenance, about ways of transportation and exposure, about size and frequency of exposure (exposure probability). Moreover, kind and velocity of the decomposition of harmful substances and their being carried away have to be taken into consideration. It is of particular importance to understand the action mechanism of the noxa, the dose response relation, the duration of an effect, and the possibility of combined effects.

6.3.1.2 Protection Objectives

It is an important part of the determination of the protection objectives to study whether the standard in question is supposed to serve exclusively or mainly the protection of humans, of other living beings, of biotopes, or of abiotic objects (e.g. buildings). Since, finally, it is humans whom the determination of protection objectives aims at (see chapter 1.2.2), it must be investigated whether the respective protection is meant to be generally valid or valid only for certain groups (occupationally exposed persons, risk groups). Particular problems arise in this context from protection objectives which, for actual or operative reasons, are sensible only on a supranational or global level.

6.3.1.3 Principles of Assessment

We have to examine the ascertained data to find out whether there is a threshold value for the noxa in consideration and for the effect resulting from an exposure. If this is the case, it has to be examined further whether this threshold value is suitable as a standard, and which safety margins should be prescribed in addition (e.g. with respect to the genetic variability of the total population or with respect to certain risk groups). If no threshold value can be recognized, it has to be decided on the basis of kind and extent of the potential danger whether the precaution principle or the minimization principle (e.g. the ALARA-principle) would be suitable as a foundation. Rational risk assessments are essential for a decision of these questions. The normative implications of these risk assessments may, in many cases, be decided by risk-risk comparisons within risk classes (cf. chapters 1.2.3 and 5). The problem of acceptability has to be solved on the basis of a risk-risk comparison as well. It is the task of political institutions to relate the results of rational risk assessment to subjective (individual or collective) risk perception and thus lead a public risk debate.

Reliability and credibility of the institutional procedure crucially depend on expertise, independence and neutrality of those participating in the procedure. All scientific disciplines relevant for issues of environment protection are to be represented by their expert knowledge. The individual representatives of the disciplines have to enjoy the confidence of the great majority of their colleagues and have to be ready to defend their statements before the scientific community.

It is the reliability of facts ascertained by the natural sciences and the credibility of the scientist who collects them or uses them which are presuppositions for a justified establishment of environmental standards. Evidence of a scientist's credibility may be found in, among other things, the scientific acknowledgement he or she has found in their specific discipline. This acknowledgement is based on scientific achievements and the expert knowledge recognizable in these achievements. In that case, an expert rates as an expert through the judgment of the scientific community. This judgment can, at any time, be verified, due to the „publicity“ of the achievements, even if it refers to an „outsider“ in some discipline.

Lacking acknowledgement on the part of the scientific community does not alone suffice to marginalize a scientist's opinion or position within the discussion or to leave it unconsidered. But the conditions of reproducibility and controllability will have to be fulfilled. Whoever does not accept or meet these conditions will be excluded from scientific discussion. Hence his or her opinion is also irrelevant in the process of establishing environmental standards.

Difficulties concerning competence might arise in defining borders of respective disciplines. Borders must not be drawn too narrowly, but not too widely either. They would be too wide if, e.g., achievements in other fields of medicine or biology were accepted as legitimation for a radiobiological expertise. It is the radiobiologist or the representatives of radiobiology qualified by scientific achievements who have to appraise the radiobiologist. There are certainly many justified objections against overrating experts in environmental discussion; but, with the present state of natural sciences, only the scientist can warrant the reliability of facts and data from the field of natural scientists.

In this context, also the „random test procedure to settle scientific-technological controversies" proposed by Maier-Leibnitz (Maier-Leibnitz 1983) should be mentioned. The core of this procedure is a controversial discussion, in writing or orally, directed by a non-involved, generally-acknowledged member of the scientific community. As long as the controversy in question is one of natural sciences only, as a rule, some „truth" will show up after a certain time, to an expert and unprejudiced observer. If, in this procedure, one specific source turns out to be tendentious and refutable more often than the average, a general loss in credibility or reliability is justified. Incidentally, one would be mistaken to believe that scientific error could be excluded for good by a procedure. To err is also part of the scientific process, and can thus persist up into the establishment of standards.

Statements of the science system are not statements which coexist in a competitive, confirming or differing way with statements of other systems, e.g. with the system of public opinion. Science claims to serve the truth, on the basis of presuppositions settled by philosophy of science. Truth is the result of a process of cognition which links the development of scientific statements strictly to conditions of reproducibility and controllability; the scientific cognition process does not tolerate any partiality except the partiality for the conditions of truth themselves. In that sense, the idea of science is to provide generalizable knowledge, the scientist is only responsible to conditions of scientific formation of truth.

This does not mean that science and the individual scientist always live up to these demands. In fact, science and scientists once and again lag behind their demands. Yet this does not constrain the claim of the science system not to be opinion among opinions or party among parties, it only means that science and each individual scientist must accept being always measured by a normative idea of science. This holds with respect to both quality of scientific work and significance of scientific statements. Quality presupposes meeting the generally accepted scientific standards, which, again, include the conditions of substantiation and control; the significance of scientific statements is measured, among other things, by the clarity of distinction between knowledge and ignorance in scientific contexts, i.e., between what can be explained with scientific strictness and what remains to be explained. Science in this sense always means clarity about what we know and what we do not (yet) know. Obscuring borders between these two states also means obscuring borders to opinions which pretend to be knowledge, it means that science might also become biased. In such cases, statements of the science system lose their rationality-securing character.

Among the rationality obligations of the science system are not only the obligation to truth but also the *obligation to responsibility*. Science which follows its normative idea – which is, in a set phrase, described by 'pure knowledge' and 'truth' – is neither one-eyed nor biased, but is also not situated beyond social contexts, since science is not only a form of knowledge formation (just formation of knowledge in the form of science) but also a social institution. In both forms, in its form of knowledge and in its institutional form, it is subject to responsibilities. Yet this also implies that science has a moral form (see Mittelstraß 1982, 11ff). Or, to put it differently: statements of the science system, which should provide a rational basis for the environmental debate, have not only a scientific dimension in a narrower sense, namely a dimension relating to standards of quality and research, but also an ethical one, which, however, should not lead to a specific ethics of science but to the general ethics of citizens.

6.3.2 Environment Council

There are several conceivable ways of fulfilling the conditions of institutionalizing the environmental discourse as described in section 6.3.1, one of which is to select departments concerned with environmental issues across relevant institutions and put them in charge of this discourse. The emphasis should be on the contributions of the science system here, though. Therefore, we propose the formation of an *environment council* and choose the German science council as a model for its constitution and procedures.

The science council was established on the basis of an agreement between the German federal and state governments in 1957 (confirmed by administrative agreements in 1975 and 1991). It works out recommendations about contents and structure of the development of universities, science and research (Art. 2). It consists of two chambers: the scientific commission, whose members are appointed by the German Federal President, and the administrative commission, whose members are appointed by federal and state governments (Art. 5). The resolutions of the plenum are prepared by the scientific commission under expert and scientific points of view, by the administrative commission under administrative and financial points of view (Art. 6). They have a relatively binding character beyond their form of recommendation, as they already include the consent of federal and state governments.

In analogy to the structure of the science council, we suggest a two-chamber system for the environment council as well. Members of the *scientific chamber* shall be scientists from the fields of natural sciences, social sciences (including economics and jurisprudence) and philosophy (practical philosophy and philosophy of science). Natural scientists represent the environmentally relevant disciplines of the natural sciences. Members of the scientific chamber have to reach a consent. In order to guarantee neutrality and independence of this body, the members of the scientific chamber are appointed by the German Federal President on the proposal of the „alliance", i.e., the science council, the „Deutsche Forschungsgemeinschaft, DFG" (a German institution for research promotion), the „Max-Planck-Gesellschaft" (another German institution for research promotion), the „Arbeitsgemeinschaft der Großforschungseinrichtungen" (working group on large-scale research installations), the „Fraunhofer-Gesellschaft" (a German institution for promotion of applied and application-oriented research) and the conference of rectors of German universities. Members of the *administrative chamber* shall be representatives of legislation and executive. The representatives of legislation are appointed according to the ratio of representation in parliament, the representatives of the executive are appointed by directive.

Cooperation of both chambers is meant to be determined by the primacy of expert knowledge; statements of the science system are placed in the first line in all deliberations and decisions. To ensure this primacy, the scientific chamber forms both permanent and ad hoc *expert commissions*. Their members are in part also members of the scientific chamber. The expert commissions work out recommendations, which are submitted to scientific chamber and administrative chamber for deliberation and which are decided upon in the plenum, the common session of both chambers, as statement of the environment council. If a common decision is not reached, the recommendation is issued by the scientific chamber alone, on its own behalf. A separate recommendation on the part of the administrative chamber is not provided for in that case.

Since the environment council is supposed to bring to bear the statements of the science system, a general obligation of both chambers to reach a consensus would be dysfunctional. The administrative chamber may participate in discussions, not least because of aspects of enforceability, and it would be quite desirable if common recommendations were reached, but participation of the administrative chamber should not make the environment council a place of compromise between scientific reason and administrative or political reason.

Evidently, the structure of the environment council differs here from that of the science council. Whereas, in the case of the science council, scientists are appointed on the basis of their scientific competence and their competence regarding *science policy*, it is scientific competence alone which is relevant in the case of the environment council. The environment council does not so much prepare and advise the decisions of the federal and state governments as bring to bear scientific expert knowledge. Recommendations of the environment council are published. They are addressed to parliaments and governments.

The environment council determines its work program annually. It is appealed to by the federal and state governments, but it can also become active on its own. There should be no possibility of an appeal by associations and other social groups, in order to avoid too low a 'threshold of appeal', which would hinder the working ability of the environment council. The environment council may, however, adopt on its own initiative suggestions from outside. This may also take the form of *hearings* (e.g. with conservation organizations, representatives of the economy and the trade unions). These hearings may take place before both chambers, but also before the scientific chamber alone or before the permanent or ad hoc commissions. Such procedures should have a 'quasi-judicial' character.

The (already existing) *radiation protection commission* and a *commission on chemicals* should be among the permanent expert commissions of the environment council.

6.3.2.1 Commission on Radiation Protection

The commission is supposed to deal with risks originating from ionizing rays. It shall give recommendations about ways of minimizing emissions and exposures to radiation, and about what exposures are tolerable under what conditions. The evaluations have to be elaborated for proper operation, for malfunction and for accidents in the installations. Also long-term safety, e.g., regarding repositories of radioactive substances, has to be considered. The commission is supposed to give its opinion and advice on all fundamental questions of radiation protection.

To this end, measurement equipment at the site of the polluters, spreading of radioactive substances in the media atmosphere, soil and water, and transportation of substances via food chains to humans, have to be evaluated with the help of respective data or models. Furthermore, dose distributions in human bodies and radiation effects in individual organs due to specific radioactive substances after incorporation and after exposure to external radiation have to be evaluated by means of biokinetic models and dosimetric concepts. Therefore, scientists of all fields of radiation protection must be represented in the commission on radiation protection, in particular scientists of the following disciplines: radiophysics, radioecology, radiobiology, medicine. The scientists have to be independent; as a rule, they should not work in institutions where they are subject to directives. The members could be appointed according to the model of the expert committees of the DFG. Election proposals may come from professional associations, large-scale research facilities, environmental institutes and universities, with the participation of the scientific chamber of the environment council; the election is carried out by the scientific chamber.

The radiation protection commission may discuss subjects proposed by the environmental council, the scientific chamber and the federal and state governments. Moreover, it may discuss subjects on its own initiative. Subjects may be of a general kind or they may refer to special facilities. The commission works out recommendations, which are addressed to the environment council. It is supposed to give the governments detailed advice in issues of radiation protection and may establish committees (permanent or ad hoc) on specific fields or issues. The chairman of such a committee has to be member of the commission. The commission has to point at open questions of radiation protection and stimulate or demand research programs necessary for a solution.

6.3.2.2 Expert Commission on Chemicals and Toxicology

The commission is supposed to deal with existing or foreseeable long-term damage potentials by chemical noxa (e.g., ozone depletion by CFCs, climatic change by CO_2 and trace gases, forest damages due to different influences which are not yet completely known, dying lakes due to acid rain and over-fertilization of surrounding fields, poisoning of soils by chemicals and heavy metals, health impairments by pharmaceutical products). It shall use its influence to have research programs about the respective correlations established and corresponding means provided (by environment ministries, by the Federal Environmental Protection Agency, by the DFG, industry, the European Union, European organizations). Its 'everyday work' is examination, assessment and documentation of current data. An integration of the MAK-commission into the work of the commission on chemicals and toxicology could be thought of. The commission collects well-founded facts about chemical behavior and biological effectiveness of substances and delegates individual problems and their solutions to respective agencies (Federal Environmental Protection Agency etc.).

It should be possible to appeal to the commission within the frame of regulations proposed for the environment council in general, for instance if some contamination is caused by a 'substance of the month'. The commission should be able to directly assess the severity of the consequences; it should, if possible, suggest countermeasures from a scientific viewpoint, not from a technological one.

(At least) the following disciplines should be represented in the commission: inorganic chemistry, organic chemistry, chemical physics, biochemistry, pharmacology, toxicology, meteorology. Disciplines should not be defined too narrowly and specialization should not be too detailed. Members of the commission should not come from environment authorities. The DFG committees could again serve as a model, not only for committees in certain classical disciplines but also in fields like environmental chemistry, environmental physics and other environmental sciences. Election proposals could come from professional associations, large-scale research installations, environmental institutes and individual universities, with the participation of the scientific chamber of the environment council, as in the case of the radiation protection commission; the election is carried out by the scientific chamber. Furthermore, the regulations formulated for the radiation protection commission hold here as well.

6.3.3 Normal Procedure: Guidelines for Establishing Environmental Standards by Environmental Authorities

The environment council will not interfere with normal tasks of federal or state environment authorities. These are tasks concerning early recognition of potential environmental damages, identification and assessment of options of acting, and the effective choice of the suitable standard or of an equivalent instrument for abatement or reduction of the environmental damage. At the same time, practical application of the standards has to be monitored and the effectiveness has to be controlled. All these tasks have to be carried out, moreover, in close cooperation with advisory committees of scientists. The sequence and structure of this process should, like the decision sequence in the environment council, orient by decision analytic criteria and proceed in the following steps:

6.3.3.1 Registration of the Problem

In the first step of the decision process, the necessity of a regulative intervention has to be ascertained and documented publicly. This step may rely on the following mechanisms:

- new scientific insights, which are systematically collected and evaluated by a review committee,
- regulation efforts in other countries, which are continuously transferred and evaluated with the help of an international network (by electronic data transfer and exchange of documents),
- improvements of retention technologies, which are permanently observed by a group within the VDI or UBA and which may be promoted by legal incentives,
- systematic early recognition of environmental damages by carefully directed research projects in universities and large-scale research installations, with feedback to the authorities,
- establishment of a registration office, where environmental damages observed by social groups are gathered and evaluated.

It seems sensible to entrust the federal environment ministry, in cooperation with the Federal Environmental Protection Agency, with the initiative authority for normal procedures. There must be a central institution with the task of collecting scientific data, of commissioning new studies, of systematically analyzing the international practice as to environmental standards, and of institutionally attending to early recognition of environmental damages. The authorities of the German states, on the other hand, should be entrusted with the implementation of standards.

6.3.3.2 Check for Conflict Potential

If, on the basis of the first step, it has to be expected that establishing a standard will lead to serious conflicts or controversies in society, the normal procedure has to be abandoned and the environment council has to be appealed to. This holds also for decisions of particular environment-political relevance.

6.3.3.3 Selection of Options

If the check for conflict potential is negative, the next step will be to investigate how far there is need for action at all (dependent on the extent of the damage and the legal possibility of intervention). If this leads to the conclusion that there are sufficient reasons for a governmental intervention, possibilities have to be considered, with the help of which the problem could be handled. The environment authority should therefore discuss possible options of taking action, assisted by a committee (internal or external) in which natural scientists, economists, technicians and social scientists are represented, and choose those options for which a positive balance between returns and expenditures is to be expected. In any case, several options have to be proposed.

6.3.3.4 Option Performance Measurement

In this step, the probable consequences of the different options have to be assessed according to scientific criteria. This assessment may, in the routine case, be elaborated by the authority itself or by its external experts, in special cases also by the expert commissions of the environment council. Finally, a matrix will be accomplished, which assesses all options of action according to the following criteria:

- risk determination for non-intervention by the government, and potential risk reductions to be expected on the basis of the proposed options (or continuous reduction functions),
- determination of costs and economic benefit to be expected for each option,
- cost-effectiveness function for individual options with respect to option combinations,
- determination of distribution effects of benefit and risk.

6.3.3.5 Transformation into Utilities

Since all measured values are presented in their natural dimensions, a transformation of these values into units of utility (positive or negative) is required. This may be done numerically, with the help of decision analytic procedures, but this is not indispensable. Since transfer into units of utility is not a purely scientific task but nevertheless requires sufficient expert knowledge, it seems appropriate to have this task accomplished either by an internal working group within the administrative body or to organize a hearing of experts. It seems furthermore sensible at this step to integrate other authorities and institutions into the deliberations, to identify possible impacts of the options on fields which are not risk-relevant, and to take them into consideration as far as possible.

6.3.3.6 Option Weighting

For this step, there should be a working group within the administrative body, which may consist of members of the environment ministry and representatives of the German states (e.g. federal and states conference). Also external experts, above all, scientists, may be invited to participate in elaborating a decision in this body. It may, if need be, organize hearings with interest groups.

6.3.3.7 Decision

Complying with the respective official regulations, the respective authority should make a decision on the basis of the scientifically established profiles of consequences and of benefit, and on the basis of the results of the hearing. This means, in the first place, to choose one of the discussed options, then, to establish level or intensity of the respective selected option (e.g., level of the standard or the tax reduction).

6.3.3.8 Reasons for the Decision, and Objection

The respective authority has to give detailed reasons for each decision in written form, basing their arguments on application of the rules as well as profiles and evidence gathered before the decision. The trade-offs between different criteria should be made clear, and each individual step of achieving the decision should be understandable for a potential reader. This increases the transparency of the decision process and counteracts exaggerated expectations of the citizens.

Within a certain period of time, the groups involved in the decision process may object. An objection can only be sustained if there is the suspicion of violation of the regulations, or the obvious ignorance of actual knowledge, or conflict of interests.

The procedure proposed here does not require a basic reform of the current administrative activities. It merely proposes a new sequential division of labor. This division of labor is supposed to ensure scientific foundation of the standards, on the one hand, and improve interdepartmental cooperation, on the other hand. Finally, an improved transparency of the decision process will probably be an important advantage in finding social acceptance.

6.3.4 Possible Institutional Objections

The proposal of creating a new, central institution of debate on environmental issues in Germany requires a (self-)critical examination of benefits and costs to be expected of such an institution. After all, institutions may, apart from their desired effects, provoke unintended and undesired side-effects through their mere existence or performance of their tasks, which have to be considered in time during planning and which, if possible, have to be avoided during practical realization.

In addition, objections against the planning of an environment council have to be expected on the part of interest groups, since the strategic scopes of politics, administration, economy and environmental associations are restricted by such an institution vested with great scientific competence – even if it is explicitly a body of deliberation and not of decision. To take this criticism seriously and to take it into account from the beginning is not only a question of scientific care but also a question of practical reason.

Arguments against an environment council can be anticipated and will probably comprise five or six different fears:

(a) Establishing an environment council could strengthen *centralistic tendencies*, while at the same time decentralized, regional and sectorial particularities of environmental policy are increasingly neglected. This objection overlooks that the environment council is by no means supposed to be responsible for all environmental issues, and that moreover, even in a central body regional aspects can be taken into consideration. The environment council is neither meant to monopolize advisory tasks nor to constrain necessary local scopes of action. Yet, it cannot be doubted that universal environmental problems also require as general as possible standards to be established, which necessitates central advisory bodies and decision instances.

(b) An environment council could encourage *bureaucratization* of processes of deliberation and decision, and this with regard to problems which many citizens experience as existential ones. This fear would not be unjustified if a big office, working full-time, gained a lead in competence, information and influence against honorary members of the environment council. The kind of institutionalization, the designing of working procedures, and the ensuring of public control will therefore from the beginning have to block, in a continuous and controlled way, the tendencies of bureaucratization genuinely immanent in such a body. This requires full transparency of work and of results of work. It will enable the environment council to not only efficiently counteract its immanent bureaucratization mechanisms but also to contribute to constraining the presently dominant but not very transparent role of bureaucracy in establishing environmental standards.

(c) Closely linked to the suspected tendencies of centralization and bureaucratization might be the suspicion that an environment council could be used for a „*neutralization*“ of social conflicts. This erroneously assumes that a body of experts could neutralize and hence weaken personal concern, individual value judgments, social potentials of conflict and the collective readiness for action of the citizens by means of the authority of scientific judgments, rational decision processes and consensual models of deliberation (to say it in a polemic way: by constructed inherent necessities). Yet stripping arguments of emotional elements does by no means necessarily impair personal commitment to environmental problems, it can also contribute to a better orientation of persons concerned and persons feeling concern.

Only a neutralization of conflicts would lead to a *depolitization* of the public environmental debate – and this will presumably be another point of criticism. Even if we disregard that interest group which wants to use the attention value of environmental problems to push through other political objectives, there are doubtless large parts of the population left to be regarded which rightly consider environmental policy to be one of the most important fields of social formation of opinion and will. It will probably be agreed upon by many that this eminently political task should not be delegated to an expert body. The task of the environment council, however, would be completely misunderstood by such criticism. This institution is not supposed to usurp the formation of political will but to base it on objective knowledge by providing advice as competent, as objective, and as non-particularistic as possible.

(d) In close connection with the fear of a possible depolitization of the environmental debate, the objection might arise that the environment council would replace the present variety and difference of scientific formation of judgments by an inadmissible *uniformity*, due to its institutional structure and function. There are at least three misunderstandings at the root of such a concern: first, today the deplorable state exists that some (natural) scientists mix their personal convictions and their professional judgments in a way which is not transparent to the public. This dissatisfying situation can and should be overcome by discussions in the environment council. Second, frank conversation of competent professionals will contribute to relativize opinions which are scientifically founded but one-sided or not yet sufficiently warranted, and lead to a judgment which best gives effect to the respective state of knowledge and the practical conclusions which can be derived from it. Finally, it has to be stressed, as a third point, that the environment council has the duty to lay bare scientific controversies and differing, but equally substantiated, judgments. This may be done in the form of labeling opinions of majorities and minorities.

In addition to that, it has, of course, to be left to the body to give that recommendation which, in a concrete case, best matches the scientific state of knowledge if not being directly derivable from it. This requires also that scientific lack of knowledge, contradictions and uncertainties are expressed openly. Notwithstanding these problems, a body of experts of different scientific disciplines should, after all, best guarantee an objective and solid judgment.

(e) Finally, it will perhaps be argued that the environment council needs plenty of time for its deliberations, but that environmental problems are often recognized all of a sudden and require quick solutions. Politicians may acquire a particularly strong public image if, in such an acute problem situation, they act quickly, effectively and convincingly. They will find this opportunity at least impaired by a body like the environment council. This, however, is only partly true, and it is true in a desired way. There is no doubt that, independently of the environment council, political decision mechanisms have to exist which allow of quick reaction in case of acute problems and conflicts, and which avoid *temporal delay*. The careful and knowledgeable formation of judgments in a body like the environment council, however, guarantees that quickly made decisions can also be corrected on the basis of improved insights. On the other hand, one important function of the proposed environment council is exactly to modify such reactive political decisions under massive pressure of time and in the light of public attention, in favor of an environmental policy which is designed over a long term, which is active, and which is not controlled by topical problems but by the scientific state of knowledge.

This is an argument which has to be emphasized in favor of the establishment of an environment council during the public debate to be expected. The point is, by combining professional reason of natural sciences, social sciences and administration, would contribute to making environmental policy better able than now to: first, avoid environmental problems in time or solve them, in a purposive and precautionary way, instead of reacting hectically on suddenly arising problems; second, draw on a permanently available, scientifically representative expertise, instead of having to rely on the ad hoc advice of more or less renowned scientists; third, have a body which can combine specialized professional knowledge with a holistic formation of judgments, instead of having to ask many individual specialists.

This institutionalized form of political advice is not meant to avoid social conflicts, to neutralize personal commitment or even to remove the political process from environmental policy, but to make advice more objective by providing a basis of information as complete as possible and a rational evaluation procedure.

6.3.5 Environment Council and Risk Communication

The political acceptance of the environment council will largely depend on whether the environment council can successfully be established, through communication with interested social groups and the public, as a competent and supra-partisan advisory body. Communication about environmental risks and their limitations take place in a political arena, in which opinions concerning environmental issues are particularly polarized, and respective political positions are linked to extreme value judgments. Since risks, by their very nature, do not allow of direct conclusions about individual outcomes (cf. chapter 1.2.2), statements about risks are hard to falsify. One example of that is the attempt to name exposure to radiation as causal explanation for anomalies or random variations of frequencies of diseases.

The environment council is thus situated in a communication environment where the public expects information not bound to interests but has no possibility to check the correctness of statements and the reliability of information sources. Therefore, the environment council is faced with the question of how it might gain authority as an instance of supra-partisan expert advice.

Of primary influence for the perception of competence and credibility of an institution is its perceived balance of achievements. This balance depends on two factors: expectations placed on the institution and perceived achievements. If expectations are high, the public will withdraw its confidence in the respective organizations even if their achievement levels are relatively high. But if expectations are low, organizations which steadily exceed the expectations by their achievements may increasingly occupy a major function, starting from a rather unimportant role. It is therefore reasonable to entrust the environment council with only few tasks in the beginning, thus not arousing too high expectations. If the environment council by its recommendations exceeds the expectations placed in it, it will soon gain in reputation and political influence. The MAK-commission is a good example of such a development.

If, therefore, the environment council wants to reach a situation where confidence is placed in its efficiency and its expert knowledge, it will have to fulfill some preconditions, which determine its role in risk communication. Among them are:

(a) *Perceived expert knowledge:* As a rule, expert authority is perceived in connection with successfully carried out tasks (e.g., correct prognoses, successful crisis management, congruence of institutionally proclaimed objectives and their realization). In cases where success is hard to assess, peripheral patterns of evaluation become important. Then, the reputation of the institution and its members and their status in the political framework serve as an orientation for an assessment of expert knowledge. At this point, a conflict situation might arise for the environment council: it should rank relatively low in the political hierarchy so as not to arouse too high expectations regarding its efficiency; but a high status within the system of institutions will be of particular advantage for acquiring enough reputation. This conflict can, for instance, be solved by the fact that the members of the environment council enjoy a high status, due to their appointment by the Federal President, while at the same time the environment council exerts a rather informal advisory function vis-à-vis the ministries.

(b) *Perceived objectivity:* As long as there are criteria according to which the truth of statements can easily be proved or disproved, there is no difficulty in perceiving objectivity. In complex situations, people again rely on peripheral indicators. They often substitute fairness (which will be treated in the next point) for objectivity. Then, the institution is thought to be objective if its proposals are situated in the center between extreme demands of groups. This interpretation of objectivity will become problematic for the environment council if this 'mean value' is not congruent with the scientifically ascertained truth. The perception of objectivity can be improved by independence of the information source, reputation of the experts, and by taking social groups into consideration. The appointment procedure and the independence of its members of political or economic interest groups is crucial for the environment council. A proportional representation of social groups in the environment council, however, is not compatible with its scientific claim.

(c) *Perceived fairness:* The readiness of an institution to examine, without reservation, all relevant points of view in the course of the decision process, does not mean that its judgment has to result in the 'mean value' of all opinions from different parties (although this is often seen as the indicator of a fair solution). What is important is rather the readiness of the institution to accept concerns and claims of all relevant groups and to integrate them into the judgment process. For this reason, it is recommendable that the environment council gives the substantiated impression of a fair consideration of all aspects put forward by relevant groups, by means of hearings and other procedures of participation.

(d) *Predictability:* With similar facts and boundary conditions, institutions should reach equal results, independent of actors and points of time. Such a continuity of formation of judgment, and hence of accomplishment of functions, increases credibility. On a metalevel, predictability creates confidence in the system in a positive sense, on the institutional level it decreases the need of control as a kind of organized distrust. The basis of predictability is the consent of the participating persons to voluntarily comply with procedural and decision rules which are acknowledged as rational and reasonable. Such predictability is achieved by binding the environment council to certain procedural and decision rules (e.g. risk-risk comparisons, see chapter 6.3.3).

(e) *Expediency:* If a certain institution is considered to be necessary for accomplishing a socially important function, it will also meet with the corresponding confidence. The environment council will have particular problems in this respect, since environmental policy seems to go on well without it. Thus, it will be important to point out that its expediency goes beyond the so far established institutional procedures in environmental policy. An objectivation alone will probably not suffice as basis for a legitimation here. Rather, the integrative character of the decision process within the environment council and its position as a „scientific clearing house“ will have to be emphasized towards the public.

The creation of a suitable organizational structure of the environment council and its appearance in public will have to pay attention to the efficiency of these five components of creating confidence. Without credibility, the environment council will not play any role in establishing environmental standards, whatever institutional form it will have. Credibility is the product of a complex structure of influences, among which efficiency is the most important component, and perceived expert knowledge, objectivity, fairness, predictability and expediency determine public reputation. Successful communication begins before imparting information; it creates the institutional and structural preconditions for information and resulting recommendations to be accepted.

ANNOTATION:

1 For the German private environmental law in general, see Gerlach 1989; Medicus 1986, 778ff; Diederichsen 1986, L 48ff; Kloepfer 1989, 225ff; Baur 1987, 317ff.

2 See Münch-Komm/Säcker 1986, § 906, Rdnr. 6.

3 For the system of claims for prevention in general, see Gerlach 1989, 177ff.

4 Münch-Komm/Säcker 1986, § 906 Rdnr. 20; Palandt-Bassenge 1990, § 903 Anm. 26 aa; Staudinger-Roth 1988, § 906 Rdnr. 114ff.

5 See Hohloch 1988, 691.

6 See the examples in Palandt-Bassenge 1990, § 906 Anm. 3b.

7 In public law, limits are binding for decisions of the administrative court. See Marburger 1986; Diederichsen 1986, L 58ff for problems of environmental standards in civil law.

8 BGHZ 46, 35; 69, 105; Palandt-Bassenge 1990, § 906 Anm. 3b.

9 Münch-Komm/Säcker 1986, § 906 Rdnr. 78ff.

10 Staudinger-Roth 1988, § 906, Rdnr. 182.

11 Marburger 1986, C 102ff.

12 BGH NJW 1983, 751f („tennis court decision") Permissions for a certain utilization, on the part of the development plan, are only seen as indicators of a customary utilization in a certain area; Staudinger-Roth 1988, § 906, Rdnr. 188.

13 So far, jurisdiction has left open the question whether the right to claim should be extended to a larger group of persons: BGHZ 92, 143, 145 („cupola furnace decision"); see further Marburger 1986, C 115 with attempts to find a solution, Münch-Komm/Säcker 1986, § 906 Rdnr. 18.

14 Kloepfer 1989, 235; Marburger 1986, C 120f; Medicus 1986, 778 (779).

15 See Marburger 1986, C 121; Medicus 1986 778 (780); BGHZ 92, 143 (152).

16 BGHZ 92, 143 (152).

17 Marburger 1986, C 122f; Hohloch 1988, 696; BGHZ 86, 356 (362); Münch-Komm/Mertens 1986, 823 Rdnr. 154f.

18 See Kloepfer 1989, 236ff; Marburger 1986, C 123ff; Münch-Komm/Säcker 1986, § 906 Rdnr. 123ff; Diederichsen 1986, C 79ff; Diederichsen 1976, 1ff; Baur 1987, 317 (320f).

19 See Diederichsen 1986, C 82, who makes public planning law responsible for these deficiencies.

20 Marburger 1986, C 124f.

21 Hager 1986, 1961 (1967), who wants to adopt the idea of pollution share liability, developed in the U.S.; Marburger 1986, C 124f, who proposes to establish compensation funds, see also Gerlach 1989, 364ff.

22 BGHZ 92, 143ff („cupola furnace decision"); see also Marburger & Herrmann 1986, 354ff, and Gerlach 1989, 246.

23 BGHZ 92, 143 (147); 70, 102 (107).

24 BGHZ 92, 143 (150), in analogy to a reversed burden of proof with respect to producer liability. For problems of reversal of burden of proof, see Adams 1986, 132 (148ff).

25 Marburger & Herrmann 1986, 354 (357).

[26] In addition, there is a multitude of further principles, see Kloepfer 1989, 72; of importance is, e.g., the principle of community.

[27] Kloepfer 1989, 74ff.

[28] Kloepfer 1989, 83ff.

[29] Kloepfer 1989, 91ff.

[30] Kloepfer 1987, 3ff.

[31] A general discussion is found in Kloepfer 1989, 99ff with further reference.

[32] For a critical discussion, see Breuer 1967/68, 636.

[33] See Kloepfer 1989, 117ff; a general discussion is found in Maurer 1988, § 9 Rdnr. 51ff.

[34] See Kloepfer 1989, 142ff.

[35] E.g. § 21 BImSchG; § 17 Abs. 35 AtG; § 12 Abs. 2 Nr. 24 WHG.

[36] See Kloepfer & Meßerschmidt 1986, 136f.

[37] Kloepfer 1975, 295 (301).

[38] See, e.g., § 52 BImSchG.

[39] Kloepfer 1989, 116.

[40] So, for instance, immission protection officers, §§ 53ff; BImSchG; water protection officers, §§ 21 a ff WHG; radiation protection officers, §§ 26ff StrlSchVO.

[41] § 30e in connection with § 30 BNatSchG; §§ 59ff LuftVG.

[42] E.g. § 62 BImSchG, § 46 AtG, § 51 WHG, § 18 AbfG.

[43] For a general discussion, see Kloepfer 1989, 159ff.

[44] For a general discussion, see Meßerschmidt 1986.

[45] Meßerschmidt 1986, 46ff.

[46] Regulated in the law on waste water levies.

[47] Benkert 1984, 132ff.

[48] Particularly by possible write-offs.

[49] E.g., reduction of fees for low-noise airplanes.

[50] See Ossenbühl 1986.

[51] Kloepfer 1989, 198.

[52] See also Sander 1988, 344ff, with examples.

[53] See Diederichsen 1986, L 58ff; Marburger 1986, C 86ff.

[54] Just see the „tennis court decision", BGH NJW 83, 751f; for a critical discussion, see Johlen 1984, 134ff.

[55] For a general discussion, see Gerlach 1989, 43ff; Gerlach 1988, 161ff; Peine 1987, 169ff; Kleinlein 1987, 8ff; Marburger 1986, C 14ff; Diederichsen 1986, L 48ff; Hohloch 1988, 702.

[56] Therefore, see also the present function of §906 BGB regarding protection against small immissions in narrower neighbor law, like noise etc., Münch-Komm/Säcker 1986, § 906 Rdnr. 13.

[57] General opinion also in civil law literature; see Diederichsen 1986, C 48f; Marburger 1986, C 15.

[58] See also Medicus 1986, 778 (785); Ritter 1987, 927 (935).

[59] Cf. section 6.2.2.1.

[60] See Baur 1987, 317 (320); Ritter 1987, 929 (939f); Marburger & Herrmann 1986, 354 (358).

[61] Baur 1987, 317 (329); Gerlach 1988, 161 (163).

[62] Cf. §§ 24, 26 VwVfG, and § 86 VwGO in administrative procedure law.

[63] So, e.g., § 5 Abs.1 Nr. 1, § 20 Abs. 2, 2; § 25 Abs. 2 BImSchG; §§ 5, 7 Abs. 2 Nr. 3, § 8 Abs. 3, 2 AtG; § 28 Abs. 3, § 45 StrlSchVO.

[64] Accordingly, the direction does not only have to serve the general interest but has also to aim at the protection of individual interests, st. Rspr. BVerwGE 55, 280 (285); 66, 307 (308)

LITERATURE:

Adams, M. (1986): Zur Aufgabe des Haftungsrechts im Umweltschutz, ZZP Marburger, 99, pp. 129

Arbeitsgemeinschaft für Umweltfragen (1986): Das Umweltgespräch: Umweltstandards – Findungs- und Entscheidungsprozeß, Schriftenreihe der Arbeitsgemeinschaft für Umweltfragen e. V., AGU–29, Bonn.

Baumol, W.J., Oates, W.E. (1975): The Theory of Environmental Policy. Externalities. Public Outlays and the Quality of Life, Englewood Cliffs, Prentice Hall.

Baur, F. (1987): Zur Entstehung des Umweltschutzrechts aus dem Sachenrecht des BGB, JZ 42.1, pp. 317

Benkert, W. (1984): Staatliche Finanzhilfen zur Förderung des Umweltschutzes, NuR 6, pp. 132

Breuer, R. (1967/68): Die hoheitliche raumgestaltende Planung, Dissertation, Bonn.

Coarse, R.H. (1960): The Problem of Social Cost, Journal of Law and Economics 3, 1–44.

Diederichsen, U. (1976): Zivilrechtliche Probleme des Umweltschutzes, in: Festschrift für Reimer Schmidt, Karlsruhe, Verlag Versicherungswirtschaft, pp. 1

Diederichsen, U. (1986): Ausbau des Individualschutzes gegen Umweltbelastungen als Aufgabe des bürgerlichen und des öffentlichen Rechts (Referat), in: 56 DJT 1986, vol. II, pp. L48

Gerlach, J.W. (1988): Die Grundstrukturen des privaten Umweltrechts im Spannungsverhältnis zum öffentlichen Recht, JZ 43.1, pp. 161

Gerlach, J. W. (1989): Privatrecht und Umweltschutz im System des Umweltrechts, Berlin, Duncker & Humblot.

Hager, G. (1986): Umweltschäden – ein Prüfstein für die Wandlungs- und Leistungsfähigkeit des Deliktrechts, NJW 39, 2.1, pp. 1961

Hohloch, G. (1988): Umweltprivatrecht, in: Handwörterbuch des Umweltrechts, vol. II, Berlin, Erich Schmidt, pp. 691

Johlen, H. (1984): Bauplanungsrecht und privatrechtlicher Immissionsschutz, BauR 15, pp. 134

Kleinlein, K. (1986): Das System des Nachbarrechts, Dissertation, Berlin. Kloepfer, M. (1975): Der Editionnvorbehalt bei Verwaltungsakten, Die Verwaltung 8, pp. 295

Kloepfer, M. (1987): Instrumente des staatlichen Umweltschutzes in der Bundesrepublik Deutschland, Jahrbuch des Umwelt- und Technikrechts 1987, pp. 3

Kloepfer, M. (1989): Umweltrecht, München, Beck.

Kloepfer, M., Meßerschmidt, K. (1986): Innere Harmonisierung des Umweltrechts, Berichte 6/86, Umweltbundesamt (ed.), Berlin.

Koslowski, P. (1983): Markt- und Demokratieversagen? Grenzen individualistischer gesellschaftlicher Entscheidungssysteme am Beispiel der Umwelt- und Kernenergiefrage, Politische Vierteljahresschrift 24, 166–187.

Maier-Leibnitz, H. (1983): Stichprobenverfahren zur Klärung wissenschaftlich-technischer Kontroversen, Die Naturwissenschaften 70, 65–69.

Marburger, P. (1986): Ausbau des Individualschutzes gegen Umweltbelastungen als Aufgabe des bürgerlichen und des öffentlichen Rechts, in: 56 DJT 1986, vol. 1, Gutachten pp. C 1

Marburger, P., Herrmann, H. (1986): Zur Verteilung der Darlegungs- und Beweislast bei der Haftung für Umweltschäden BGHZ 92, 143, JuS 26, pp. 354

Maurer, H. (1988): Allgemeines Verwaltungsrecht, 6. ed., München, Beck.

Medicus, D. (1986): Zivilrecht und Umweltschutz, JZ 41.2, pp. 778

Meßerschmidt, K. (1986): Umweltabgaben als Rechtsprobleme, Berlin, Duncker & Humblot.

Mittelstraß, J. (1982): Wissenschaft als Lebensform. Reden über philosophische Orientierungen in Wissenschaft und Universität, Frankfurt a.M., Suhrkamp. Mohr, E. (1990): Marktwirtschaftliche Instrumente für Umweltstandards, Vortragsmanuskript für die Arbeitsgruppe „Umweltstandards" der Akademie der Wissenschaften zu Berlin.

Münchener Kommentar (1986): Bürgerliches Gesetzbuch, vol. 3, 2. Halbband, vol. 4, 2. ed. München, Beck.

O'Riordan, T. (1988): The Politics of Environmental Regulation in Great Britain, Environment 30, No. 8, 4–9, 39–44.

Ossenbühl, F. (1986): Umweltpflege durch behördliche Warnungen und Empfehlungen, Köln, C. Heymann.

Palandt (1990): Bürgerliches Gesetzbuch, 49. ed., München, Beck.

Peine, F.-J. (1987): Öffentliches und Privates Nachbarrecht, JuS 27, pp. 169

Renn, O., Kals, J. (1990): Technische Risikoanalyse und unternehmerisches Handeln, in: Risiko und Wagnis: Die Herausforderung der industriellen Welt, M. Schüz (ed.), Pfullingen, Neske, 60–80.

Ritter, E.-H. (1987): Umweltpolitik und Rechtsentwicklung, NVwZ6, pp. 929.

Sander, H. (1988): Selbstbeschränkung, in: Handwörterbuch des Umweltrechts, vol. II, Berlin, Erich Schmidt, pp. 344.

Schürmann, H.J. (1978): Ökonomische Ansätze zu einer rationalen Umweltpolitik und wirtschaftspolitische Konsequenzen, München, Oldenbourg.

Sen, A. (1967): Isolation, Assurance and the Social Rate of Discount, Quarterly Journal of Economics 81, 112–124.

Siebert, H. (1987a): Economics of the Environment, Theory and Policy, 2nd ed., Berlin, Springer.

Siebert, H. (1987b): Umwelt als knappes Gut, in: Umwelt-Wirtschaft-Gesellschaft. Wege zu einem Grundverständnis, R. Wildenmann (ed.), Stuttgart, Staatsministerium Baden-Württemberg, 77–87.

Smith, V.K. (1986): A Conceptual Overview of the Foundations of Benefit-Cost Analysis, in: Benefits Assessment: The State of the Art, J.D. Bentkover, V.T. Covello, & J. Mumpower (eds.), Dordrecht, Reidel, 13–34.

Staudinger (1989): Kommentar zum Bürgerlichen Gesetzbuch, Drittes Buch, 12. ed., Berlin, Schweitzer/de Gruyter.

Steger, U., unter Mitarbeit von Feess-Dor, E., Prätonus, G. (1990): Übernahme von Umweltkosten und -risiken durch die Industrie, in: Risiko und Wagnis: Die Herausforderung der industriellen Welt, M. Schüz (ed.), Pfullingen, Neske, 194–209.

Tietenberg, T. (1984): Environmental and Natural Resource Economics, Glenview, I11, Scott and Foresman.

Tullock, G. (1976): Private Wants – Public Means. An Economic Analysis of the Desirable Scope of Government, New York, Basic.

Wicke, L. (1990): Der ökonomische Wert der Ökologie, in: Risiko und Wagnis:

Die Herausforderung der industriellen Welt, M. Schüz (ed.), Pfullingen, Neske, 210–228.

Züricher, M. (1989): Marktorientierte Instrumente der Umweltpolitik, Studie der Robert Madler AG im Auftrag der Freising-Demokratischen Partei der Schweiz, Zürich, Hektographiertes Manuskript.

CONCLUSION

The scientific and social relevance of the results of this study goes far beyond the studied environmental problems. Environmental standards are regulative directions to cope with consequences of technologies or, more generally speaking, with consequences of actions in their effect on the environment. Such regulative directions are necessary, since the carrying capacity of the system environment is no longer unlimited as to consequences of the manifestations of life of a growing population.

The three-step process of defining objectives, ascertaining scientific regular correlations (above all by the natural sciences) and making a decision within the framework of an interdisciplinary accounting discourse, which is elaborated in this study, is of general validity. Assessment of consequences of technologies has to be better understood in accounting discourses, it must not merely consist in a scientific ascertainment of facts.

The study gains important insights as to starting point, objective and procedure of such an accounting discourse. Starting point is the existing situation, objective is by no means paradise but an improvement of the existing situation in the interest of humane forms of life. Judgments linking the different dimensions of problems and preferences can only be found on the basis of pragmatic consistency of these forms of life. The orientations of these forms of life, in their turn, can only claim generalizability if they are determined by rational ways of proceeding. Their rationality is the best warrant for their social efficiency. This is, at the same time, the only way open to democratic societies to cope with problems which arise from the increasing burden imposed on the system Earth by humans and their activities.